OPERATIONS
RESEARCH
APPLICATIONS

The Operations Research Series

Series Editor: A. Ravi Ravindran
Dept. of Industrial & Manufacturing Engineering
The Pennsylvania State University, USA

OPERATIONS
RESEARCH
APPLICATIONS

EDITED BY A. RAVI RAVINDRAN

CRC Press
Taylor & Francis Group
Boca Raton London New York

CRC Press is an imprint of the
Taylor & Francis Group, an **informa** business

CRC Press
Taylor & Francis Group
6000 Broken Sound Parkway NW, Suite 300
Boca Raton, FL 33487-2742

First issued in paperback 2019

© 2009 by Taylor & Francis Group, LLC
CRC Press is an imprint of Taylor & Francis Group, an Informa business

No claim to original U.S. Government works

ISBN-13: 978-1-4200-9186-1 (hbk)
ISBN-13: 978-0-367-38647-4 (pbk)

Visit the Taylor & Francis Web site at
http://www.taylorandfrancis.com

and the CRC Press Web site at
http://www.crcpress.com

Dedication

This book is dedicated to the memory of Professor G. V. Loganathan, who was killed in the Virginia Tech campus tragedy on April 16, 2007.

Contents

Preface

Operations research (OR), which began as an interdisciplinary activity to solve complex problems in the military during World War II, has grown in the past 50 years to a full-fledged academic discipline. Now OR is viewed as a body of established mathematical models and methods to solve complex management problems. OR provides a quantitative analysis of the problem from which the management can make an objective decision. OR has drawn upon skills from mathematics, engineering, business, computer science, economics, and statistics to contribute to a wide variety of applications in business, industry, government, and military. OR methodologies and their applications continue to grow and flourish in a number of decision-making fields.

The objective of this book is to provide a comprehensive overview of OR applications in practice in a single volume. This book is not an OR textbook or a research monograph. The intent is that this book becomes the first resource a practitioner would reach for when faced with an OR problem or application. The key features of this book are as follows:

- Single source guide to OR applications
- Comprehensive resource, but concise
- Coverage of functional applications of OR
- Quick reference guide to students, researchers, and practitioners
- Coverage of industry-specific applications of OR
- References to computer software availability
- Designed and edited with nonexperts in mind

This book contains 12 chapters that cover not only OR applications in the functional areas of business but also industry-specific areas. Each chapter in this book is written by leading authorities in the field and is devoted to a specific application area listed as follows:

- Airlines
- E-commerce
- Energy systems
- Finance
- Military
- Production systems
- Project management
- Quality control
- Reliability
- Supply chain management
- Water resources

This book ends with a chapter on the future of OR applications. This book will be an ideal reference book for OR practitioners in business, industry, government, and academia.

It can also serve as a supplemental text in undergraduate and graduate OR courses in the universities. Readers may also be interested in the companion book titled *Operations Research Methodologies*, which contains a comprehensive review of the OR models and methods that are used in the applications discussed here.

A. Ravi Ravindran
University Park, Pennsylvania

Acknowledgments

First and foremost I would like to thank the authors, who have worked diligently in writing the various handbook chapters that are comprehensive, concise, and easy to read, bridging the gap between theory and practice. The development and evolution of this handbook have also benefited substantially from the advice and counsel of my colleagues and friends in academia and industry, who are too numerous to acknowledge individually. They helped me identify the key topics to be included in the handbook, suggested chapter authors, and served as reviewers of the manuscripts.

I express my sincere appreciation to Atul Rangarajan, an industrial engineering doctoral student at Penn State University, for serving as my editorial assistant and for his careful review of the page proofs returned by the authors. Several other graduate students also helped me with the handbook work, in particular, Ufuk Bilsel, Ajay Natarajan, Richard Titus, Vijay Wadhwa, and Tao Yang. Special thanks go to Professor Prabha Sharma at the Indian Institute of Technology, Kanpur, for her careful review of several chapter manuscripts. I also acknowledge the pleasant personality and excellent typing skills of Sharon Frazier during the entire book project.

I thank Cindy Carelli, Senior Acquisitions Editor, and Jessica Vakili, project coordinator at CRC Press, for their help from inception to publication of the handbook. Finally, I wish to thank my dear wife, Bhuvana, for her patience, understanding, and support when I was focused completely on the handbook work.

A. Ravi Ravindran

Editor

A. Ravi Ravindran, Ph.D., is a professor and the past department head of Industrial and Manufacturing Engineering at the Pennsylvania State University. Formerly, he was a faculty member at the School of Industrial Engineering at Purdue University for 13 years and at the University of Oklahoma for 15 years. At Oklahoma, he served as the director of the School of Industrial Engineering for 8 years and as the associate provost of the university for 7 years, with responsibility for budget, personnel, and space for the academic area. He holds a B.S. in electrical engineering with honors from the Birla Institute of Technology and Science, Pilani, India. His graduate degrees are from the University of California, Berkeley, where he received an M.S. and a Ph.D. in industrial engineering and operations research.

Dr. Ravindran's area of specialization is operations research with research interests in multiple criteria decision-making, financial engineering, health planning, and supply chain optimization. He has published two major textbooks (*Operations Research: Principles and Practice* and *Engineering Optimization: Methods and Applications*) and more than 100 journal articles on operations research. He is a fellow of the Institute of Industrial Engineers. In 2001, he was recognized by the Institute of Industrial Engineers with the Albert G. Holzman Distinguished Educator Award for significant contributions to the industrial engineering profession by an educator. He has won several Best Teacher awards from IE students. He has been a consultant to AT&T, General Motors, General Electric, IBM, Kimberly Clark, Cellular Telecommunication Industry Association, and the U.S. Air Force. He currently serves as the Operations Research Series editor for Taylor & Francis/CRC Press.

Contributors

Adedeji B. Badiru
Air Force Institute of Technology
Dayton, Ohio

P. Balasubramanian
Theme Work Analytics
Bangalore, India

Qianmei Feng
University of Houston
Houston, Texas

Bobbie L. Foote
U.S. Military Academy (Retd.)
West Point, New York

Aliza R. Heching
IBM T. J. Watson Research Center
Yorktown Heights, New York

C. Randy Hudson
Oak Ridge National Laboratory
Oak Ridge, Tennessee

Kailash C. Kapur
University of Washington
Seattle, Washington

Alan J. King
IBM T. J. Watson Research Center
Yorktown Heights, New York

Lawrence M. Leemis
College of William and Mary
Williamsburg, Virginia

G. V. Loganathan
Virginia Tech
Blacksburg, Virginia

Katta G. Murty
University of Michigan
Ann Arbor, Michigan

Giuseppe Paleologo
IBM T. J. Watson Research Center
Yorktown Heights, New York

Sowmyanarayanan Sadagopan
Indian Institute of Information Technology
Bangalore, India

Jane L. Snowdon
IBM T. J. Watson Research Center
Yorktown Heights, New York

Marlin U. Thomas
Air Force Institute of Technology
Wright-Patterson AFB, Ohio

Donald P. Warsing
North Carolina State University
Raleigh, North Carolina

Jeffery D. Weir
Air Force Institute of Technology
Wright-Patterson AFB, Ohio

History of Operations Research

A. Ravi Ravindran
Pennsylvania State University

Origin of Operations Research

To understand what operations research (OR) is today, one must know something of its history and evolution. Although particular models and techniques of OR can be traced back to much earlier origins, it is generally agreed that the discipline began during World War II. Many strategic and tactical problems associated with the Allied military effort were simply too complicated to expect adequate solutions from any one individual, or even a single discipline. In response to these complex problems, groups of scientists with diverse educational backgrounds were assembled as special units within the armed forces. These teams of scientists started working together, applying their interdisciplinary knowledge and training to solve such problems as deployment of radars, anti-aircraft fire control, deployment of ships to minimize losses from enemy submarines, and strategies for air defense. Each of the three wings of Britain's armed forces had such interdisciplinary research teams working on military management problems. As these teams were generally assigned to the commanders in charge of military operations, they were called *operational research* (OR) *teams*. The nature of their research came to be known as *operational research* or *operations research*.

The work of these OR teams was very successful and their solutions were effective in military management. This led to the use of such scientific teams in other Allied nations, in particular the United States, France, and Canada. At the end of the war, many of the scientists who worked in the military operational research units returned to civilian life in universities and industries. They started applying the OR methodology to solve complex management problems in industries. Petroleum companies were the first to make use of OR models for solving large-scale production and distribution problems. In the universities advancements in OR techniques were made that led to the further development and applications of OR. Much of the postwar development of OR took place in the United States.

An important factor in the rapid growth of operations research was the introduction of electronic computers in the early 1950s. The computer became an invaluable tool to the *operations researchers*, enabling them to solve large problems in the business world.

The Operations Research Society of America (ORSA) was formed in 1952 to serve the professional needs of these operations research scientists. Due to the application of OR in industries, a new term called management science (MS) came into being. In 1953, a national society called The Institute of Management Sciences (TIMS) was formed in the United States to promote scientific knowledge in the understanding and practice of management. The journals of these two societies, *Operations Research* and *Management Science*, as well as the joint conferences of their members, helped to draw together the many diverse results into some semblance of a coherent body of knowledge. In 1995, the two societies, ORSA and TIMS, merged to form the Institute of Operations Research and Management Sciences (INFORMS).

Another factor that accelerated the growth of operations research was the introduction of OR/MS courses in the curricula of many universities and colleges in the United States. Graduate programs leading to advanced degrees at the master's and doctorate levels were introduced in major American universities. By the mid-1960s many theoretical advances in OR techniques had been made, which included linear programming, network analysis, integer programming, nonlinear programming, dynamic programming, inventory theory, queueing theory, and simulation. Simultaneously, new applications of OR emerged in service organizations such as banks, health care, communications, libraries, and transportation. In addition, OR came to be used in local, state, and federal governments in their planning and policy-making activities.

It is interesting to note that the modern perception of OR as a body of established models and techniques—that is, a discipline in itself—is quite different from the original concept of OR as an *activity*, which was preformed by interdisciplinary teams. An evolution of this kind is to be expected in any emerging field of scientific inquiry. In the initial formative years, there are no experts, no traditions, no literature. As problems are successfully solved, the body of specific knowledge grows to a point where it begins to require specialization even to know what has been previously accomplished. The pioneering efforts of one generation become the standard practice of the next. Still, it ought to be remembered that at least a portion of the record of success of OR can be attributed to its ecumenical nature.

Meaning of Operations Research

From the historical and philosophical summary just presented, it should be apparent that the term "operations research" has a number of quite distinct variations of meaning. To some, OR is that certain body of problems, techniques, and solutions that has been accumulated under the name of OR over the past 50 years and we apply OR when we recognize a problem of that certain genre. To others, it is an activity or process, which by its very nature is applied. It would also be counterproductive to attempt to make distinctions between "operations research" and the "systems approach." For all practical purposes, they are the same.

How then can we define operations research? The Operational Research Society of Great Britain has adopted the following definition:

> Operational research is the application of the methods of science to complex problems arising in the direction and management of large systems of men, machines, materials and money in industry, business, government, and defense. The distinctive approach is to develop a scientific model of the system, incorporating measurement of factors such as chance and risk, with which to predict and compare the outcomes of alternative decisions, strategies or controls. The purpose is to help management determine its policy and actions scientifically.

The Operations Research Society of America has offered a shorter, but similar, description:

> Operations research is concerned with scientifically deciding how to best design and operate man–machine systems, usually under conditions requiring the allocation of scarce resources.

In general, most of the definitions of OR emphasize its methodology, namely its unique approach to problem solving, which may be due to the use of interdisciplinary teams or due to the application of scientific and mathematical models. In other words, each problem may be analyzed differently, though the same basic approach of operations research is employed. As more research went into the development of OR, the researchers were able to classify to some extent many of the important management problems that arise in practice. Examples of such problems are those relating to allocation, inventory, network, queuing, replacement, scheduling, and so on. The theoretical research in OR concentrated on developing appropriate mathematical models and techniques for analyzing these problems under different conditions. Thus, whenever a management problem is identified as belonging to a particular class, all the models and techniques available for that class can be used to study that problem. In this context, one could view OR as a collection of mathematical models and techniques to solve complex management problems. Hence, it is very common to find OR courses in universities emphasizing different mathematical techniques of operations research such as mathematical programming, queueing theory, network analysis, dynamic programming, inventory models, simulation, and so on.

For more on the early activities in operations research, see Refs. 1–5. Readers interested in the timeline of major contributions in the history of OR/MS are referred to the excellent review article by Gass [6].

References

1. Haley, K.B., War and peace: the first 25 years of OR in Great Britain, *Operations Research*, 50, Jan.–Feb. 2002.
2. Miser, H.J., The easy chair: what OR/MS workers should know about the early formative years of their profession, *Interfaces*, 30, March–April 2000.
3. Trefethen, F.N., A history of operations research, in *Operations Research for Management*, J.F. McCloskey and F.N. Trefethen, Eds., Johns Hopkins Press, Baltimore, MD, 1954.
4. Horner, P., History in the making, *ORMS Today*, 29, 30–39, 2002.
5. Ravindran, A., Phillips, D.T., and Solberg, J.J., *Operations Research: Principles and Practice*, Second Edition, John Wiley & Sons, New York, 1987 (Chapter 1).
6. Gass, S.I., Great moments in histORy, *ORMS Today*, 29, 31–37, 2002.

1

Project Management

Adedeji B. Badiru
Air Force Institute of Technology

1.1 Introduction

Project management techniques continue to be a major avenue to accomplishing goals and objectives in various organizations ranging from government, business, and industry to academia. The techniques of project management can be divided into three major tracks as summarized below:

- Qualitative managerial principles
- Computational decision models
- Computer implementation tools.

This chapter focuses on computational network techniques for project management. Network techniques emerged as a formal body of knowledge for project management during World War II, which ushered in a golden era for operations research and its quantitative modeling techniques.

Badiru (1996) defines project management as the process of managing, allocating, and timing resources to achieve a given objective expeditiously. The phases of project management are:

1. Planning
2. Organizing
3. Scheduling
4. Controlling.

The network techniques covered in this chapter are primarily for the scheduling phase, although network planning is sometimes included in the project planning phase too. Everyone in every organization needs project management to accomplish objectives. Consequently, the need for project management will continue to grow as organizations seek better ways to satisfy the constraints on the following:

- Schedule constraints (time limitation)
- Cost constraints (budget limitation)
- Performance constraints (quality limitation).

The different tools of project management may change over time. Some tools will come and go over time. But the basic need of using network analysis to manage projects will always be high. The network of activities in a project forms the basis for scheduling the project. The critical path method (CPM) and the program evaluation and review technique (PERT) are the two most popular techniques for project network analysis. The precedence diagramming method (PDM) has gained popularity in recent years because of the move toward concurrent engineering. A project network is the graphical representation of the contents and objectives of the project. The basic project network analysis of CPM and PERT is typically implemented in three phases (network planning phase, network scheduling phase, and network control phase), which has the following advantages:

- Advantages for communication
  It clarifies project objectives.
  It establishes the specifications for project performance.
  It provides a starting point for more detailed task analysis.
  It presents a documentation of the project plan.
  It serves as a visual communication tool.
- Advantages for control
  It presents a measure for evaluating project performance.
  It helps determine what corrective actions are needed.
  It gives a clear message of what is expected.
  It encourages team interactions.
- Advantages for team interaction
  It offers a mechanism for a quick introduction to the project.
  It specifies functional interfaces on the project.
  It facilitates ease of application.
  It creates synergy between elements of the project.

Network planning is sometimes referred to as activity planning. This involves the identification of the relevant activities for the project. The required activities and their precedence relationships are determined in the planning phase. Precedence requirements may be determined on the basis of technological, procedural, or imposed constraints. The activities are then represented in the form of a network diagram. The two popular models for network drawing are the activity-on-arrow (AOA) and the activity-on-node (AON) conventions. In the AOA approach, arrows are used to represent activities, whereas nodes represent starting and ending points of activities. In the AON approach, nodes represent activities, whereas arrows represent precedence relationships. Time, cost, and resource requirement estimates are developed for each activity during the network planning phase. Time estimates may be based on the following:

1. Historical records
2. Time standards
3. Forecasting
4. Regression functions
5. Experiential estimates.

Network scheduling is performed by using forward pass and backward pass computational procedures. These computations give the earliest and latest starting and finishing times for each activity. The slack time or float associated with each activity is determined in the computations. The activity path with the minimum slack in the network is used to determine the critical activities. This path also determines the duration of the project. Resource allocation and time-cost tradeoffs are other functions performed during network scheduling.

Network control involves tracking the progress of a project on the basis of the network schedule and taking corrective actions when needed; an evaluation of actual performance versus expected performance determines deficiencies in the project.

1.2 Critical Path Method

Precedence relationships in a CPM network fall into the three major categories listed below:

1. Technical precedence
2. Procedural precedence
3. Imposed precedence

Technical precedence requirements are caused by the technical relationships among activities in a project. For example, in conventional construction, walls must be erected before the roof can be installed. Procedural precedence requirements are determined by policies and procedures. Such policies and procedures are often subjective, with no concrete justification. Imposed precedence requirements can be classified as resource-imposed, project-imposed, or environment-imposed. For example, resource shortages may require that one task be scheduled before another. The current status of a project (e.g., percent completion) may determine that one activity be performed before another. The environment of a project, for example, weather changes or the effects of concurrent projects, may determine the precedence relationships of the activities in a project.

The primary goal of a CPM analysis of a project is the determination of the "critical path." The critical path determines the minimum completion time for a project. The computational analysis involves forward pass and backward pass procedures. The forward pass determines the earliest start time and the earliest completion time for each activity in the network. The backward pass determines the latest start time and the latest completion time for each activity. Figure 1.1 shows an example of an activity network using the activity-on-node convention. Conventionally the network is drawn from left to right. If this convention is followed, there is no need to use arrows to indicate the directional flow in the activity network. The notations used for activity A in the network are explained below:

A: Activity identification
ES: Earliest starting time
EC: Earliest completion time
LS: Latest starting time
LC: Latest completion time
t: Activity duration

During the forward pass analysis of the network, it is assumed that each activity will begin at its earliest starting time. An activity can begin as soon as the last of its predecessors is finished. The completion of the forward pass determines the earliest completion time of the project. The backward pass analysis is a reverse of the forward pass. It begins at the latest project completion time and ends at the latest starting time of the first activity in the project network. The rules for implementing the forward pass and backward pass analyses in CPM are presented below. These rules are implemented iteratively until the ES, EC, LS, and LC have been calculated for all nodes in the network.

Rule 1: Unless otherwise stated, the starting time of a project is set equal to time zero. That is, the first node in the network diagram has an earliest start time of zero.

Rule 2: The earliest start time (ES) for any activity is equal to the maximum of the earliest completion times (EC) of the immediate predecessors of the activity. That is,

$$ES = Maximum\ \{Immediately\ Preceding\ ECs\}$$

Rule 3: The earliest completion time (EC) of an activity is the activity's earliest start time plus its estimated duration. That is,

$$ES = ES + (Activity\ Time)$$

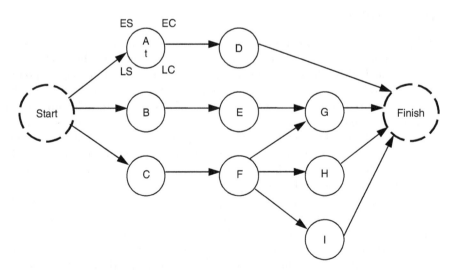

FIGURE 1.1   Example of activity network.

Rule 4: The earliest completion time of a project is equal to the earliest completion time of the very last node in the network. That is,

$$EC\ of\ Project = EC\ of\ last\ activity$$

Rule 5: Unless the latest completion time (LC) of a project is explicitly specified, it is set equal to the earliest completion time of the project. This is called the *zero-project-slack* assumption. That is,

$$LC\ of\ Project = EC\ of\ Project$$

Rule 6: If a desired deadline is specified for the project, then

$$LC\ of\ Project = Specified\ Deadline$$

It should be noted that a latest completion time or deadline may sometimes be specified for a project based on contractual agreements.

Rule 7: The latest completion time (LC) for an activity is the smallest of the latest start times of the activity's immediate successors. That is,

$$LC = Minimum\ \{Immediately\ Succeeding\ LS's\}$$

Rule 8: The latest start time for an activity is the latest completion time minus the activity time. That is,

$$LS = LC - (Activity\ Time)$$

1.2.1 CPM Example

Table 1.1 presents the data for an illustrative project. This network and its extensions will be used for other computational examples in this chapter. The AON network for the example is given in Figure 1.2. Dummy activities are included in the network to designate single starting and ending points for the project.

TABLE 1.1 Data for Sample Project for CPM Analysis

Activity	Predecessor	Duration (Days)
A	–	2
B	–	6
C	–	4
D	A	3
E	C	5
F	A	4
G	B,D,E	2

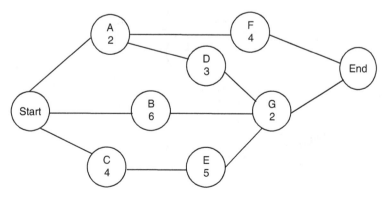

FIGURE 1.2 Project network for illustrative example.

1.2.2   Forward Pass

The forward pass calculations are shown in Figure 1.3. Zero is entered as the ES for the initial node. As the initial node for the example is a dummy node, its duration is zero. Thus, EC for the starting node is equal to its ES. The ES values for the immediate successors of the starting node are set equal to the EC of the START node and the resulting EC values are computed. Each node is treated as the "start" node for its successor or successors. However, if an activity has more than one predecessor, the maximum of the ECs of the preceding activities is used as the activity's starting time. This happens in the case of activity G, whose ES is determined as Max $\{6,5,9\} = 9$. The earliest project completion time for the example is 11 days. Note that this is the maximum of the immediately preceding earliest completion times: Max $\{6,11\} = 11$. As the dummy ending node has no duration, its earliest completion time is set equal to its earliest start time of 11 days.

1.2.3   Backward Pass

The backward pass computations establish the latest start time (LS) and latest completion time (LC) for each node in the network. The results of the backward pass computations are shown in Figure 1.4. As no deadline is specified, the latest completion time of the project is set equal to the earliest completion time. By backtracking and using the network analysis rules presented earlier, the latest completion and start times are determined for each node. Note that in the case of activity A with two successors, the latest completion time is determined as the minimum of the immediately succeeding latest start times. That

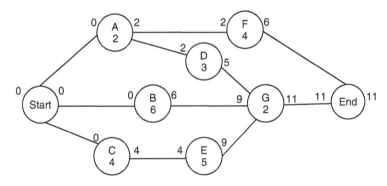

FIGURE 1.3   Forward pass analysis for CPM example.

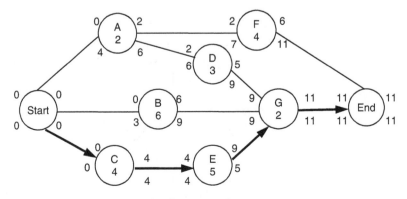

FIGURE 1.4   Backward pass analysis for CPM example.

is, Min{6,7} = 6. A similar situation occurs for the dummy starting node. In that case, the latest completion time of the dummy start node is Min {0,3,4} = 0. As this dummy node has no duration, the latest starting time of the project is set equal to the node's latest completion time. Thus, the project starts at time 0 and is expected to be completed by time 11.

Within a project network, there are usually several possible paths and a number of activities that must be performed sequentially and some activities that may be performed concurrently. If an activity has ES and LS times that are not equal, then the actual start and completion times of that activity may be flexible. The amount of flexibility an activity possesses is called a slack. The slack time is used to determine the critical activities in the network as discussed below.

1.2.4 Determination of Critical Activities

The critical path is defined as the path with the least slack in the network diagram. All the activities on the critical path are said to be critical activities. These activities can create bottlenecks in the network if they are delayed. The critical path is also the longest path in the network diagram. In some networks, particularly large ones, it is possible to have multiple critical paths. If there is a large number of paths in the network, it may be very difficult to visually identify all the critical paths.

The slack time of an activity is also referred to as its *float*. There are four basic types of activity slack as described below:

- *Total Slack* (TS). Total slack is defined as the amount of time an activity may be delayed from its earliest starting time without delaying the latest completion time of the project. The total slack time of an activity is the difference between the latest completion time and the earliest completion time of the activity, or the difference between the latest starting time and the earliest starting time of the activity.

$$TS = LC - EC$$

  Or

$$TS = LS - ES$$

  Total slack is the measure that is used to determine the critical activities in a project network. The critical activities are identified as those having the minimum total slack in the network diagram. If there is only one critical path in the network, then all the critical activities will be on that one path.

- *Free Slack* (FS). Free slack is the amount of time an activity may be delayed from its earliest starting time without delaying the starting time of any of its immediate successors. Activity free slack is calculated as the difference between the minimum earliest starting time of the activity's successors and the earliest completion time of the activity.

$$FS = Min\{Succeeding\ ES's\} - EC$$

- *Interfering Slack* (IS). Interfering slack or interfering float is the amount of time by which an activity interferes with (or obstructs) its successors when its total slack is fully used. This is rarely used in practice. The interfering float is computed as the difference between the total slack and the free slack.

$$IS = TS - FS$$

- *Independent Float* (IF). Independent float or independent slack is the amount of float that an activity will always have regardless of the completion times of its predecessors or the starting times of its successors. Independent float is computed as:

$$\text{IF} = \text{Max}\{0, (\text{ES}_j - \text{LC}_i - t)\}$$

where ES_j is the earliest starting time of the preceding activity, LC_i is the latest completion time of the succeeding activity, and t is the duration of the activity whose independent float is being calculated. Independent float takes a pessimistic view of the situation of an activity. It evaluates the situation whereby the activity is pressured from either side, that is, when its predecessors are delayed as late as possible while its successors are to be started as early as possible. Independent float is useful for conservative planning purposes, but it is not used much in practice. Despite its low level of use, independent float does have practical implications for better project management. Activities can be buffered with independent floats as a way to handle contingencies.

In Figure 1.4 the total slack and the free slack for activity A are calculated, respectively, as:

$$\text{TS} = 6 - 2 = 4 \text{ days}$$
$$\text{FS} = \text{Min}\{2, 2\} - 2 = 2 - 2 = 0$$

Similarly, the total slack and the free slack for activity F are:

$$\text{TS} = 11 - 6 = 5 \text{ days}$$
$$\text{FS} = \text{Min}\{11\} - 6 = 11 - 6 = 5 \text{ days}$$

Table 1.2 presents a tabulation of the results of the CPM example. The table contains the earliest and latest times for each activity as well as the total and free slacks. The results indicate that the minimum total slack in the network is zero. Thus, activities C, E, and G are identified as the critical activities. The critical path is highlighted in Figure 1.4 and consists of the following sequence of activities:

$$\textbf{Start} \rightarrow \textbf{C} \rightarrow \textbf{E} \rightarrow \textbf{G} \rightarrow \textbf{End}$$

The total slack for the overall project itself is equal to the total slack observed on the critical path. The minimum slack in most networks will be zero as the ending LC is set equal to the ending EC. If a deadline is specified for a project, then we would set the project's latest completion time to the specified deadline. In that case, the minimum total slack in the network would be given by:

$$\text{TS}_{\text{Min}} = \text{Project Deadline} - \text{EC of the last node}$$

TABLE 1.2   Result of CPM Analysis for Sample Project

Activity	Duration	ES	EC	LS	LC	TS	FS	Criticality
A	2	0	2	4	6	4	0	–
B	6	0	6	3	9	3	3	–
C	4	0	4	0	4	0	0	Critical
D	3	2	5	6	9	4	4	–
E	5	4	9	4	9	0	0	Critical
F	4	2	6	7	11	5	5	–
G	2	9	11	9	11	0	0	Critical

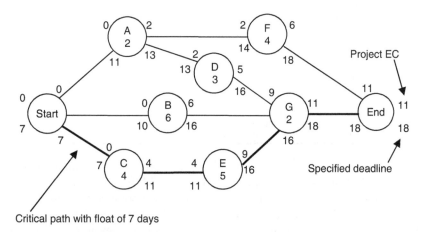

FIGURE 1.5 CPM network with deadline.

This minimum total slack will then appear as the total slack for each activity on the critical path. If a specified deadline is lower than the EC at the finish node, then the project will start out with a negative slack. That means that it will be behind schedule before it even starts. It may then become necessary to expedite some activities (i.e., crashing) to overcome the negative slack. Figure 1.5 shows an example with a specified project deadline. In this case, the deadline of 18 days comes after the earliest completion time of the last node in the network.

1.2.5 Using Forward Pass to Determine the Critical Path

The critical path in CPM analysis can be determined from the forward pass only. This can be helpful in cases where it is desired to quickly identify the critical activities without performing all the other calculations needed to obtain the latest starting times, the latest completion times, and total slacks. The steps for determining the critical path from the forward pass only are:

1. Complete the forward pass in the usual manner.
2. Identify the last node in the network as a critical activity.
3. Work backward from the last node. Whenever a merge node occurs, the critical path will be along the path where the earliest completion time (EC) of the predecessor is equal to the earliest start time (ES) of the current node.
4. Continue the backtracking from each critical activity until the project starting node is reached. Note that if there is a single starting node or a single ending node in the network, then that node will always be on the critical path.

1.2.6 Subcritical Paths

In a large network, there may be paths that are near critical. Such paths require almost as much attention as the critical path as they have a high potential of becoming critical when changes occur in the network. Analysis of subcritical paths may help in the classification of tasks into ABC categories on the basis of Pareto analysis. Pareto analysis separates the "vital" few activities from the "trivial" many activities. This permits a more efficient allocation of resources. The principle of Pareto analysis originated from the work of Italian economist Vilfredo Pareto (1848–1923). Pareto discovered from his studies that most of the wealth in his country was held by a few individuals.

TABLE 1.3   Analysis of Sub-Critical Paths

Path No.	Activities on Path	Total Slack	λ (%)	λ'
1	A,C,G,H	0	100	10
2	B,D,E	1	97.56	9.78
3	F,I	5	87.81	8.90
4	J,K,L	9	78.05	8.03
5	O,P,Q,R	10	75.61	7.81
6	M,S,T	25	39.02	4.51
7	N,AA,BB,U	30	26.83	3.42
8	V,W,X	32	21.95	2.98
9	Y,CC,EE	35	17.14	2.54
10	DD,Z,FF	41	0	1.00

For project control purposes, the Pareto principle states that 80% of the bottlenecks are caused by only 20% of the tasks. This principle is applicable to many management processes. For example, in cost analysis, one can infer that 80% of the total cost is associated with only 20% of the cost items. Similarly, 20% of an automobile's parts cause 80% of the maintenance problems. In personnel management, about 20% of the employees account for about 80% of the absenteeism. For critical path analysis, 20% of the network activities will take up 80% of our control efforts. The ABC classification based on Pareto analysis divides items into three priority categories: A (most important), B (moderately important), and C (least important). Appropriate percentages (e.g., 20%, 25%, 55%) may be assigned to the categories.

With Pareto analysis, attention can be shifted from focusing only on the critical path to managing critical and near-critical tasks. The level of criticality of each path may be assessed by the steps below:

1. Sort paths in increasing order of total slack.
2. Partition the sorted paths into groups based on the magnitudes of their total slacks.
3. Sort the activities within each group in increasing order of their earliest starting times.
4. Assign the highest level of criticality to the first group of activities (e.g., 100%). This first group represents the usual critical path.
5. Calculate the relative criticality indices for the other groups in decreasing order of criticality.

Define the following variables:
α_1 = the minimum total slack in the network
α_2 = the maximum total slack in the network
β = total slack for the path whose criticality is to be calculated.

Compute the path's criticality as:

$$\lambda = \frac{\alpha_2 - \beta}{\alpha_2 - \alpha_1}(100\%)$$

The above procedure yields relative criticality levels between 0% and 100%. Table 1.3 presents a hypothetical example of path criticality indices. The criticality level may be converted to a scale between 1 (least critical) and 10 (most critical) by the expression below:

$$\lambda' = 1 + 0.09\lambda$$

1.2.7   Gantt Charts

When the results of a CPM analysis are fitted to a calendar time, the project plan becomes a schedule. The Gantt chart is one of the most widely used tools for presenting a project

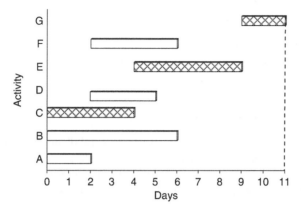

FIGURE 1.6 Gantt chart based on earliest starting times.

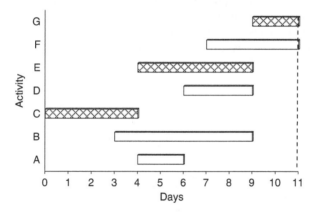

FIGURE 1.7 Gantt chart based on latest starting times.

schedule. A Gantt chart can show planned and actual progress of activities. The time scale is indicated along the horizontal axis, while horizontal bars or lines representing activities are ordered along the vertical axis. As a project progresses, markers are made on the activity bars to indicate actual work accomplished. Gantt charts must be updated periodically to indicate project status. Figure 1.6 presents the Gantt chart using the earliest starting (ES) times from Table 1.2. Figure 1.7 presents the Gantt chart for the example based on the latest starting (LS) times. Critical activities are indicated by the shaded bars.

Figure 1.6 shows the Gantt chart for the example based on earliest starting time. From the CPM network computations, it is noted that activity F can be delayed from day two until day seven (i.e., TS = 5) without delaying the overall project. Likewise, A, D, or both may be delayed by a combined total of 4 days (TS = 4) without delaying the overall project. If all the 4 days of slack are used up by A, then D cannot be delayed. If A is delayed by 1 day, then D can only be delayed by up to 3 days without causing a delay of G, which determines project completion. CPM computations also reveal that activity B may be delayed up to 3 days without affecting the project completion time.

In Figure 1.7, the activities are scheduled by their latest completion times. This represents a pessimistic case where activity slack times are fully used. No activity in this schedule can be delayed without delaying the project. In Figure 1.7, only one activity is scheduled over the first 3 days. This may be compared to the schedule in Figure 1.6, which has three starting activities. The schedule in Figure 1.7 may be useful if there is a situation that permits only a few activities to be scheduled in the early stages of the project. Such situations may involve shortage of project personnel, lack of initial budget, time for project initiation, time for personnel training, allowance for learning period, or general resource constraints. Scheduling of activities based on ES times indicates an optimistic view. Scheduling on the basis of LS times represents a pessimistic approach.

1.2.8 Gantt Chart Variations

The basic Gantt chart does not show the precedence relationships among activities. The chart can be modified to show these relationships by coding appropriate bars, as shown by the cross-hatched bars in Figure 1.8. Other simple legends can be added to show which bars are related by precedence linking. Figure 1.9 shows a Gantt chart that presents a comparison of planned and actual schedules. Note that two tasks are in progress at the current time indicated in the figure. One of the ongoing tasks is an unplanned task. Figure 1.10 shows a Gantt chart on which important milestones have been indicated. Figure 1.11 shows a Gantt chart in which bars represent a combination of related tasks. Tasks may be combined for scheduling purposes or for conveying functional relationships required in a project. Figure 1.12 presents a Gantt chart of project phases. Each phase is further divided into

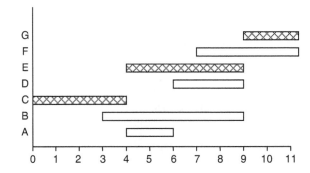

FIGURE 1.8   Coding of bars that are related by precedence linking.

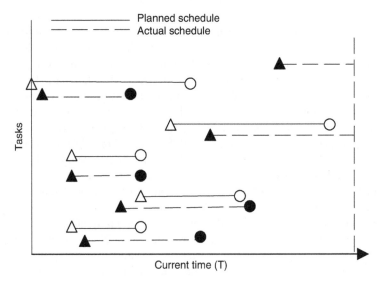

FIGURE 1.9   Progress monitoring Gantt chart.

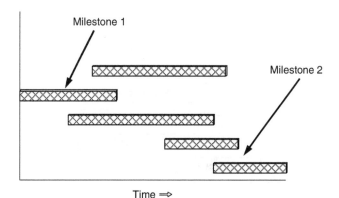

FIGURE 1.10   Milestone Gantt chart.

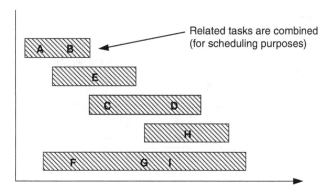

FIGURE 1.11   Task combination Gantt chart.

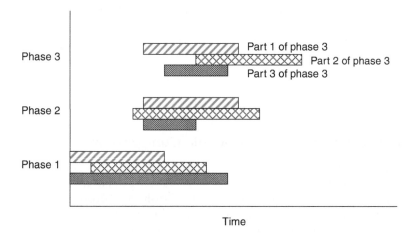

FIGURE 1.12   Phase-based Gantt chart.

parts. Figure 1.13 shows a Gantt chart for multiple projects. Multiple project charts are useful for evaluating resource allocation strategies. Resource loading over multiple projects may be needed for capital budgeting and cash flow analysis decisions. Figure 1.14 shows a project slippage chart that is useful for project tracking and control. Other variations of the basic Gantt chart may be developed for specific needs.

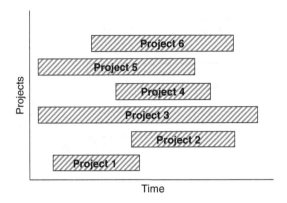

FIGURE 1.13   Multiple projects Gantt chart.

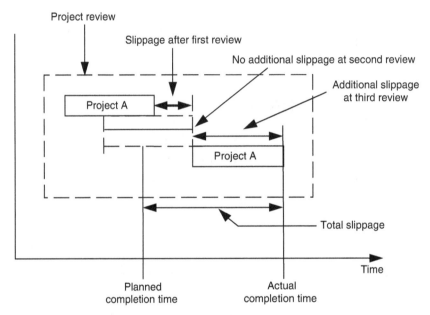

FIGURE 1.14   Project slippage tracking Gantt chart.

1.2.9   Activity Crashing and Schedule Compression

Schedule compression refers to reducing the length of a project network. This is often accomplished by crashing activities. Crashing, sometimes referred to as expediting, reduces activity durations, thereby reducing project duration. Crashing is done as a tradeoff between shorter task duration and higher task cost. It must be determined whether the total cost savings realized from reducing the project duration is enough to justify the higher costs associated with reducing individual task durations. If there is a delay penalty associated with a project, it may be possible to reduce the total project cost even though individual task costs are increased by crashing. If the cost savings on delay penalty is higher than the incremental cost of reducing the project duration, then crashing is justified. Under conventional crashing, the more the duration of a project is compressed, the higher the total cost of the project. The objective is to determine at what point to terminate further crashing in a network. Normal task duration refers to the time required to perform a task under normal

TABLE 1.4 Normal and Crash Time and Cost Data

Activity	Normal Duration	Normal Cost	Crash Duration	Crash Cost	Crashing Ratio
A	2	$210	2	$210	0
B	6	400	4	600	100
C	4	500	3	750	250
D	3	540	2	600	60
E	5	750	3	950	100
F	4	275	3	310	35
G	2	100	1	125	25
		$2775		$3545	

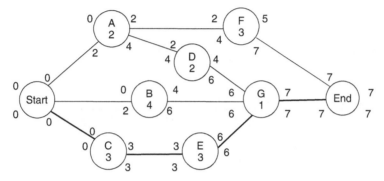

FIGURE 1.15 Example of fully crashed CPM network.

circumstances. Crash task duration refers to the reduced time required to perform a task when additional resources are allocated to it.

If each activity is assigned a range of time and cost estimates, then several combinations of time and cost values will be associated with the overall project. Iterative procedures are used to determine the best time and cost combination for a project. Time-cost trade-off analysis may be conducted, for example, to determine the marginal cost of reducing the duration of the project by one time unit. Table 1.4 presents an extension of the data for the earlier example to include normal and crash times as well as normal and crash costs for each activity. The normal duration of the project is 11 days, as seen earlier, and the normal cost is $2775.

If all the activities are reduced to their respective crash durations, the total crash cost of the project will be $3545. In that case, the crash time is found by CPM analysis to be 7 days. The CPM network for the fully crashed project is shown in Figure 1.15. Note that activities C, E, and G remain critical. Sometimes, the crashing of activities may result in a new critical path. The Gantt chart in Figure 1.16 shows a schedule of the crashed project using the ES times. In practice, one would not crash all activities in a network. Rather, some heuristic would be used to determine which activity should be crashed and by how much. One approach is to crash only the critical activities or those activities with the best ratios of incremental cost versus time reduction. The last column in Table 1.4 presents the respective ratios for the activities in our example. The crashing ratios are computed as:

$$r = \frac{\text{Crash Cost} - \text{Normal Cost}}{\text{Normal Duration} - \text{Crash Duration}}$$

This method of computing the crashing ratio gives crashing priority to the activity with the lowest cost slope. It is a commonly used approach in CPM networks.

Activity G offers the lowest cost per unit time reduction of $25. If our approach is to crash only one activity at a time, we may decide to crash activity G first and evaluate the increase in project cost versus the reduction in project duration. The process can then be

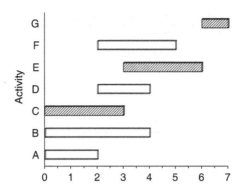

FIGURE 1.16   Gantt chart of fully crashed CPM network.

TABLE 1.5   Selected Crashing Options for CPM Example

Option No.	Activities Crashed	Network Duration	Time Reduction	Incremental Cost	Total Cost
1.	None	11	–	–	2775
2.	G	10	1	25	2800
3.	G,F	10	0	35	2835
4.	G,F,D	10	0	60	2895
5.	G,F,D,B	10	0	200	3095
6.	G,F,D,B,E	8	2	200	3295
7.	G,F,D,B,E,C	7	1	250	3545

repeated for the next best candidate for crashing, which is activity F in this case. After F has been crashed, activity D can then be crashed.

This approach is repeated iteratively in order of activity preference until no further reduction in project duration can be achieved or until the total project cost exceeds a specified limit.

A more comprehensive analysis is to evaluate all possible combinations of the activities that can be crashed. However, such a complete enumeration would be prohibitive, as there would be a total of 2^c crashed networks to evaluate, where c is the number of activities that can be crashed out of the n activities in the network ($c <= n$). For our example, only 6 out of the 7 activities in the sample network can be crashed. Thus, a complete enumeration will involve $2^6 = 64$ alternate networks. Table 1.5 shows 7 of the 64 crashing options. Activity G, which offers the best crashing ratio, reduces the project duration by only 1 day. Even though activities F, D, and B are crashed by a total of 4 days at an incremental cost of $295, they do not generate any reduction in project duration. Activity E is crashed by 2 days and it generates a reduction of 2 days in project duration. Activity C, which is crashed by 1 day, generates a further reduction of 1 day in the project duration. It should be noted that the activities that generate reductions in project duration are the ones that were identified earlier as the critical activities.

Figure 1.17 shows the crashed project duration versus the crashing options, while Figure 1.18 shows a plot of the total project cost after crashing versus the selected crashing options. As more activities are crashed, the project duration decreases while the total project cost increases. If full enumeration were performed, Figure 1.17 would contain additional points between the minimum possible project duration of 7 days (fully crashed) and the normal project duration of 11 days (no crashing). Similarly, the plot for total project cost (Figure 1.18) would contain additional points between the normal cost of $2775 and the crash cost of $3545.

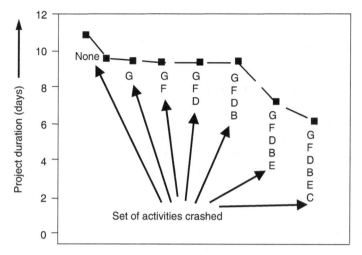

FIGURE 1.17   Duration as a function of crashing options.

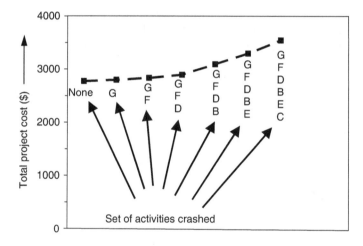

FIGURE 1.18   Project cost as a function of crashing options.

Several other approaches exist for determining which activities to crash in a project network. Two alternate approaches are presented below for computing the crashing ratio, r.

Let r = Criticality Index

or

Let $r = \dfrac{\text{Crash Cost} - \text{Normal Cost}}{(\text{Normal Duration} - \text{Crash Duration})(\text{Criticality Index})}$.

The first approach uses a *critical index* criterion, giving crashing priority to the activity with the highest probability of being on the critical path. In deterministic networks, this refers to the critical activities. In stochastic networks, an activity is expected to fall on the critical path only a percentage of the time. The second approach is a combination of the approach used for the illustrative example and the criticality index approach. It reflects the process of selecting the least-cost expected value. The denominator of the expression represents the expected number of days by which the critical path can be shortened. For different project networks, different crashing approaches should be considered, and the one that best fits the nature of the network should be selected.

1.3   PERT Network Analysis

Program evaluation review technique (PERT) is an extension of CPM that incorporates variability in activity durations into project network analysis. PERT has been used extensively and successfully in practice. In real life, activities are often prone to uncertainties that determine the actual durations of the activities. In CPM, activity durations are assumed to be deterministic. In PERT, the potential uncertainties in activity durations are accounted for by using three time estimates for each activity. The three time estimates represent the spread of the estimated activity duration. The greater the uncertainty of an activity the wider the range of the estimates.

1.3.1   PERT Estimates and Formulas

PERT uses three time estimates (optimistic, most likely, and pessimistic) to compute the expected duration and variance for each activity. The PERT formulas are based on a simplification of the expressions for the mean and variance of a beta distribution. The approximation formula for the mean is a simple weighted average of the three time estimates, with the end points assumed to be equally likely and the mode four times as likely. The approximation formula for PERT is based on the recognition that most of the observations from a distribution will lie within plus or minus three standard deviations, or a spread of six standard deviations. This leads to the simple method of setting the PERT formula for standard deviation equal to one-sixth of the estimated duration range. While there is no theoretical validation for these approximation approaches, the PERT formulas do facilitate ease of use. The formulas are presented below:

$$t_e = \frac{a + 4m + b}{6}$$

$$s = \frac{(b - a)}{6}$$

where
a = optimistic time estimate
m = most likely time estimate
b = pessimistic time estimate $a < m < b$
t_e = expected time for the activity
s^2 = variance of the duration of the activity

After obtaining the estimate of the duration for each activity, the network analysis is carried out in the same manner previously illustrated for the CPM approach. The major steps in PERT analysis are summarized below:

1. Obtain three time estimates a, m, and b for each activity.
2. Compute the expected duration for each activity by using the formula for t_e.
3. Compute the variance of the duration of each activity from the formula for s^2.
4. Compute the expected project duration, T_e. As in the case of CPM, the duration of a project in PERT analysis is the sum the durations of the activities on the critical path.

5. Compute the variance of the project duration as the sum of the variances of the activities on the critical path. The variance of the project duration is denoted by S^2. It should be recalled that CPM cannot compute the variance of the project duration, as variances of activity durations are not computed.

6. If there are two or more critical paths in the network, choose the one with the largest variance to determine the project duration and the variance of the project duration. Thus, PERT is pessimistic with respect to the variance of project duration when there are multiple critical paths in the network. For some networks, it may be necessary to perform a mean-variance analysis to determine the relative importance of the multiple paths by plotting the expected project duration versus the path duration variance.

7. If desired, compute the probability of completing the project within a specified time period. This is not possible under CPM.

1.3.2 Modeling of Activity Times

In practice, a question often arises as to how to obtain good estimates of a, m, and b. Several approaches can be used to obtain the time estimates for PERT. Some of the approaches are:

- Estimates furnished by an experienced person
- Estimates extracted from standard time data
- Estimates obtained from historical data
- Estimates obtained from simple regression or forecasting
- Estimates generated by simulation
- Estimates derived from heuristic assumptions
- Estimates dictated by customer requirements.

The pitfall of using estimates furnished by an individual is that they may be inconsistent, as they are limited by the experience and personal bias of the person providing them. Individuals responsible for furnishing time estimates are usually not experts in estimation, and they generally have difficulty in providing accurate PERT time estimates. There is often a tendency to select values of $a, m,$ and b that are optimistically skewed. This is because a conservatively large value is typically assigned to b by inexperienced individuals.

The use of time standards, on the other hand, may not reflect the changes occurring in the current operating environment due to new technology, work simplification, new personnel, and so on. The use of historical data and forecasting is very popular because estimates can be verified and validated by actual records. In the case of regression and forecasting, there is the danger of extrapolation beyond the data range used for fitting the regression and forecasting models. If the sample size in a historical data set is sufficient and the data can be assumed to reasonably represent prevailing operating conditions, the three PERT estimates can be computed as follows:

$$\hat{a} = \bar{t} - kR$$
$$\hat{m} = \bar{t}$$
$$\hat{b} = \bar{t} + kR$$

where
R = range of the sample data
$\bar{t}$ = arithmetic average of the sample data

$k = 3/d_2$

d_2 = an adjustment factor for estimating the standard deviation of a population

If $kR > \bar{t}$, then set $a = 0$ and $b = 2\bar{t}$. The factor d_2 is widely tabulated in the quality control literature as a function of the number of sample points, n. Selected values of d_2 are presented below.

n	5	10	15	20	25	30	40	50	75	100
d_2	2.326	3.078	3.472	3.735	3.931	4.086	4.322	4.498	4.806	5.015

As mentioned earlier, activity times can be determined from historical data. The procedure involves three steps:

1. Appropriate organization of the historical data into histograms.
2. Determination of a distribution that reasonably fits the shape of the histogram.
3. Testing of the goodness-of-fit of the hypothesized distribution by using an appropriate statistical model. The chi-square test and the Kolmogrov-Smirnov (K-S) test are two popular methods for testing goodness-of-fit. Most statistical texts present the details of how to carry out goodness-of-fit tests.

1.3.3   Beta Distribution

PERT analysis assumes that the probabilistic properties of activity duration can be modeled by the beta probability density function. The beta distribution is defined by two end points and two shape parameters. The beta distribution was chosen by the original developers of PERT as a reasonable distribution to model activity times because it has finite end points and can assume a variety of shapes based on different shape parameters. While the true distribution of activity time will rarely ever be known, the beta distribution serves as an acceptable model. Figure 1.19 shows examples of alternate shapes of the standard beta

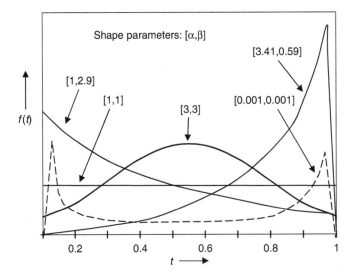

FIGURE 1.19   Alternate shapes of the beta distribution.

distribution between zero and one. The uniform distribution between 0 and 1 is a special case of the beta distribution with both shape parameters equal to one.

The standard beta distribution is defined over the interval 0 to 1, while the general beta distribution is defined over any interval a to b. The general beta probability density function is given by:

$$f(t) = \frac{\Gamma(\alpha + \beta)}{\Gamma(\alpha)\Gamma(\beta)} \cdot \frac{1}{(b-a)^{\alpha+\beta-1}} \cdot (t-a)^{\alpha-1}(b-t)^{\beta-1}$$

$$\text{for } a \leq t \leq b \quad \text{and} \quad \alpha > 0, \beta > 0$$

where
$a =$ lower end point of the distribution
$b =$ upper end point of the distribution
α and β are the shape parameters for the distribution.

The mean, variance, and mode of the general beta distribution are defined as:

$$\mu = a + (b-a)\frac{\alpha}{\alpha + \beta}$$

$$\sigma^2 = (b-a)^2 \frac{\alpha\beta}{(\alpha + \beta + 1)(\alpha + \beta)^2}$$

$$m = \frac{a(\beta - 1) + b(\alpha - 1)}{\alpha + \beta - 2}$$

The general beta distribution can be transformed into a standardized distribution by changing its domain from $[a, b]$ to the unit interval $[0, 1]$. This is accomplished by using the relationship $t_s = a + (b-a)t_s$, where t_s is the standard beta random variable between 0 and 1. This yields the standardized beta distribution, given by:

$$f(t) = \frac{\Gamma(\alpha + \beta)}{\Gamma(\alpha)\Gamma(\beta)} t^{\alpha-1}(1-t)^{\beta-1}; \quad 0 < t < 1; \quad \alpha, \beta > 0$$

$$= 0; \quad \text{elsewhere}$$

with mean, variance, and mode defined as:

$$\mu = \frac{\alpha}{\alpha + \beta}$$

$$\sigma^2 = \frac{\alpha\beta}{(\alpha + \beta + 1)(\alpha + \beta)^2}$$

$$m = \frac{a(\beta - 1) + b(\alpha - 1)}{\alpha + \beta - 2}$$

1.3.4   Triangular Distribution

The triangular probability density function has been used as an alternative to the beta distribution for modeling activity times. The triangular density has three essential parameters: a

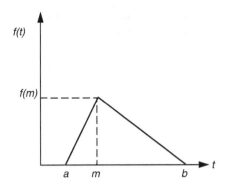

FIGURE 1.20   Triangular probability density function.

minimum value (a), a mode (m) and a maximum (b). It is defined mathematically as:

$$f(t) = \frac{2(t - a)}{(m - a)(b - a)}; \quad a \le t \le m$$

$$= \frac{2(b - t)}{(b - m)(b - a)}; \quad m \le t \le b$$

with mean and variance defined, respectively, as:

$$\mu = \frac{a + m + b}{3}$$

$$\sigma^2 = \frac{a(a - m) + b(b - a) + m(m - b)}{18}$$

Figure 1.20 presents a graphical representation of the triangular density function. The three time estimates of PERT can be inserted into the expression for the mean of the triangular distribution to obtain an estimate of the expected activity duration. Note that in the conventional PERT formula, the mode (m) is assumed to carry four times as much weight as either a or b when calculating the expected activity duration. By contrast, under the triangular distribution, the three time estimates are assumed to carry equal weights.

1.3.5   Uniform Distribution

For cases where only two time estimates instead of three are to be used for network analysis, the uniform density function may be assumed for activity times. This is acceptable for situations where extreme limits of an activity duration can be estimated and it can be assumed that the intermediate values are equally likely to occur. The uniform distribution is defined mathematically as:

$$f(t) = \frac{1}{b - a}; \quad a \le t \le b$$

$$= 0; \qquad \text{otherwise}$$

with mean and variance defined, respectively, as:

$$\mu = \frac{a + b}{2}$$

$$\sigma^2 = \frac{(b - a)^2}{12}$$

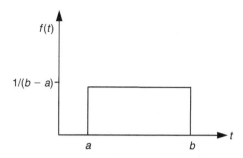

FIGURE 1.21   Uniform probability density function.

Figure 1.21 presents a graphical representation of the uniform distribution, for which the expected activity duration is computed as the average of the upper and lower limits of the distribution. The appeal of using only two time estimates a and b is that the estimation error due to subjectivity can be reduced and the estimation task simplified.

Other distributions that have been explored for activity time modeling include the normal distribution, lognormal distribution, truncated exponential distribution, and Weibull distribution. Once the expected activity durations have been computed, the analysis of the activity network is carried out just as in the case of single-estimate CPM network analysis.

1.4   Statistical Analysis of Project Duration

Regardless of the distribution assumed for activity durations, the central limit theorem suggests that the distribution of the project duration will be approximately normally distributed. The theorem states that the distribution of averages obtained from any probability density function will be approximately normally distributed if the sample size is large and the averages are independent. In mathematical terms, the theorem is stated as follows.

1.4.1   Central Limit Theorem

Let X_1, $X_2, \ldots, X_N$ be independent and identically distributed random variables. Then the sum of the random variables is normally distributed for large values of N. The sum is defined as:

$$T = X_1 + X_2 + \cdots + X_N$$

In activity network analysis, T represents the total project length as determined by the sum of the durations of the activities of the critical path. The mean and variance of T are expressed as:

$$\mu = \sum_{i=1}^{N} E[X_i]$$

$$\sigma^2 = \sum_{i=1}^{N} V[X_i]$$

where
$E[X_i] =$ expected value of random variable X_i
$V[X_i] =$ variance of random variable X_i.

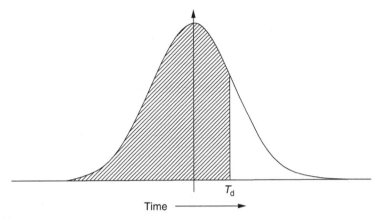

FIGURE 1.22 Area under the normal curve.

When applying the central limit theorem to activity networks, one should note that the assumption of independent activity times may not always be satisfied. Because of precedence relationships and other interdependencies of activities, some activity durations may not be independent.

1.4.2 Probability Calculation

If the project duration T_e can be assumed to be approximately normally distributed based on the central limit theorem, then the probability of meeting a specified deadline T_d can be computed by finding the area under the standard normal curve to the left of T_d. Figure 1.22 shows an example of a normal distribution describing the project duration.

Using the familiar transformation formula below, a relationship between the standard normal random variable z and the project duration variable can be obtained:

$$z = \frac{T_d - T_e}{S}$$

where
T_d = specified deadline
T_e = expected project duration based on network analysis
S = standard deviation of the project duration.

The probability of completing a project by the deadline T_d is then computed as:

$$P(T \leq T_d) = P\left(z \leq \frac{T_d - T_e}{S}\right)$$

The probability is obtained from the standard normal table. Examples presented below illustrate the procedure for probability calculations in PERT.

1.4.3 PERT Network Example

Suppose we have the project data presented in Table 1.6. The expected activity durations and variances as calculated by the PERT formulas are shown in the last two columns of the table. Figure 1.23 shows the PERT network. Activities C, E, and G are shown to be critical, and the project completion time is 11 time units.

TABLE 1.6   Data for PERT Network Example

Activity	Predecessors	a	m	b	t_e	s^2
A	–	1	2	4	2.17	0.2500
B	–	5	6	7	6.00	0.1111
C	–	2	4	5	3.83	0.2500
D	A	1	3	4	2.83	0.2500
E	C	4	5	7	5.17	0.2500
F	A	3	4	5	4.00	0.1111
G	B,D,E	1	2	3	2.00	0.1111

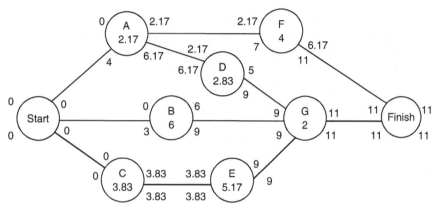

FIGURE 1.23   PERT network example.

The probability of completing the project on or before a deadline of 10 time units (i.e., $T_d = 10$) is calculated as shown below:

$$T_e = 11$$

$$S^2 = V[C] + V[E] + V[G]$$

$$= 0.25 + 0.25 + 0.1111$$

$$= 0.6111$$

$$S = \sqrt{0.6111}$$

$$= 0.7817$$

$$P(T \le T_d) = P(T \le 10)$$

$$= P\left(z \le \frac{10 - T_e}{S}\right)$$

$$= P\left(z \le \frac{10 - 11}{0.7817}\right)$$

$$= P(z \le -1.2793)$$

$$= 1 - P(z \le 1.2793)$$

$$= 1 - 0.8997$$

$$= 0.1003$$

Thus, there is just over 10% probability of finishing the project within 10 days. By contrast, the probability of finishing the project in 13 days is calculated as:

$$P(T \leq 13) = P\left(z \leq \frac{13 - 11}{0.7817}\right)$$

$$= P(z \leq 2.5585)$$

$$= 0.9948$$

This implies that there is over 99% probability of finishing the project within 13 days. Note that the probability of finishing the project in exactly 13 days will be zero. That is, $P(T = T_{\mathrm{d}}) = 0$. If we desire the probability that the project can be completed within a certain lower limit (T_{L}) and a certain upper limit (T_{U}), the computation will proceed as follows: Let $T_{\mathrm{L}} = 9$ and $T_{\mathrm{U}} = 11.5$. Then,

$$P(T_{\mathrm{L}} \leq T \leq T_{\mathrm{U}}) = P(9 \leq T \leq 11.5)$$

$$= P(T \leq 11.5) - P(T \leq 9)$$

$$= P\left(z \leq \frac{11.5 - 11}{0.7817}\right) - P\left(z \leq \frac{9 - 11}{0.7817}\right)$$

$$= P(z \leq 0.6396) - P(z \leq -2.5585)$$

$$= P(z \leq 0.6396) - [1 - P(z \leq 2.5585)]$$

$$= 0.7389 - [1 - 0.9948]$$

$$= 0.7389 - 0.0052$$

$$= 0.7337$$

That is, there is 73.4% chance of finishing the project within the specified range of duration.

1.5   Precedence Diagramming Method

The precedence diagramming method (PDM) was developed in the early 1960s as an extension of PERT/CPM network analysis. PDM permits mutually dependent activities to be performed partially in parallel instead of serially. The usual finish-to-start dependencies between activities are relaxed to allow activities to overlap. This facilitates schedule compression. An example is the requirement that concrete should be allowed to dry for a number of days before drilling holes for handrails. That is, drilling cannot start until so many days after the completion of concrete work. This is a finish-to-start constraint. The time between the finishing time of the first activity and the starting time of the second activity is called the lead–lag requirement between the two activities. Figure 1.24 shows the basic lead–lag relationships between activity A and activity B. The terminology presented in Figure 1.24 is explained as follows.

SS$_{\mathbf{AB}}$ (Start-to-Start) lead: Activity B cannot start until activity A has been in progress for at least SS time units.

FF$_{\mathbf{AB}}$ (Finish-to-Finish) lead: Activity B cannot finish until at least FF time units after the completion of activity A.

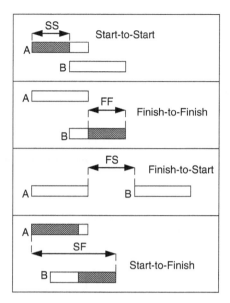

FIGURE 1.24 Lead–lag relationships in PDM.

FS$_{AB}$ (Finish-to-Start) lead: Activity B cannot start until at least FS time units after the completion of activity A. Note that PERT/CPM approaches use FS$_{AB}$ = 0 for network analysis.

SF$_{AB}$ (Start-to-Finish) lead: This specifies that there must be at least SF time units between the start of activity A and the completion of activity B.

The leads or lags may, alternately, be expressed in percentages rather than time units. For example, we may specify that 25% of the work content of activity A must be completed before activity B can start. If the percentage of work completed is used for determining lead–lag constraints, then a reliable procedure must be used for estimating the percent completion. If the project work is broken up properly using work breakdown structure (WBS), it will be much easier to estimate percent completion by evaluating the work completed at the elementary task level. The lead–lag relationships may also be specified in terms of at most relationships instead of at least relationships. For example, we may have at most FF lag requirement between the finishing time of one activity and the finishing time of another activity. Splitting of activities often simplifies the implementation of PDM, as will be shown later with some examples. Some of the factors that will determine whether or not an activity can be split are technical limitations affecting splitting of a task, morale of the person working on the split task, set-up times required to restart split tasks, difficulty involved in managing resources for split tasks, loss of consistency of work, and management policy about splitting jobs.

Figure 1.25 presents a simple CPM network consisting of three activities. The activities are to be performed serially and each has an expected duration of 10 days. The conventional CPM network analysis indicates that the duration of the network is 30 days. The earliest times and the latest times are as shown in the figure.

The Gantt chart for the example is shown in Figure 1.26. For a comparison, Figure 1.27 shows the same network but with some lead–lag constraints. For example, there is an SS constraint of 2 days and an FF constraint of 2 days between activities A and B. Thus, activity B can start as early as 2 days after activity A starts, but it cannot finish until 2 days after the completion of A. In other words, at least 2 days must separate the finishing

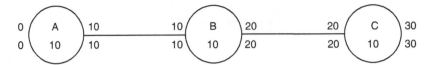

FIGURE 1.25   Serial activities in CPM network.

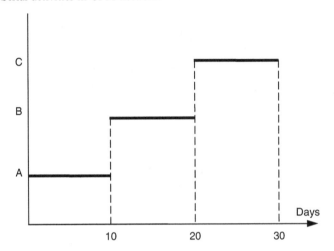

FIGURE 1.26   Gantt chart of serial activities in CPM example.

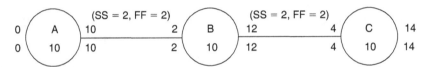

FIGURE 1.27   PDM network example.

time of A and the finishing time of B. A similar precedence relationship exists between activity B and activity C. The earliest and latest times obtained by considering the lag constraints are indicated in Figure 1.27.

The calculations show that if B is started just 2 days after A is started, it can be completed as early as 12 days as opposed to the 20 days obtained in the case of conventional CPM. Similarly, activity C is completed at time 14, which is considerably less than the 30 days calculated by conventional CPM. The lead–lag constraints allow us to compress or overlap activities. Depending on the nature of the tasks involved, an activity does not have to wait until its predecessor finishes before it can start. Figure 1.28 shows the Gantt chart for the example incorporating the lead–lag constraints. It should be noted that a portion of a succeeding activity can be performed simultaneously with a portion of the preceding activity.

A portion of an activity that overlaps with a portion of another activity may be viewed as a distinct portion of the required work. Thus, partial completion of an activity may be evaluated. Figure 1.29 shows how each of the three activities is partitioned into contiguous parts. Even though there is no physical break or termination of work in any activity, the distinct parts (beginning and ending) can still be identified. This means that there is no physical splitting of the work content of any activity. The distinct parts are determined on the basis of the amount of work that must be completed before or after another activity, as dictated by the lead–lag relationships. In Figure 1.29, activity A is partitioned into parts A_1 and A_2. The duration of A_1 is 2 days because there is an SS = 2 relationship between

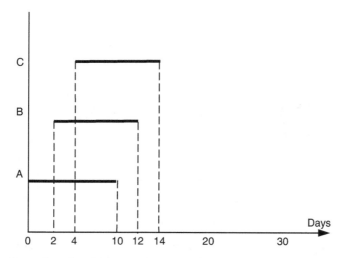

FIGURE 1.28   Gantt chart for PDM example.

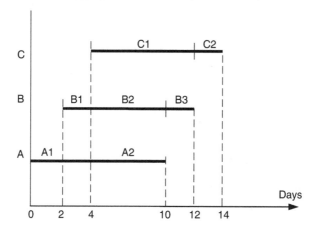

FIGURE 1.29   Partitioning of activities in PDM example.

activity A and activity B. As the original duration of A is 10 days, the duration of A_2 is then calculated to be $10 - 2 = 8$ days.

Likewise, activity B is partitioned into parts B_1, B_2, and B_3. The duration of B_1 is 2 days because there is an $SS = 2$ relationship between activity B and activity C. The duration of B_3 is also 2 days because there is an $FF = 2$ relationship between activity A and activity B. As the original duration of B is 10 days, the duration of B_2 is calculated to be $10 - (2 + 2) = 6$ days. In a similar fashion, activity c is partitioned into C_1 and C_2. The duration of C_2 is 2 days because there is an $FF = 2$ relationship between activity B and activity C. As the original duration of C is 10 days, the duration of C_1 is then calculated to be $10 - 2 = 8$ days. Figure 1.30 shows a conventional CPM network drawn for the three activities after they are partitioned into distinct parts. The conventional forward and backward passes reveal that all the activity parts are performed serially and no physical splitting of activities has been performed. Note that there are three critical paths in Figure 1.30, each with a length of 14 days. It should also be noted that the distinct parts of each activity are performed contiguously.

Figure 1.31 shows an alternate example of three serial activities. The conventional CPM analysis shows that the duration of the network is 30 days. When lead–lag constraints are

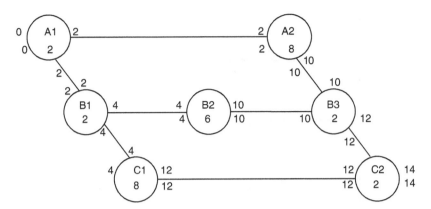

FIGURE 1.30   CPM network of partitioned activities.

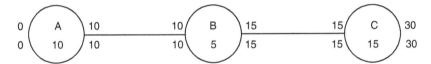

FIGURE 1.31   Another CPM example of serial activities.

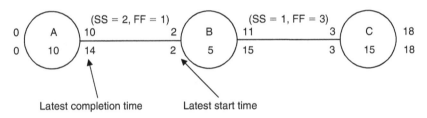

FIGURE 1.32   Compressed PDM network.

introduced into the network as shown in Figure 1.32, the network duration is compressed to 18 days.

In the forward pass computations in Figure 1.32, note that the earliest completion time of B is time 11, because there is an FF $= 1$ restriction between activity A and activity B. As A finishes at time 10, B cannot finish until at least time 11. Even though the earliest starting time of B is time 2 and its duration is 5 days, its earliest completion time cannot be earlier than time 11. Also note that C can start as early as time 3 because there is an SS $= 1$ relationship between B and C. Thus, given a duration of 15 days for C, the earliest completion time of the network is $3 + 15 = 18$ days. The difference between the earliest completion time of C and the earliest completion time of B is $18 - 11 = 7$ days, which satisfies the FF $= 3$ relationship between B and C.

In the backward pass, the latest completion time of B is 15 (i.e., $18 - 3 = 15$), as there is an FF $= 3$ relationship between activity B and activity C. The latest start time for B is time 2 (i.e., $3 - 1 = 2$), as there is an SS $= 1$ relationship between activity B and activity C. If we are not careful, we may erroneously set the latest start time of B to 10 (i.e., $15 - 5 = 10$). But that would violate the SS $= 1$ restriction between B and C. The latest completion time of A is found to be 14 (i.e., $15 - 1 = 14$), as there is an FF $= 1$ relationship between A and B. All the earliest times and latest times at each node must be evaluated to ensure that they conform to all the lead–lag constraints. When computing earliest start or earliest completion times, the smallest possible value that satisfies the lead–lag constraints should

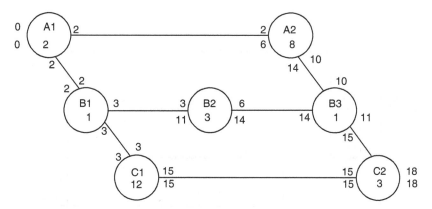

FIGURE 1.33 CPM expansion of second PDM example.

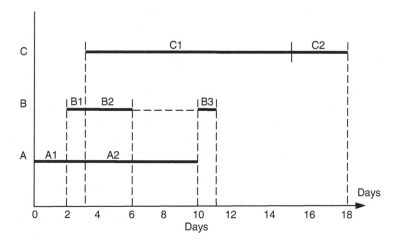

FIGURE 1.34 Compressed PDM schedule based on ES times.

be used. By the same reasoning, when computing the latest start or latest completion times, the largest possible value that satisfies the lead–lag constraints should be used.

Manual evaluations of the lead–lag precedence network analysis can become very tedious for large networks. A computer tool may be needed to implement PDM. If manual analysis must be done for PDM computations, it is suggested that the network be partitioned into more manageable segments. The segments may then be linked after the computations are completed. The expanded CPM network in Figure 1.33 was developed on the basis of the precedence network in Figure 1.32. It is seen that activity A is partitioned into two parts, activity B is partitioned into three parts, and activity C is partitioned into two parts. The forward and backward passes show that only the first parts of activities A and B are on the critical path, whereas both parts of activity C are on the critical path.

Figure 1.34 shows the corresponding earliest-start Gantt chart for the expanded network. Looking at the earliest start times, one can see that activity B is physically split at the boundary of B_2 and B_3 in such a way that B_3 is separated from B_2 by 4 days. This implies that work on activity B is temporarily stopped at time 6 after B_2 is finished and is not started again until time 10. Note that despite the 4-day delay in starting B_3, the entire project is not delayed. This is because B_3, the last part of activity B, is not on the critical path. In fact, B_3 has a total slack of 4 days. In a situation like this, the duration of activity

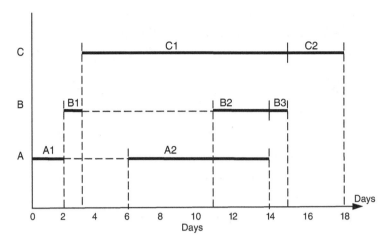

FIGURE 1.35 Compressed PDM schedule based on LS times.

B can actually be increased from 5 days to 9 days without any adverse effect on the project duration. It should be recognized, however, that increasing the duration of an activity may have negative implications for project cost, personnel productivity, and morale.

If physical splitting of activities is not permitted, then the best option available in Figure 1.34 is to stretch the duration of B_3 so as to fill up the gap from time 6 to time 10. An alternative is to delay the starting time of B_1 until time 4 so as to use up the 4-day slack right at the beginning of activity B. Unfortunately, delaying the starting time of B_1 by 4 days will delay the overall project by 4 days, as B_1 is on the critical path as shown in Figure 1.33. The project analyst will need to evaluate the appropriate tradeoffs between splitting of activities, delaying activities, increasing activity durations, and incurring higher project costs. The prevailing project scenario should be considered when making such trade-off decisions. Figure 1.35 shows the Gantt chart for the compressed PDM schedule based on latest start times. In this case, it will be necessary to split both activities A and B even though the total project duration remains the same at 18 days. If activity splitting is to be avoided, then we can increase the duration of activity A from 10 to 14 days and the duration of B from 5 to 13 days without adversely affecting the entire project duration. The benefit of precedence diagramming is that the ability to overlap activities facilitates flexibility in manipulating individual activity times and reducing project duration.

1.5.1 Reverse Criticality in PDM Networks

Care must be exercised when working with PDM networks because of the potential for misuse or misinterpretation. Because of the lead and lag requirements, activities that do not have any slacks may appear to have generous slacks. Also, "reverse critical" activities may occur in PDM. Reverse critical activities are activities that can cause a decrease in project duration when their durations are increased. This may happen when the critical path enters the completion of an activity through a finish lead–lag constraint. Also, if a "finish-to-finish" dependency and a "start-to-start" dependency are connected to a reverse critical task, a reduction in the duration of the task may actually lead to an increase in the project duration. Figure 1.36 illustrates this anomalous situation. The finish-to-finish constraint between A and B requires that B should finish no earlier than 20 days. If the duration of task B is reduced from 10 days to 5 days, the start-to-start constraint between

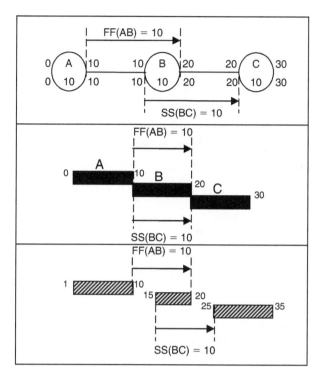

FIGURE 1.36 Reverse critical activity in PDM network.

B and C forces the starting time of C to be shifted forward by 5 days, thereby resulting in a 5-day increase in the project duration.

The preceding anomalies can occur without being noticed in large PDM networks. One safeguard against their adverse effects is to make only one activity change at a time and document the resulting effect on the network structure and duration. The following categorizations are used for the unusual characteristics of activities in PDM networks.

Normal Critical (NC): This refers to an activity for which the project duration shifts in the same direction as the shift in the duration of the activity.

Reverse Critical (RC): This refers to an activity for which the project duration shifts in the reverse direction to the shift in the duration of the activity.

Bi-Critical (BC): This refers to an activity for which the project duration increases as a result of any shift in the duration of the activity.

Start Critical (SC): This refers to an activity for which the project duration shifts in the direction of the shift in the start time of the activity, but is unaffected (neutral) by a shift in the overall duration of the activity.

Finish Critical (FC): This refers to an activity for which the project duration shifts in the direction of the shift in the finish time of the activity, but is unaffected (neutral) by a shift in the overall duration of the activity.

Mid Normal Critical (MNC): This refers to an activity whose mid-portion is normal critical.

Mid Reverse Critical (MRC): This refers to an activity whose mid-portion is reverse critical.

Mid Bi-Critical (MBC): This refers to an activity whose mid-portion is bi-critical.

A computer-based decision support system can facilitate the integration and consistent usage of all the relevant information in a complex scheduling environment. Task precedence relaxation assessment and resource-constrained heuristic scheduling constitute an example of a problem suitable for computer implementation. Examples of pseudocoded heuristic rules for computer implementation are shown below:

IF: *Logistical conditions are satisfied*
THEN: *Perform the selected scheduling action*

IF: condition A is satisfied and
            condition B is false and
            evidence C is present or
            observation D is available
THEN: precedence belongs in class X

IF: *precedence belongs to class X*
THEN: *activate heuristic scheduling procedure Y*

The function of the computer model will be to aid a decision maker in developing a task sequence that fits the needs of concurrent scheduling. Based on user input, the model will determine the type of task precedence, establish precedence relaxation strategy, implement task scheduling heuristic, match the schedule to resource availability, and present a recommended task sequence to the user. The user can perform "what-if" analysis by making changes in the input data and conducting sensitivity analysis. The computer implementation can be achieved in an interactive environment as shown in Figure 1.37. The user will provide task definitions and resource availabilities with appropriate precedence requirements. At each stage, the user is prompted to consider potential points for precedence relaxation. Wherever schedule options exist, they will be presented to the user for consideration and approval. The user will have the opportunity to make final decisions about task sequence.

1.6   Software Tools for Project Management

There are numerous commercial software packages available for project management. Because of the dynamic changes in software choices on the market, it will not be effective to include a survey of the available software in an archival publication of this nature. Any such review may be outdated before the book is even published. To get the latest on commercial software capabilities and choices, it will be necessary to consult one of the frequently published trade magazines that carry software reviews. Examples of such magazines are *PC Week*, *PC World*, and *Software*. Other professional publications also carry project management software review occasionally. Examples of such publications are *Industrial Engineering Magazine*, *OR/MS Today Magazine*, *Manufacturing Engineering Magazine*, *PM Network Magazine*, and so on. Practitioners can easily get the most current software information through the Internet. Prospective buyers are often overwhelmed by the range of products available. However, there are important factors to consider when selecting a software package. Some of the factors are presented in this section.

The proliferation of project management software has created an atmosphere whereby every project analyst wants to have and use a software package for every project situation. However, not every project situation deserves the use of project management software. An analyst should first determine whether or not the use of software is justified. If this

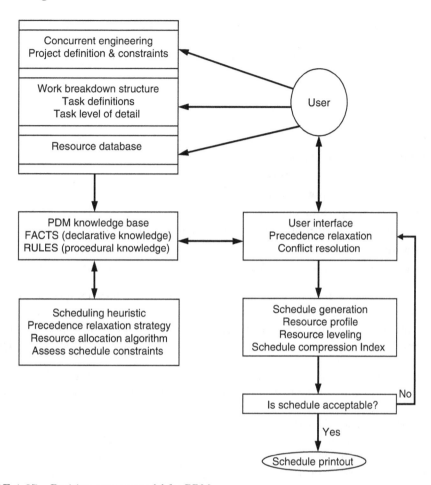

FIGURE 1.37 Decision support model for PDM.

evaluation is affirmative, then the analyst will need to determine which specific package out of the many that are available should be used. Some of the important factors that may indicate the need for project management software are:

1. Multiple projects are to be managed concurrently.
2. A typical project contains more than 20 tasks.
3. The project scheduling environment is very complex.
4. More than five resource types are involved in each project.
5. There is a need to perform complementing numerical analysis to support project control functions.
6. The generation of graphics (e.g., Gantt, PERT charts) are needed to facilitate project communication.
7. Cost analysis is to be performed on a frequent basis.
8. It is necessary to generate forecasts from historical project data.
9. Automated reporting is important to the organization.
10. Computerization is one of the goals of the organization.

A close examination of a project will reveal whether it fits the criteria for using project management software. Only very small and narrowly focused projects will not need the help of software for effective management. Some of the important factors for software selection are summarized below:

1. Cost
2. Need
3. Complexity of project scope and plan
4. Diversity of resource types
5. Frequency of progress tracking
6. Need for versatile report generation
7. Ease of use and learning curve requirement
8. Supporting analytical tools (e.g., optimization add-on, economic analysis)
9. Hardware requirements
10. General characteristics (e.g., software version, vendor accessibility, technical support).

Some of the above factors may be more important than others in specific project situations. A careful overall analysis should be done by the project analyst. With more and more new programs and updates appearing in the market, a crucial aspect of the project management function is keeping up with what is available and making a good judgment in selecting a software tool.

The project management software market continues to be very competitive. Many packages that were originally developed for specific and narrow applications, such as data analysis and conventional decision support, now offer project management capabilities. For example, SAS/OR, a statistical analysis software package that is popular on large computers, has a PC-based version that handles project management. Similarly, AutoCAD, the popular computer-aided design software, now has a project management option within it. The option, called AutoProject, has very good graphics and integrated drafting capabilities.

Some of the most popular project management software packages include InstaPlan 5000, Artemis Project, Microsoft Project for Windows, Plantrac II, Advanced Project Workbench, Qwiknet Professional, Super Project Plus, Time Line, Project Scheduler, Primavera, Texim, ViewPoint, PROMIS, Topdown Project Planner, Harvard Project Manager, PCS (Project Control System), PAC III, VISION, Control Project, SAS/OR, Autoproject, Visual Planner, Project/2, Timesheet, Task Monitor, Quick Schedule Plus, Suretrak Project Scheduler, Supertime, On Target, Great Gantt, Pro Tracs, Autoplan, AMS Time Machine, Mac-Project II, Micro Trak, Checkpoint, Maestro II, Cascade, OpenPlan, Vue, and Cosmos. So prolific are the software offerings that the developers are running out of innovative productive names. At this moment, new products are being introduced while some are being phased out. Prospective buyers of project management software should consult vendors for the latest products. No blanket software recommendation can be offered in this chapter as product profiles change quickly and frequently.

1.6.1 Computer Simulation Software

Special purpose software tools have found a place in project management. Some of these tools are simulation packages, statistical analysis packages, optimization programs, report writers, and others. Computer simulation is a versatile tool that has a potential for enhancing project planning and control analysis. Computer simulation is a tool that can be effectively utilized to enhance project planning, scheduling, and control. At any given time, only a small segment of a project network will be available for direct observation and analysis. The major portion of the project either will have been in the past or will be expected in the future. Such unobservable portions of the project can be studied by simulation.

Using the historical information from previous segments of a project and the prevailing events in the project environment, projections can be made about future expectations of the project. Outputs of simulation can alert management to real and potential problems. The information provided by simulation can be very helpful in projecting selection decisions. Simulation-based project analysis may involve the following components:

- Activity time modeling
- Simulation of project schedule
- What-if analysis and statistical modeling
- Management decisions and sensitivity analysis.

1.7 Conclusion

This chapter has presented the basic techniques of activity network analysis for project management. In business and industry, project management is rapidly becoming one of the major tools used to accomplish goals. Engineers, managers, and OR professionals are increasingly required to participate on teams in complex projects. The technique of network analysis is frequently utilized as a part of the quantitative assessment of such projects. The computational approaches contained in this chapter can aid project analysts in developing effective project schedules and determining the best way to exercise project control whenever needed.

Further Reading

1. Badiru, A.B., *Project Management in Manufacturing and High Technology Operations*, 2nd edition, John Wiley & Sons, New York, 1996.
2. Badiru, A.B. and P. S. Pulat, *Comprehensive Project Management: Integrating Optimization Models, Management Principles, and Computers*, Prentice-Hall, Englewood Cliffs, NJ, 1995.

2

Quality Control

Qianmei Feng
University of Houston

Kailash C. Kapur
University of Washington

2.1 Introduction

Quality has been defined in different ways by various experts and the operational definition has even changed over time. The best way is to start from the beginning and look at the origin and meaning of the word. Quality, in Latin *qualitas*, comes from the word *qualis*, meaning "how constituted" and signifying "such as a thing really is." The Merriam-Webster dictionary defines quality as "...peculiar and essential character...a distinguishing attribute...." Thus, a product has several or infinite qualities. Juran and Gryna [1] looked at multiple elements of fitness of use based on various quality characteristics (or qualities), such as technological characteristics (strength, dimensions, current, weight, ph values), psychological characteristics (beauty, taste, and many other sensory characteristics), time-oriented characteristics (reliability, availability, maintainability, safety, and security), cost (purchase price, life cycle cost), and product development cycle. Deming also discussed several faces of quality; the three corners of quality relate to various quality characteristics and focus on evaluation of quality from the viewpoint of the customer [2]. The American Society for Quality defines quality as the "totality of features and characteristics of a product or service that bear on its ability to satisfy a user's given needs" [3]. Thus the quality of a process or product is defined and evaluated by the customer. Any process has many processes before it that are typically called the suppliers and has many processes after it that are its customers. Thus, anything (the next process, environment, user) the present process affects is its customer.

Quality

⇓   ⇑

Customer satisfaction

⇓   ⇑

Voice of customer

⇓   ⇑

Substitute characteristics

⇓   ⇑

Target values

FIGURE 2.1   Quality, customer satisfaction, and target values.

One of the most important tasks in any quality program is to understand and evaluate the needs and expectations of the customer and to then provide products and services that meet or exceed those needs and expectations. Shewhart states this as follows: "The first step of the engineer in trying to satisfy these wants is, therefore, that of translating as nearly as possible these wants into the physical characteristics of the thing manufactured to satisfy these wants. In taking this step, intuition and judgment play an important role as well as the broad knowledge of human element involved in the wants of individuals. The second step of the engineer is to set up ways and means of obtaining a product which will differ from the arbitrary set standards for these quality characteristics by no more than may be left to chance" [4]. One of the objectives of quality function deployment (QFD) is exactly to achieve this first step proposed by Shewhart. Mizuno and Akao have developed the necessary philosophy, system, and methodology to achieve this step [5]. QFD is a means to translate the "voice of the customer" into substitute quality characteristics, design configurations, design parameters, and technological characteristics that can be deployed (horizontally) through the whole organization: marketing, product planning, design, engineering, purchasing, manufacturing, assembly, sales, and service [5,6]. Products have several characteristics, and an "ideal" state or value of these characteristics must be determined from the customer's viewpoint. This ideal state is called the target value (Figure 2.1). QFD is a methodology to develop target values for substitute quality characteristics that satisfy the requirements of the customer. The purpose of statistical process control is to accomplish the second step mentioned by Shewhart, and his pioneering book [4] developed the methodology for this purpose.

2.2   Quality Control and Product Life Cycle

Dynamic competition has become a key concept in world-class design and manufacturing. Competition is forcing us to provide products and services with less variation than our competitors. Quality effort in many organizations relies on a combination of audits, process and product inspections, and statistical methods using control charts and sampling inspection. These techniques are used to control the manufacturing process and meet specifications in the production environment. We must now concentrate on achieving high quality by starting at the beginning of the product life cycle. It is not enough for a product to work well when manufactured according to engineering specifications. A product must be designed for manufacturability and must be insensitive to variability present in the production environment and in the field when used by the customer. Reduced variation translates into greater

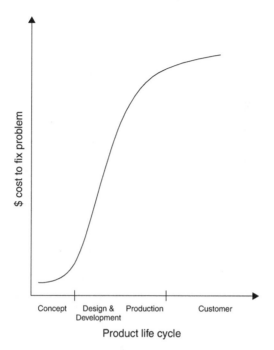

FIGURE 2.2 Cost to fix problems vs. product life cycle.

repeatability, reliability, and ultimately cost savings to both the producer and the consumer and thus the whole society.

Figure 2.2 shows how the cost to fix or solve problems increases as we move downstream in the product life cycle. Another way to emphasize the early and proactive activities related to quality is to evaluate Figure 2.3, which shows that approximately 90% of the life cycle cost is determined by the concept and development phases of the life cycle.

Different quality improvement methods should be applied for the different phases of the product life cycle. During the phase of design and development of products, design of experiments should be utilized to select the optimal combination of input components and minimize the variation of the output quality characteristic, as shown in Figure 2.4. Sometimes, we use offline quality engineering to refer to the quality improvement efforts including experiment design and robust parameter design, because these efforts are made off the production line.

As opposed to offline quality engineering, online quality control refers to the techniques employed during the manufacturing process of products. Statistical quality control (SQC) is a primary online control technique for monitoring the manufacturing process or any other process with key quality characteristics of interest. Before we use statistical quality control to monitor the manufacturing process, the optimal values of the mean and standard deviation of the quality characteristics are determined by minimizing the variability of the quality characteristics through experimental design and process adjustment techniques. Consequently, the major goal of SQC (SPC) is to monitor the manufacturing process, keep the values of mean and standard deviation stable, and finally reduce variability.

The manufactured products need to be inspected or tested before they reach the customer. Closely related to the inspection of output products, acceptance sampling is defined as the inspection and classification of samples from a lot randomly and decision about disposition of the lot.

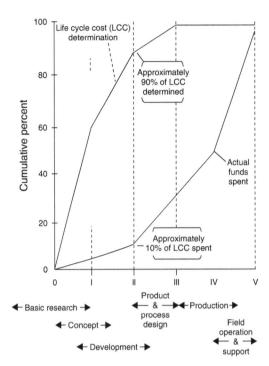

FIGURE 2.3 Life cycle cost and production life cycle.

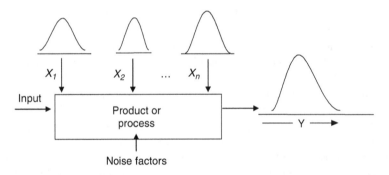

FIGURE 2.4 Offline quality engineering.

At different phases of the product life cycle, the modeling approaches in operations research and management science (OR/MS) are widely used. For example, as a collection of statistical and optimization methods, the response surface methodology (RSM) is a specialized experimental design technique in off-line quality engineering. The basic idea of RSM is to fit a response model for the output variable and then explore various settings of the input design variables with the purpose to maximize or minimize the response [7]. For online quality control and inspection efforts, various optimization modeling approaches are involved for process adjustment, economic design of control charts, and optimization of specifications, and so on [8,9].

Six Sigma methodology is one of the most distinctive applications of quality improvement effort with the involvement of OR/MS approaches. As a collection of optimization, statistical, engineering, and management methods, Six Sigma methodology emerged in manufacturing industry, and has been widely practiced in many industries including energy, health care, transportation, and financial services. The next section will describe the basic ideas of Six Sigma, and its relationship with the quality control effort.

2.3 New Trends and Relationship to Six Sigma

Industrial, manufacturing, and service organizations are interested in improving their processes by decreasing the process variation because the competitive environment leaves little room for error. Based on the ideal or target value of the quality characteristic from the viewpoint of the customer, the traditional evaluation of quality is based on average measures of the process/product and their deviation from the target value. But customers judge the quality of process/product not only on the average measure, but also by the variance in each transaction with the process or use of the product. Customers want consistent, reliable, and predictable processes that deliver best-in-class levels of quality. This is what the Six Sigma process strives to achieve. Six Sigma has been applied by many manufacturing companies such as GE, Motorola, and service industries including health care systems.

Six Sigma is a customer-focused, data-driven, and robust methodology, which is well rooted in mathematics and statistics [10–12]. A typical process for Six Sigma quality improvement has six phases: Define, Measure, Analyze, Improve, Control, and Technology Transfer, denoted by (D)MAIC(T). Traditionally, a four-phase process, MAIC is often referred in the literature [13]. We extend it to the six-phase process, (D)MAIC(T). We want to emphasize the importance of the define (D) phase as the first phase for problem definition and project selection and technology transfer (T) as the never-ending phase for continuous applications of the Six Sigma technology to other parts of the organization to maximize the rate of return on the investment in developing this technology.

The process of (D)MAIC(T) stays on track by establishing deliverables at each phase, and by creating engineering models over time to reduce the process variation. Each of the six phases answers some target questions, and this continuously improves the implementation and the effectiveness of the methodology [8].

Define—What problem needs to be solved?

Measure—What is the current capability of the process?

Analyze—What are the root causes for the process variability?

Improve—How to improve the process capability?

Control—What controls can be put in place to sustain the improvement?

Technology Transfer—Where else can these improvements be applied?

In each phase, there are several steps that need to be implemented. For each step, many quality improvement methods, tools, and techniques are used. We next describe the six phases in more detail.

Phase 0: Define (D)

Once an organization decides to launch a Six Sigma process improvement project, they need to first define the improvement activities. Usually the following two steps are taken in the define phase:

Step 0.1: *Identify and prioritize customer requirements.* Methods such as benchmarking surveys, spider charts, and customer needs mapping must be put in place to ensure that the customer requirements are properly identified. The critical to quality (CTQ) characteristics are defined from the viewpoint of customers, which are also called external CTQs. We need to translate the external CTQs into internal CTQs that are key process requirements. This translation is the foremost step in the measure phase.

Step 0.2: *Select projects.* Based on customer requirements, the target project is selected by analyzing the gap between the current process performance and the requirement of customers. Specifically, we need to develop a charter for the project, including project scope, expectations, resources, milestones, and the core processes.

Phase 1: Measure (M)

Six Sigma is a data-driven approach that requires quantifying and benchmarking the process using actual data. In this phase, the performance or process capability of the process for the CTQ characteristics are evaluated.

Step 1.1: *Select CTQ characteristics.* This step uses tools such as QFD and FMECA to translate the external CTQs established in the define phase into internal requirements denoted by Y's. Some of the objectives for this step are:

- Define, construct, and interpret the QFDs.
- Participate in a customer needs mapping session.
- Apply failure mode, effect, and criticality analysis (FMECA) to the process of selecting CTQ characteristics.
- Identify CTQs and internal Y's.

Step 1.2: *Define performance standards.* After identifying the product requirements, Y's, measurement standards for the Y's are defined in this step. QFD, FMECA, as well as process mapping can be used to establish internal measurement standards.

Step 1.3: *Validate measurement system.* We need to learn how to validate measurement systems and determine the repeatability and reproducibility of these systems using tools such as gage R&R. This determination provides for separation of variability into components and thus into targeted improvement actions.

Phase 2: Analyze (A)

Once the project is understood and the baseline performance is documented, it is time to do an analysis of the process. In this phase, the Six Sigma approach applies statistical tools to validate the root causes of problems. The objective is to understand the process in sufficient detail so that we are able to formulate options for improvement.

Step 2.1: *Establish product capability.* This step determines the current product capability, associated confidence levels, and sample size by process capability analysis, described in Section 2.5. The typical definition for process capability index, C_{pk}, is $C_{\text{pk}} = \min\left\{\frac{USL - \hat{\mu}}{3\hat{\sigma}}, \frac{\hat{\mu} - LSL}{3\hat{\sigma}}\right\}$, where USL is the upper specification limit, LSL is the lower specification limit, $\hat{\mu}$ is the point estimator of the mean, and $\hat{\sigma}$ is the point estimator of the standard deviation. If the process is centered at the middle of the specifications, which is also interpreted as the target value, that is, $\hat{\mu} = \frac{USL + LSL}{2} = y_0$, then 6σ process means that $C_{pk} = 2$.

Step 2.2: *Define performance objectives.* The performance objectives are defined to establish a balance between improving customer satisfaction and available resources. We should distinguish between courses of actions necessary to improve process capability versus technology capability.

Step 2.3: *Identify variation sources.* This step begins to identify the causal variables that affect the product requirements, or the responses of the process. Some of these causal variables might be used to control the responses Y's. Experimental design and analysis should be applied for the identification of variation sources.

Phase 3: Improve (I)

In the improvement phase, ideas and solutions are implemented to initialize the change. Experiments are designed and analyzed to find the best solution using optimization approaches.

Step 3.1: *Discover variable relationships.* In the previous step, the causal variables X's are identified with a possible prioritization as to their importance in controlling the Y's. In this step, we explore the impact of each vital X on the responses Y's. A system transfer function (STF) is developed as an empirical model relating the Y's and the vital X's.

Step 3.2: *Establish operating tolerances.* After understanding the functional relationship between the vital X's and the responses Y's, we need to establish the operating tolerances of the X's that optimize the performance of the Y's. Mathematically, we develop a variance transmission equation (VTE) that transfers the variances of the vital X's to variances of Y's.

Step 3.3: *Optimize variable settings.* The STF and VTE will be used to determine the key operating parameters and tolerances to achieve the desired performance of the Y's. Optimization models are developed to determine the optimum values for both means and variances for these vital X's.

Phase 4: Control (C)

The key to the overall success of the Six Sigma methodology is its sustainability. Performance tracking mechanisms and measurements are put in place to assure that the process remains on the new course.

Step 4.1: *Validate measurement system.* The measurement system tools first applied in Step 1.3 will now be used for the X's.

Step 4.2: *Implement process controls.* Statistical process control is a critical element in maintaining a Six Sigma level. Control charting is the major tool used to control the vital few X's. Special causes of process variations are identified through the use of control charts, and corrective actions are implemented to reduce variations.

Step 4.3: *Documentation of the improvement.* We should understand that the project is not complete until the changes are documented in the appropriate quality management system, such as QS9000/ISO9000. A translation package and plan should be developed for possible technology transfer.

Phase ∞: Technology Transfer (T)

Using the infinity number, we convey the meaning that transferring technology is a never-ending phase for achieving Six Sigma quality. Ideas and knowledge developed in one part of the organization can be transferred to other parts of the organization. In addition, the methods and solutions developed for one product or process can be applied to other similar products or processes. With technology transfer, the Six Sigma approach starts to create phenomenal returns.

2.4   Statistical Process Control

A traditional approach to manufacturing and addressing quality is to depend on production to make the product and on quality control to inspect the final product and screen out the items that do not meet the requirements of the next customer. This detection strategy related to after-the-fact inspection is mostly uneconomical, as the wasteful production has

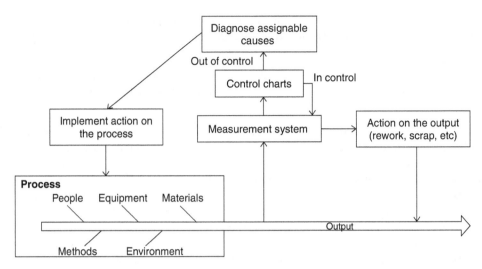

FIGURE 2.5 A process control system.

already been produced. A better strategy is to avoid waste by not producing the unacceptable output in the first place and thus focus more on prevention. Statistical process control (SPC) is an effective prevention strategy to manufacture products that will meet the requirements of the customer [14]. We will discuss the following topics in this section.

1. What is a process control system?
2. What are different types of variation and how do they affect the process output? We will discuss two types of variation based on common or system causes and special or assignable causes.
3. How can a control chart or statistical methods tell us whether a problem is due to special causes or common causes?
4. What is meant by a process being in statistical control?
5. What are control charts and how are they used?
6. What benefits can be expected from using control charts?

2.4.1   Process Control System

A process control system can be described as a feedback system as shown in Figure 2.5. Four elements of that system are important to the discussions that will follow:

1. The process—by the process, we mean the whole combination of people, equipment, input materials, methods, and environment that work together to produce output. The total performance of the process—the quality of its output and its productive efficiency—depends on the way the process has been designed and built, and on the way it is operated. The rest of the process control system is useful only if it contributes to improved performance of the process.
2. Information about performance—much information about the actual performance of the process can be learned by studying the process output. In a broad sense, process output includes not only the products that are produced, but also any intermediate "outputs" that describe the operating state of the process, such as

temperatures, cycle times, and the like. If this information is gathered and interpreted correctly, it can show whether action is necessary to correct the process or the just-produced product. If timely and appropriate actions are not taken, however, any information-gathering effort is wasted.

3. Action on the process—action on the process is future-oriented, as it is taken when necessary to prevent the production of nonconforming products. This action might consist of changes in the operations (e.g., operator training, changes to the incoming materials, etc.) or the more basic elements of the process itself (e.g., the equipment—which may need rehabilitation, or the design of the process as a whole—which may be vulnerable to changes in shop temperature or humidity).

4. Action on the output—action on the output is past-oriented, because it involves detecting out-of-specification output already produced. Unfortunately, if current output does not consistently meet customer requirements, it may be necessary to sort all products and to scrap or rework any nonconforming items. This must continue until the necessary corrective action on the process has been taken and verified, or until the product specifications have been changed.

It is obvious that inspection followed by action only on the output is a poor substitute for effective first-time process performance. Therefore, the discussions that follow focus on gathering process information and analyzing it so that action can be taken to correct the process itself.

Process control plays a very important role during the effort for process improvement. When we try to control a process, analysis and improvement naturally result; and when we try to make an improvement, we naturally come to understand the importance of control. We can only make a breakthrough when we have achieved control. Without process control, we do not know where to improve, and we cannot have standards and use control charts. Improvement can only be achieved through process analysis.

2.4.2 Sources of Variation

Usually, the sources of variability in a process are classified into two types: chance causes and assignable causes of variation. Chance causes, or common causes, are the sources of inherent variability, which cannot be removed easily from the process without fundamental changes in the process itself. Assignable causes, or special causes, arise in somewhat unpredictable fashion, such as operator error, material defects, or machine failure. The variability due to assignable causes is comparatively larger than chance causes, and can cause the process to go out of control. Table 2.1 compares the two sources of variation, including some examples.

TABLE 2.1   Sources of Variation

Common or Chance Causes	Special or Assignable Causes
1. Consist of many individual causes	1. Consist of one or just a few individual causes
2. Any one chance cause results in only a minute amount of variation. (However, many of chance causes together result in a substantial amount of variation)	2. Any one assignable cause can result in a large amount of variation
3. As a practical matter, chance variation cannot be economically eliminated—the process may have to be changed to reduce variability	3. The presence of assignable variation can be detected (by control charts) and action to eliminate the causes is usually economically justified
4. Examples: • Slight variations in raw materials • Slight vibrations of a machine • Lack of human perfection in reading instruments or setting controls	4. Examples: • Batch of defective raw materials • Faulty setup • Untrained operator

2.4.3   Use of Control Charts for Problem Identification

Control charts by themselves do not correct problems. They indicate that something is wrong and it is up to you to take the corrective action. Assignable causes are the factors that cause the process to go out of control. They are due to change in the condition of manpower, materials, machines, or methods or a combination of all of these.

Assignable causes relating to manpower:

- New or wrong man on the job
- Careless workmanship and attitudes
- Improper instructions
- Domestic, personal problems

Assignable causes relating to materials:

- Improper work handling
- Stock too hard or too soft
- Wrong dimensions
- Contamination, dirt, etc.
- Improper flow of materials

Assignable causes relating to machines or methods:

- Dull tools
- Poor housekeeping
- Machine adjustment
- Improper machine tools, jigs, fixtures
- Improper speeds, feeds, etc.
- Improper adjustments and maintenance
- Worn or improperly located locators

When assignable causes are present, as shown in Figure 2.6, the probability of nonconformance may increase, and the process quality deteriorates significantly. The eventual goal of SPC is to improve the process quality by reducing variability in the process. As one of the primary SPC techniques, the control chart can effectively detect the variation due to the assignable causes and reduce process variability if the identified assignable causes can be eliminated from the process.

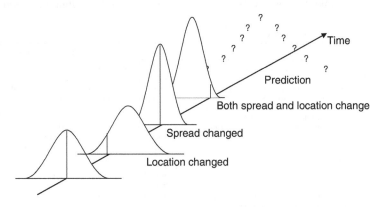

FIGURE 2.6   Unstable and unpredictable process with presence of assignable causes.

2.4.4 Statistical Control

Statistically, SPC techniques aim to detect changes over time in the parameters (e.g., mean and standard deviation) of the underlying distribution for the process. In general, the statistical process control problem can be described as below [15]. Let X denote a random variable for a quality characteristic with the probability density function, $f(x; \theta)$, where θ is a set of parameters. If the process is operating with $\theta = \theta_0$, it is said to be in statistical control; otherwise, it is out of control. The value of θ_0 is not necessarily equal to the target (or ideal) value of the process. Due to the effort of experimental design and process adjustment techniques, a process is assumed to start with the in-control state [7,16,17]. After a random length of time, variability in the process will possibly cause deterioration or shift of the process. This shift can be reflected by a change in θ from the value of θ_0, and the process is said to be out of control. Therefore, the basic goal of control charts is to detect changes in θ that can occur over time.

A process is said to be operating in statistical control when the only source of variation is common causes. The status of statistical control is obtained by eliminating special causes of excessive variation one by one.

Process capability is determined by the total variation that comes from common causes. A process must first be brought into statistical control, and then its capability to meet specifications can be assessed. We will discuss the details of process capability analysis in the next section.

2.4.5 Control Charts

The basic concept of control charts was proposed by Walter A. Shewhart of the Bell Telephone Laboratories in the 1920s, which indicates the formal beginning of statistical quality control. The effective use of the control chart involves a series of process improvement activities. For a process variable of interest, one must observe data from a process over time, or monitor the process, and apply a control chart to detect process changes. When the control chart signals the possible presence of an assignable cause, efforts should be made to diagnose the assignable causes and implement corrective actions to remove the assignable causes so as to reduce variability and improve the process quality. The long history of control charting application in many industries has proven its effectiveness for improving productivity, preventing defects and providing information about diagnostic and process capability.

A typical control chart is given in Figure 2.7. The basic model for Shewhart control charts consist of a center line, an upper control limit (UCL) and a lower control limit (LCL) [18].

$$\text{UCL} = \mu_s + L\sigma_s$$
$$\text{Center line} = \mu_s \qquad (2.1)$$
$$\text{LCL} = \mu_s - L\sigma_s$$

where μ_s and σ_s are the mean and standard deviation of the sample statistic, such as sample mean (X-bar chart), sample range (R chart), and sample proportion defective (p chart). $L\sigma_s$ is the distance of the control limits from the center line and L is most often set at three. To construct a control chart, one also needs to specify the sample size and sampling frequency. The common wisdom is to take smaller samples at short intervals or larger samples at longer intervals, so that the sampling effort can be allocated economically. An important concept related to the sampling scheme is the rational subgroup approach, recommended by Shewhart. To maximize detection of assignable causes between samples, the rational subgroup approach takes samples in a way that the within-sample variability is only due to common causes, while the between-sample variability should indicate assignable causes in

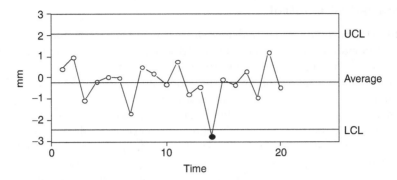

FIGURE 2.7 A typical control chart (x-bar chart).

TABLE 2.2 The Most Commonly Used Shewhart Control Charts

Symbol	Description	Sample Size
Variable Charts		
X-bar and R	The average (mean) and range of measurements in a sample	Must be constant
X-bar and S	The average (mean) and standard deviation of measurements in a sample	May be variable
Attributes Charts		
p	The percent of defective (nonconforming) units in a sample	May be variable
np	The number of defective (nonconforming) units in a sample	Must be constant
c	The number of defects in a sample	Must be constant
u	The number of defects per unit	May be variable

the process. Further discussion of the rational subgroup approach can be found in Refs. [19] and [20].

An out-of-control signal is given when a sample statistic falls beyond the control limits, or a nonrandom pattern presents. Western Electric rules are used to identify the nonrandom pattern in the process. According to Western Electric rules [21], a process is considered out of control if any of the following occur:

1. One or more points outside 3σ limits.
2. Two of three consecutive points outside 2σ limits.
3. Four of five consecutive points beyond the 1σ limits.
4. A run of eight consecutive points on one side of the center line.

More decision rules or sensitizing rules can be found in the textbook by Montgomery.

The measurements of quality characteristics are typically classified as attributes or variables. Continuous measurements, such as length, thickness, or voltage, are variable data. Discrete measurements, such as number of defective units, number of nonconformities per unit, are attributes. The most commonly used Shewhart control charts for both attributes and variables are summarized in Table 2.2.

Control Charts for Variables

When a quality characteristic is measured as variables, it is necessary to monitor the process mean and standard deviation. For grouped data, we use $\overline{X}$ charts to detect the process mean shift (between-group variability), and R charts or S charts to monitor the process variation (within-group variability). The control limits of each chart are constructed based on the Shewhart model in Equation 2.1. When we use $\overline{X}$, R, and S charts, we assume that the underlying distribution of the quality characteristic is normal and the observations exhibit no correlation over time. If the quality characteristic is extremely nonnormal or the observations are autocorrelated, other control charts such as the exponentially weighted moving average control charts (EWMA) or time series model (ARIMA) may be used instead.

In practice, the parameters of the underlying distribution of a quality characteristic are not known. We need to estimate the process mean and standard deviation based on the preliminary data. It can be shown that an unbiased estimate of the standard deviation is $\hat{\sigma} = s/c_4$, where s is the sample standard deviation. A more convenient approach in quality control application is the range method, where the range of the sample, R, is used to estimate the standard deviation, and it is obtained as $\hat{\sigma} = R/d_2$. The resulting control charts using different estimators of standard deviation are the R chart and the S chart, respectively.

$\overline{X}$ and R Charts

When the sample size is not very large ($n < 10$), $\overline{X}$ and R charts are widely used to monitor variable quality characteristics due to their simplicity of application. To use the basic Shewhart model in Equation 2.1 for $\overline{X}$ and R charts, we need to estimate $\mu_{\overline{x}}$ and $\sigma_{\overline{x}}$, μ_R, and σ_R first.

It is obvious that we can use the grand average to estimate $\mu_{\overline{x}}$ and μ_R, that is, $\hat{\mu}_{\overline{x}} = \overline{\overline{x}}$ and $\hat{\mu}_R = \overline{R}$. Using the range method, we have $\hat{\sigma}_{\overline{x}} = \hat{\sigma}/\sqrt{n} = \overline{R}/(d_2\sqrt{n})$ and $\hat{\sigma}_R = d_3\hat{\sigma} = d_3\overline{R}/d_2$. The control limits for $\overline{X}$ and R charts are

$$\text{LCL} = \overline{\overline{x}} - A_2\overline{R} \qquad \text{LCL} = D_3\overline{R}$$
$$\text{CL} = \overline{\overline{x}} \qquad \text{and} \qquad \text{CL} = \overline{R}$$
$$\text{ULC} = \overline{\overline{x}} + A_2\overline{R} \qquad \text{UCL} = D_4\overline{R}$$

respectively, where $A_2 = \frac{3}{d_2\sqrt{n}}$, $D_3 = 1 - \frac{3d_3}{d_2}$, and $D_4 = 1 + \frac{3d_3}{d_2}$. The values of d_2, d_3, A_2, D_3, and D_4 can be obtained from most books on control charts for n up to 25 [22,20]. Normally, the preliminary data used to establish the control limits are 20–25 samples with sample size 3–5. The established control limits are then used to check if the preliminary samples are in control. The R chart (or S chart) should be checked first to ensure the process variability is in statistical control, and then the $\overline{X}$ chart is checked for the process mean shift. Once a set of reliable control limits is constructed, they can be used for process monitoring.

$\overline{X}$ and S Charts

When the sample size is relatively large ($n > 10$), or the sample size is variable, the $\overline{X}$ and S charts are preferred to $\overline{X}$ and R charts. To construct the control limits, we need to estimate the mean and standard deviation of $\overline{X}$ and S, that is, $\mu_{\overline{x}}$ and $\sigma_{\overline{x}}$, μ_S and σ_S first. We have $\hat{\mu}_{\overline{x}} = \overline{\overline{x}}$ and $\hat{\mu}_S = \overline{S}$. Using $\hat{\sigma} = s/c_4$, we have $\hat{\sigma}_{\overline{x}} = \hat{\sigma}/\sqrt{n} = \overline{s}/(c_4\sqrt{n})$ and $\hat{\sigma}_s = \overline{s}\sqrt{1 - c_4^2}/c_4$. Therefore, the control limits for $\overline{X}$ and S charts are

$$\text{LCL} = \overline{\overline{x}} - A_3\overline{S} \qquad \text{LCL} = B_3\overline{S}$$
$$\text{CL} = \overline{\overline{x}} \qquad \text{and} \qquad \text{CL} = \overline{S}$$
$$\text{ULC} = \overline{\overline{x}} + A_3\overline{S} \qquad \text{UCL} = B_4\overline{S}$$

respectively, where $A_3 = \frac{3}{c_4\sqrt{n}}$, $B_3 = 1 - \frac{3}{c_4}\sqrt{1 - c_4^2}$, and $B_4 = 1 + \frac{3}{c_4}\sqrt{1 - c_4^2}$. The values of c_4, A_3, B_3, and B_4 can be obtained from most books on control charts for n up to 25.

Control Charts for Attributes

When quality characteristics are expressed as attribute data, such as defective or conforming items, control charts for attributes are established. Attribute charts can handle multiple quality characteristics jointly because the unit is classified as defective if it fails to meet the specification on one or more characteristics. The inspection of samples for attribute charts is usually cheaper due to less precision requirement. Attribute charts are particularly useful in quality improvement efforts where numerical data are not easily obtained, such as service, industrial, and health care systems. In the context of quality control, the attribute data include a proportion of defective items and a number of defects on items. A defective unit may have one or more defects that are a result of nonconformance to standard on one or more quality characteristics. Nevertheless, a unit with several defects may not necessarily be classified as a defective unit. It requires two different types of attribute charts: control charts for proportion defective (p chart and np chart), and control charts for number of defects (c chart and u chart).

p Chart and np Chart

The proportion that is defective is defined as the ratio of the number of defective units to the total number of units in a population. We usually assume that the number of defective units in a sample is a binomial variable; that is, each unit in the sample is produced independently and the probability that a unit is defective is constant, p. Using preliminary samples, we can estimate the defective rate, that is, $\bar{p} = \sum_{i=1}^{m} D_i/mn$, where D_i is the number of defective units in sample i, n is the sample size, and m is the number of samples taken. The formulas used to calculate control limits are then given as

$$\mathrm{UCL}_{\hat{p}} = \bar{p} + 3\sqrt{\frac{\bar{p}(1-\bar{p})}{n}}$$

$$\mathrm{Centerline} = \bar{p}$$

$$\mathrm{LCL}_{\hat{p}} = \bar{p} - 3\sqrt{\frac{\bar{p}(1-\bar{p})}{n}}.$$

Sometimes, it may be easier to interpret the number that is defective instead of the proportion that is defective. That is why the np chart came into use:

$$\mathrm{UCL} = n\bar{p} + 3\sqrt{n\bar{p}(1-\bar{p})}$$

$$\mathrm{Centerline} = n\bar{p}$$

$$\mathrm{LCL} = n\bar{p} - 3\sqrt{n\bar{p}(1-\bar{p})}.$$

The developed trial control limits are then used to check if the preliminary data are in statistical control, and the assignable causes may be identified and removed if a point is out of control. As the process improves, we expect a downward trend in the p or np control chart.

c Chart and u Chart

Control charts for monitoring the number of defects per sample are constructed based on Poisson distribution. With this assumption of reference distribution, the probability of occurrence of a defect at any area is small and constant, the potential area for defects is infinitely large, and defects occurs randomly and independently. If the average occurrence rate per sample is a constant c, we know that both the mean and variance of the Poisson distribution are the constant c. Therefore, the parameters in the c chart for the number of defects are

$$\text{LCL} = c - 3\sqrt{c}$$

$$\text{CL} = c$$

$$\text{UCL} = c + 3\sqrt{c}$$

where c can be estimated by the average number of defects in a preliminary sample. To satisfy the assumption of constant rate of occurrence, the sample size is required to be constant.

For variable sample size, the u chart should be used instead of the c chart. Compared to the c chart that is used to monitor the number of defects per sample, the u chart is designed to check the average number of defects per inspection unit. Usually, a sample may contain one or more inspection units. For example, in a textile finishing plant, dyed cloth is inspected for defects per $50 \, \text{m}^2$, which is one inspection unit. A roll of cloth of $500 \, \text{m}^2$ is one sample with 10 inspection units. Different rolls of cloth may have various areas, hence variable sample sizes. As a result, it is not appropriate to use the c chart, because the occurrence rate of defects in each sample is not a constant. The alternative is to monitor the average number of defects per inspection unit in a sample, $u_i = c_i / n_i$. In this way, the parameters in the u chart are given as

$$\text{LCL} = \overline{u} - 3\sqrt{\frac{\overline{u}}{n}}$$

$$\text{CL} = \overline{u}$$

$$\text{UCL} = \overline{u} + 3\sqrt{\frac{\overline{u}}{n}}$$

where $\overline{u} = \sum_{i=1}^{m} u_i / m$, is an estimation of the average number of defects in an inspection unit. For variable sample size, the upper and lower control limits vary for different n.

To effectively detect small process shifts (on the order of 1.5σ or less), the cumulative sum (CUSUM) control chart and the exponentially weighted moving average (EWMA) control chart may be used instead of Shewhart control charts. In addition, there are many situations where we need to simultaneously monitor two or more correlated quality characteristics. The control charts for multivariate quality characteristics will also be discussed in the following.

2.4.6 Benefits of Control Charts

In this section, we summarize some of the important benefits that can come from using control charts.

- Control charts are simple and effective tools to achieve statistical control. They lend themselves to being maintained at the job station by the operator. They give the people closest to the operation reliable information on when action should be taken—and on when action should not be taken.
- When a process is in statistical control, its performance to specification will be predictable. Thus, both producer and customer can rely on consistent quality levels, and both can rely on stable costs of achieving that quality level.

- After a process is in statistical control, its performance can be further improved to reduce variation. The expected effects of proposed improvements in the system can be anticipated, and the actual effects of even relatively subtle changes can be identified through the control chart data. Such process improvements will:
  - Increase the percentage of output that meets customer expectations (improve quality),
  - Decrease the output requiring scrap or rework (improve cost per good unit produced), and
  - Increase the total yield of acceptable output through the process (improve effective capacity).
- Control charts provide a common language for communications about the performance of a process—between the two or three shifts that operate a process; between line production (operator, supervisor) and support activities (maintenance, material control, process engineering, quality control); between different stations in the process; between supplier and user; between the manufacturing/ assembly plant and the design engineering activity.
- Control charts, by distinguishing special from common causes of variation, give a good indication of whether any problems are likely to be correctable locally or will require management action. This minimizes the confusion, frustration, and excessive cost of misdirected problem-solving efforts.

2.5 Process Capability Studies

As discussed earlier, statistical control of a process is arrived at by eliminating special causes of excessive variation one by one. Process capability is determined by the total variation that comes from common causes. Therefore, a process must first be brought into statistical control and then its capability to meet specifications can be assessed.

The process capability to meet specifications is usually measured by process capability indices that link process parameters to product design specifications. Using a single number, process capability indices measure the degree to which the stable process can meet the specifications [23,24]. If we denote the lower specification limit as LSL and the upper specification limit as USL, the process capability index C_p is defined as:

$$C_p = \frac{\text{USL} - \text{LSL}}{6\sigma}$$

which measures the potential process capability. To measure the actual process capability, we use C_{pk} that is defined as:

$$C_{pk} = \min\left(\frac{\text{USL} - \mu}{3\sigma}, \frac{\mu - \text{LSL}}{3\sigma}\right)$$

The measure of C_{pk} takes the process centering into account by choosing the one side C_p for the specification limit closest to the process mean. The estimations of C_p and C_{pk} are obtained by replacing μ and σ using the estimates $\hat{\mu}$ and $\hat{\sigma}$. To consider the variability in terms of both standard deviation and mean, another process capability index C_{pm} is defined as

$$\hat{C}_{pm} = \frac{\text{USL} - \text{LSL}}{6\hat{\tau}}$$

where $\hat{\tau}$ is an estimator of the expected square deviation from the target, T, and is given by

$$\tau^2 = E\left[(x - T)^2\right] = \sigma^2 + (\mu - T)^2$$

Therefore, if we know the estimate of C_p, we can estimate C_{pm} as:

$$\hat{C}_{pm} = \frac{\hat{C}_p}{\sqrt{1 + \left(\dfrac{\hat{\mu} - T}{\hat{\sigma}}\right)^2}}$$

In addition to process capability indices, capability can also be described in terms of the distance of the process mean from the specification limits in standard deviation units, Z, that is

$$Z_U = \frac{\text{USL} - \hat{\mu}}{\hat{\sigma}}, \quad \text{and} \quad Z_L = \frac{\hat{\mu} - \text{LSL}}{\hat{\sigma}}$$

Z values can be used with a table of standard normal distribution to estimate the proportion of process fallout for a normally distributed and statistically controlled process. The Z value can also be converted to the capability index, C_{pk}:

$$C_{pk} = \frac{Z_{\min}}{3} = \frac{1}{3} \min\left(Z_U, Z_L\right)$$

A process with $Z_{\min} = 3$, which could be described as having $\hat{\mu} \pm 3\hat{\sigma}$ capability, would have $C_{pk} = 1.00$. If $Z_{\min} = 4$, the process would have $\hat{\mu} \pm 4\hat{\sigma}$ capability and $C_{pk} = 1.33$.

Example 2.1

For a process with $\hat{\mu} = 0.738$, $\hat{\sigma} = 0.0725$, USL $= 0.9$, and LSL $= 0.5$,

- Since the process has two-sided specification limits,

$$Z_{\min} = \min\left(\frac{\text{USL} - \hat{\mu}}{\hat{\sigma}}, \frac{\hat{\mu} - \text{LSL}}{\hat{\sigma}}\right)$$

$$= \min\left(\frac{0.9 - 0.738}{0.0725}, \frac{0.738 - 0.5}{0.0725}\right) = \min(2.23, 3.28) = 2.23$$

and the proportion of process fallout would be:

$$p = 1 - \Phi(2.23) + \Phi(-3.28) = 0.0129 + 0.0005 = 0.0134$$

The process capability index would be:

$$C_{pk} = \frac{Z_{\min}}{3} = 0.74$$

- If the process could be adjusted toward the center of the specification, the proportion of process fallout might be reduced, even with no change in σ:

$$Z_{\min} = \min\left(\frac{\text{USL} - \hat{\mu}}{\hat{\sigma}}, \frac{\hat{\mu} - \text{LSL}}{\hat{\sigma}}\right) = \min\left(\frac{0.9 - 0.7}{0.0725}, \frac{0.7 - 0.5}{0.0725}\right) = 2.76$$

and the proportion of process fallout would be:

$$p = 2\Phi(-2.76) = 0.0058$$

The process capability index would be:

$$C_{pk} = \frac{Z_{\min}}{3} = 0.92$$

- To improve the actual process performance in the long run, the variation from common causes must be reduced. If the capability criterion is $\hat{\mu} \pm 4\hat{\sigma}$ ($Z_{\min} \geq 4$), the process standard deviation for a centered process would be:

$$\sigma_{new} = \frac{\text{USL} - \hat{\mu}}{Z_{\min}} = \frac{0.9 - 0.7}{4} = 0.05$$

Therefore, actions should be taken to reduce the process standard deviation from 0.0725 to 0.05, about 31%.   ∎

At this point, the process has been brought into statistical control and its capability has been described in terms of process capability index or $Z_{\min}$. The next step is to evaluate the process capability in terms of meeting customer requirements. The fundamental goal is never-ending improvement in process performance. In the near term, however, priorities must be set as to which processes should receive attention first. This is essentially an economic decision. The circumstances vary from case to case, depending on the nature of the particular process in question. While each such decision could be resolved individually, it is often helpful to use broader guidelines to set priorities and promote consistency of improvement efforts. For instance, certain procedures require $C_{pk} > 1.33$, and further specify $C_{pk} = 1.50$ for new processes. These requirements are intended to assure a minimum performance level that is consistent among characteristics, products, and manufacturing sources.

Whether in response to a capability criterion that has not been met, or to the continuing need for improvement of cost and quality performance even beyond minimum capability requirement, the action required is the same: improve the process performance by reducing the variation that comes from common causes. This means taking management action to improve the system.

2.6   Advanced Control Charts

The major disadvantage of the Shewhart control chart is that it uses the information in the last plotted point and ignores information given by the sequence of points. This makes it insensitive to small shifts. One effective way is to use

- cumulative sum (CUSUM) control charts
- exponentially weighted moving average (EWMA) control charts

2.6.1   Cumulative Sum Control Charts

CUSUM charts incorporate all the information in the sequence of sample values by plotting the CUSUM of deviations of the sample values from a target value, defined as

$$C_i = \sum_{j=1}^{i} (\overline{x}_j - T)$$

A significant trend developed in C_i is an indication of the process mean shift. Therefore, CUSUM control charts would be more effective than Shewhart charts to detect small process shifts. Two statistics are used to accumulate deviations from the target T:

$$C_i^+ = \max[0, \quad x_i - (T + K) + C_{i-1}^+]$$
$$C_i^- = \max[0, \quad (T - K) - x_i + C_{i-1}^-]$$

where $C_0^+ = C_0^- = 0$, and K is the slack value, and it is often chosen about halfway between the target value and the process mean after shift. If either C^+ or C^- exceeds the decision interval H (a common choice is $H = 5\sigma$), the process is considered to be out of control.

2.6.2 Exponentially Weighted Moving Average Control Charts

As discussed earlier, we use Western Electric rules to increase the sensitivity of Shewhart control charts to detect nonrandom patterns or small shifts in a process. A different approach to highlight small shifts is to use a time average over past and present data values as an indicator of recent performance. Roberts [25] introduced the EWMA as such an indicator, that is, past data values are remembered with geometrically decreasing weight. For example, we denote the present and past values of a quality characteristic x by x_t, x_{t-1}, $x_{t-2}, \ldots$; then the EWMA y_t with discount factor q is

$$y_t = a(x_t + qx_{t-1} + q^2 x_{t-2} + \cdots)$$

where a is a constant that makes the weights add up to 1 and it equals to $1 - q$. In the practice of process monitoring, the constant $1 - q$ is given the distinguishing symbol λ. Using λ, the EWMA can be expressed as $y_t = \lambda x_t + (1 - \lambda)y_{t-1}$, which is a more convenient formula for updating the value of EWMA at each new observation. It is observed from the formula that a larger value of λ results in weights that die out more quickly and place more emphasis on recent observations. Therefore, a smaller value of λ is recommended to detect small process shifts, usually $\lambda = 0.05$, 0.10, or 0.20.

An EWMA control chart with appropriate limits is used to monitor the value of EWMA. If the process is in statistical control with a process mean of μ and a standard deviation of σ, the mean of the EWMA would be μ, and the standard deviation of the EWMA would be $\sigma \left(\frac{\lambda}{2-\lambda} \right)^{1/2}$. Thus, given a value of λ, three-sigma or other appropriate limits can be constructed to monitor the value of EWMA.

2.6.3 Other Advanced Control Charts

The successful use of Shewhart control charts and the CUSUM and EWMA control charts have led to the development of many new techniques over the last 20 years. A brief summary of these techniques and references to more complete descriptions are provided in this section.

The competitive global market expects smaller defect rates and higher quality level that requires 100% inspection of output products. The recent advancement of sensing techniques and computer capacity makes 100% inspection more feasible. Due to the reduced intervals between sampling of the 100% inspection, the complete observations will be correlated over time. However, one of the assumptions for Shewhart control charts is the independence between observations over time. When the observations are autocorrelated, Shewhart control charts will give misleading results in the form of many false alarms. ARIMA are used to remove autocorrelation from data, and then control charts are applied to the residuals. Further discussion on SPC with autocorrelated process data can be found in Refs. [16,20].

It is often necessary to simultaneously monitor or control two or more related quality characteristics. Using individual control charts to monitor the independent variables separately can be very misleading. Multivariate SPC control charts were developed based on multivariate normal distribution by Hotelling [26]. More discussion on multivariate SPC can be found in Refs. [20,26].

The use of control charts requires the selection of sample size, sampling frequency or interval between samples, and the control limits for the charts. The selection of these parameters has economic consequences in that the cost of sampling, the cost of false alarms, and the cost of removing assignable causes will affect the choice of the parameters. Therefore, the economic design of control charts has received attention in research and practice. Related discussion can be found in Refs. [8,27]. Other research issues and ideas in SPC can be found in a review paper by Woodall and Montgomery [28].

2.7   Limitations of Acceptance Sampling

As one of the earliest methods of quality control, acceptance sampling is closely related to inspection of output of a process, or testing of a product. Acceptance sampling is defined as the inspection and classification of samples from a lot randomly and decision about disposition of the lot. At the beginning of the concept of quality conformance back in the 1930s, the acceptance sample took the whole effort of quality improvement. The most widely used plans are given by the Military Standard tables (MIL STD 105A), which were developed during World War II. The last revision (MIL STD 105E) was issued in 1989, but cancelled in 1991. The standard was adopted by the American Society for Quality as ANSI/ASQ A1.4.

Due to its less proactive nature in terms of quality improvement, acceptance sampling is less emphasized in current quality control systems. Usually, methods of lot sentencing include no inspection, 100% inspection, and acceptance sampling. Some of the problems with acceptance sampling were articulated by Dr. W. Edwards Deming [2], who pointed out that this procedure, while minimizing the inspection cost, does not minimize the total cost to the producer. To minimize the total cost to the producer, Deming indicated that inspection should be performed either 100% or not at all, which is called Deming's "All or None Rule." In addition, acceptance sampling has several disadvantages compared to 100% inspection [20]:

- There are risks of accepting "bad" lots and rejecting "good" lots.
- Less information is usually generated about the product or process.
- Acceptance sampling requires planning and documentation of the acceptance sampling procedure.

2.8   Conclusions

The quality of a system is defined and evaluated by the customer [29]. A system has many qualities and we can develop a utility or customer satisfaction measure based on all of these qualities. The design process substitutes the voice of the customer with engineering or technological characteristics. Quality function deployment (QFD) plays an important role in the development of those characteristics. Quality engineering principles can be used to develop ideal values or targets for these characteristics. The methodology of robust design is an integral part of the quality process [30–32]. Quality should be an integral part of all the elements of the enterprise, which means that it is distributed throughout the enterprise and also all of these elements must be integrated together as shown in Figure 2.8.

Statistical quality control is a primary technique for monitoring the manufacturing process or any other process with key quality characteristics of interests. Before we use statistical quality control to monitor the manufacturing process, the optimal values of the mean and

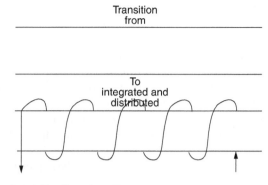

FIGURE 2.8 Integrated and distributed quality management.

standard deviation of the quality characteristic are determined by minimizing the variability of the quality characteristic through experimental design and process adjustment techniques. Consequently, the major goal of SQC (SPC) is to monitor the manufacturing process, keep the values of mean and standard deviation stable, and finally reduce variability. When the process is in control, all the assignable causes are not present and consequently the probability to produce a nonconforming unit is very small. When the process changes to out of control, the probability of nonconformance may increase, and the process quality deteriorates significantly. As one of the primary SPC techniques, the control charts we discussed in this chapter can effectively detect the variation due to the assignable causes and reduce process variability if the identified assignable causes can be eliminated from the process.

The philosophy of Deming, Juran, and other quality gurus implies that the responsibility for quality spans the entire organization. It is critical that the management in any enterprise recognize that quality improvement must be a total, company-wide activity, and that every organizational unit must actively participate. Statistical quality control techniques are the common language of communication about quality problems that enables all organizational units to solve problems rapidly and efficiently.

References

1. Juran, J.M. and Gryna, F.M. *Quality Planning and Analysis: From Product Development Through Usage*, 3rd edition. McGraw-Hill, New York, 1993.
2. Deming, W.E., *Quality, Productivity, and Competitive Position*, Massachusetts Institute of Technology, Center for Advanced Engineering Study, Cambridge, MA, 1982.
3. American Society for Quality Control, *ASQC Glossary and Tables for Statistical Quality Control*, Milwaukee, 1983.
4. Shewhart, W.A., *Economic Control of Quality of a Manufactured Product*, D. Van Nostrand Company, New York, 1931.
5. Mizuno, S. and Akao, Y. (Eds), *Quality Function Deployment: A Company-Wide Quality Approach*, JUSE Press, Tokyo, 1978.
6. Akao, Y. (Ed), *Quality Function Deployment: Integrating Customer Requirements into Product Design*, translated by Mazur, G.H., Productivity Press, Cambridge, MA, 1990.
7. Montgomery, D.C., *Design and Analysis of Experiments*, 5th edition, John Wiley & Sons, New York, 2001.

8. Kapur, K.C. and Feng, Q., Integrated optimization models and strategies for the improvement of the Six Sigma process. *International Journal of Six Sigma and Competitive Advantage*, 1(2), 2005.

9. Montgomery, D.C., The economic design of control charts: A review and literature survey, *Journal of Quality Technology*, 14, 75–87, 1980.

10. Breyfogle, F.W., *Implementing Six Sigma: Smarter Solutions Using Statistical Methods*, 2nd edition, John Wiley & Sons, New York, 2003.

11. Pyzdek, T., *The Six Sigma Handbook, Revised and Expanded: The Complete Guide for Greenbelts, Blackbelts, and Managers at All Levels*, 2nd edition, McGraw-Hill, New York, 2003.

12. Yang, K. and El-Haik, B., *Design for Six Sigma: A Roadmap for Product Development*, McGraw-Hill, New York, 2003.

13. De Feo, J.A. and Barnard, W.W., *Juran Institute's Six Sigma Breakthrough and Beyond: Quality Performance Breakthrough Methods*, McGraw-Hill, New York, 2004.

14. Duncan, A.J., *Quality Control and Industrial Statistics*, 5th edition, Irwin, Homewood, IL, 1986.

15. Stoumbos, Z.G., Reynolds, M.R., Ryan, T.P., and Woodall, W.H., The state of statistical process control as we proceed into the 21st century, *Journal of the American Statistical Association*, 95, 992–998, 2000.

16. Box, G. and Luceno, A., *Statistical Control by Monitoring and Feedback Adjustment*, John Wiley & Sons, New York, 1997.

17. Hicks, C.R. and Turner, K.V., *Fundamental Concepts in the Design of Experiments*, 5th edition, Oxford University Press, New York, 1999.

18. ASTM Publication STP-15D, *Manual on the Presentation of Data and Control Chart Analysis*, 1916 Race Street, Philadelphia, PA, 1976.

19. Alwan, L.C., Statistical Process Analysis, McGraw-Hill, New York, 2000.

20. Montgomery, D.C., *Introduction to Statistical Quality Control*, 5th edition, John Wiley & Sons, New York, 2005.

21. Western Electric, *Statistical Quality Control Handbook*, Western Electric Corporation, Indianapolis, IN, 1956.

22. Chandra, M.J., *Statistical Quality Control*, CRC Press LLC, Boca Raton, FL, 2001.

23. Kane, V.E., Process capability indices, *Journal of Quality Technology*, 18, 1986.

24. Kotz, S. and Lovelace, C.R., *Process Capability Indices in Theory and Practice*, Arnold, London, 1998.

25. Roberts, S.W., *Control Chart Tests Based on Geometric Moving Averages, Technometrics*, 42(1), 97–102 1959.

26. Hotelling, H., Multivariate Quality Control, In: *Techniques of Statistical Analysis*, Edited by Eisenhart, C., Hastay, M.W., and Wallis, W.A., McGraw-Hill, New York, 1947.

27. Duncan, A.J., The economic design of charts used to maintain current control of a process, *Journal of the American Statistical Association*, 51, 228–242, 1956.

28. Woodall, W.H. and Montgomery, D.C., Research issues and ideas in statistical process control, *Journal of Quality Technology*, 31(4), 376–386, 1999.

29. Kapur, K.C., An integrated customer-focused approach for quality and reliability, *International Journal of Reliability, Quality and Safety Engineering*, 5(2), 101–113, 1998.

30. Phadke, M.S., *Quality Engineering Using Robust Design*, Prentice-Hall, Englewood Cliffs, NJ, 1989.

31. Taguchi, G., *Introduction to Quality Engineering*, Asia Productivity Organization, Tokyo, 1986.

32. Taguchi, G., *System of Experimental Design, Volume I and II*, UNIPUB/Krauss International, White Plains, New York, 1987.

Appendix

Constants and Formulas for Constructing Control Charts

Subgroup Size n	$\overline{X}$ and R Charts				$\overline{X}$ and S Charts			
	Chart for Averages $\overline{X}$	Chart for Ranges (R)			Chart for Averages $\overline{X}$	Chart for Standard Deviations (S)		
	Factors for Control Limits A_2	Devisors for Estimate of Standard Deviation d_2	Factors for Control Limits D_3	Factors for Control Limits D_4	Factors for Control Limits A_3	Divisors for Estimator of Standard Deviation c_4	Factors for Control Limits B_3	Factors for Control Limits B_4
2	1.880	1.128	–	3.267	2.659	0.7979	–	3.267
3	1.023	1.693	–	2.574	1.954	0.8862	–	2.568
4	0.729	2.059	–	2.282	1.628	0.9213	–	2.266
5	0.577	2.326	–	2.114	1.427	0.9400	–	2.089
6	0.483	2.534	–	2.004	1.287	0.9515	0.030	1.970
7	0.419	2.704	0.076	1.924	1.182	0.9594	0.118	1.882
8	0.373	2.847	0.136	1.864	1.099	0.9650	0.185	1.815
9	0.337	2.970	0.184	1.816	1.032	0.9693	0.239	1.761
10	0.308	3.078	0.223	1.777	0.975	0.9727	0.284	1.716

$$\text{LCL} = \overline{\overline{x}} - A_2\overline{R} \qquad \text{LCL} = D_3\overline{R} \qquad \text{LCL} = \overline{\overline{x}} - A_3\overline{S} \qquad \text{LCL} = B_3\overline{S}$$

$$\text{CL} = \overline{\overline{x}} \quad \text{and} \quad \text{CL} = \overline{R} \qquad \text{CL} = \overline{\overline{x}} \quad \text{and} \quad \text{CL} = \overline{S}$$

$$\text{ULC} = \overline{\overline{x}} + A_2\overline{R} \qquad \text{UCL} = D_4\overline{R} \qquad \text{ULC} = \overline{\overline{x}} + A_3\overline{S} \qquad \text{UCL} = B_4\overline{S}$$

$$\hat{\sigma} = R/d_2 \qquad\qquad\qquad \hat{\sigma} = s/c_4$$

Guide for Selection of Charts for Attributes:

	Nonconforming Units	Nonconformities
Number of Nonconformities (Simple, but needs constant sample size)	np	c
Proportion (More complex, but adjusts to understandable proportion, and can cope with varying sample sizes)	p	u

- *p chart* for proportion of units nonconforming, from samples not necessarily of constant size: (If n varies, use $\overline{n}$ or individual n_i.)

$$\text{UCL}_{\hat{p}} = \overline{p} + 3\sqrt{\frac{\overline{p}(1 - \overline{p})}{n}}$$

$$\text{Centerline} = \overline{p}$$

$$\text{LCL}_{\hat{p}} = \overline{p} - 3\sqrt{\frac{\overline{p}(1 - \overline{p})}{n}}$$

- *np chart* for number of units nonconforming, from samples of constant size:

$$\text{UCL} = n\overline{p} + 3\sqrt{n\overline{p}(1 - \overline{p})}$$

$$\text{Centerline} = n\overline{p}$$

$$\text{LCL} = n\overline{p} - 3\sqrt{n\overline{p}(1 - \overline{p})}$$

- *c chart* for number of nonconformities, from samples of constant size:

$$\text{LCL} = c - 3\sqrt{c}$$

$$\text{CL} = c$$

$$\text{UCL} = c + 3\sqrt{c}$$

- *u chart* for number of nonconformities per unit, from samples not necessarily of constant size: (If n varies, use $\bar{n}$ or individual n_i.)

$$\text{LCL} = \bar{u} - 3\sqrt{\frac{\bar{u}}{n}}$$

$$\text{CL} = \bar{u}$$

$$\text{UCL} = \bar{u} + 3\sqrt{\frac{\bar{u}}{n}}$$

3

Reliability

Lawrence M. Leemis
The College of William and Mary

3.1   Introduction

Reliability theory involves the mathematical modeling of systems, typically comprised of components, with respect to their ability to perform their intended function over time. Reliability theory can be used as a *predictive* tool, as in the case of a new product introduction; it can also be used as a *descriptive* tool, as in the case of finding a weakness in an existing system design.

Reliability theory is based on probability theory. The reliability of a component or system at one particular point in time is a real number between 0 and 1 that represents the probability that the component or system is functioning at that time. Reliability theory also involves statistical methods. Estimating component and system reliability is often performed by analyzing a data set of lifetimes.

Recent tragedies, such as the space shuttle accidents, nuclear power plant accidents, and aircraft catastrophes, highlight the importance of reliability in design. This chapter describes probabilistic models for reliability in design and statistical techniques that can be applied to a data set of lifetimes. Although the majority of the illustrations given here come from engineering problems, the techniques described here may also be applied to problems in actuarial science and biostatistics.

Reliability engineers concern themselves primarily with lifetimes of inanimate objects, such as switches, microprocessors, or gears. They usually regard a complex system as a collection of components when performing an analysis. These components are arranged in a structure that allows the system state to be determined as a function of the component states. Interest in reliability and quality control has been revived by a more competitive international market and increased consumer expectations. A product or service that has a reputation for high reliability will have consumer goodwill and, if appropriately priced, will gain in market share.

Literature on reliability tends to use *failure*, actuarial literature tends to use *death*, and point process literature tends to use *epoch*, to describe the event at the termination of a lifetime. Likewise, reliability literature tends to use a *system*, *component*, or *item*, actuarial literature uses an *individual*, and biostatistical literature tends to use an *organism* as the object of a study. To avoid switching terms, *failure* of an *item* will be used as much as possible throughout this chapter, as the emphasis is on reliability. The concept of failure time (or lifetime or survival time) is quite generic, and the models and statistical methods presented here apply to any nonnegative random variable (e.g., the response time at a computer terminal).

The remainder of this chapter is organized as follows. Sections 3.2 through 3.4 contain probability models for lifetimes, and the subsequent remaining sections contain methods related to data collection and inference.

Mathematical models for describing the arrangement of components in a system are introduced in Section 3.2. Two of the simplest arrangements of components are series and parallel systems. The notion of the reliability of a component and system at a particular time is also introduced in this section. As shown in Section 3.3, the concept of reliability generalizes to a survivor function when time dependence is introduced. In particular, four different representations for the distribution of the failure time of an item are considered: the survivor, density, hazard, and cumulative hazard functions.

Several popular parametric models for the lifetime distribution of an item are investigated in Section 3.4. The *exponential distribution* is examined first due to its importance as the only continuous distribution with the memoryless property, which implies that a used item that is functioning has the same conditional failure distribution as a new item. Just as the normal distribution plays a central role in classical statistics due to the central limit theorem, the exponential distribution is central to the study of the distribution of lifetimes, as it is the only continuous distribution with a constant hazard function. The more flexible Weibull distribution is also outlined in this section.

The emphasis changes from developing probabilistic models for lifetimes to analyzing lifetime data sets in Section 3.5. One problem associated with these data sets is that of censored data. Data are censored when only a bound on the lifetime is known. This would be the case, for example, when conducting an experiment with light bulbs, and half of the light bulbs are still operating at the end of the experiment. This section surveys methods for fitting parametric distributions to data sets. Maximum likelihood parameter estimates are emphasized because they have certain desirable statistical properties. Section 3.6 reviews a nonparametric method for estimating the survivor function of an item from a censored data set: the Kaplan–Meier product-limit estimate. Once a parametric model has been chosen to represent the failure time for a particular item, the adequacy of the model should be assessed. Section 3.7 considers the Kolmogorov–Smirnov goodness-of-fit test for assessing how well a fitted lifetime distribution models the lifetime of the item. The test uses the largest vertical distance between the fitted and empirical survivor functions as the test statistic.

We have avoided references throughout the chapter to improve readability. There are thousands of journal articles and over 100 textbooks on reliability theory and applications. As this is not a review of the current state-of-the-art in reliability theory, we cite only a few key comprehensive texts for further reading. A classic, early reference is Barlow and

Proschan (1981). Meeker and Escobar (1998) is a more recent comprehensive textbook on reliability. The analysis of survival data is also considered by Kalbfleisch and Prentice (2002) and Lawless (2003). This chapter assumes that the reader has a familiarity with calculus-based probability and statistical inference techniques.

3.2 Reliability in System Design

This section introduces mathematical techniques for expressing the arrangement of components in a system and for determining the reliability of the associated system. We assume that an item consists of n components, arranged into a system. We first consider a system's *structural properties* associated with the arrangement of the components into a system, and then consider the system's *probabilistic properties* associated with determining the system reliability. *Structure functions* are used to map the states of the individual components to the state of the system. *Reliability functions* are used to determine the system reliability at a particular point in time, given the component reliabilities at that time.

3.2.1 Structure Functions

A *structure function* is used to describe the way that the n components are related to form a system. The structure function defines the system state as a function of the component states. In addition, it is assumed that both the components and the system can either be functioning or failed. Although this *binary assumption* may be unrealistic for certain types of components or systems, it makes the mathematics involved more tractable. The functioning and failed states for both components and systems will be denoted by 1 and 0, respectively, as in the following definition.

DEFINITION 3.1 The state of component i, denoted by x_i is

$$x_i = \begin{cases} 0 & \text{if component } i \text{ has failed} \\ 1 & \text{if component } i \text{ is functioning} \end{cases}$$

for $i = 1, 2, \ldots, n$.

These n values can be written as a system state vector, $\boldsymbol{x} = (x_1, x_2, \ldots, x_n)$. As there are n components, there are 2^n different values that the system state vector can assume, and $\binom{n}{j}$ of these vectors correspond to exactly j functioning components, $j = 0, 1, \ldots, n$. The structure function, $\phi(\boldsymbol{x})$, maps the system state vector $\boldsymbol{x}$ to 0 or 1, yielding the state of the system.

DEFINITION 3.2 The *structure function* ϕ is

$$\phi(\boldsymbol{x}) = \begin{cases} 0 & \text{if the system has failed when the state vector is } \boldsymbol{x} \\ 1 & \text{if the system is functioning when the state vector is } \boldsymbol{x} \end{cases}$$

The most common system structures are the series and parallel systems, which are defined in the examples that follow. The series system is the worst possible way to arrange components in a system; the parallel system is the best possible way to arrange components in a system.

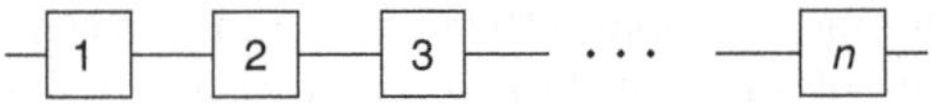

FIGURE 3.1 A series system.

Example 3.1

A *series system* functions when all its components function. Thus $\phi(\boldsymbol{x})$ assumes the value 1 when $x_1 = x_2 = \cdots = x_n = 1$, and 0 otherwise. Therefore,

$$\phi(\boldsymbol{x}) = \begin{cases} 0 & \text{if there exists an } i \text{ such that } x_i = 0 \\ 1 & \text{if } x_i = 1 \text{ for all } i = 1, 2, \ldots, n \end{cases}$$

$$= \min\{x_i, x_2, \ldots, x_n\}$$

$$= \prod_{i=1}^{n} x_i$$

These three different ways of expressing the value of the structure function are equivalent, although the third is preferred because of its compactness. Systems that function only when all their components function should be modeled as series systems.

Block diagrams are useful for visualizing a system of components. The block diagram corresponding to a series system of n components is shown in Figure 3.1. A block diagram is a graphic device for expressing the arrangement of the components to form a system. If a path can be traced through functioning components from left to right on a block diagram, then the system functions. The boxes represent the system components, and either component numbers or reliabilities are placed inside the boxes. ∎

Example 3.2

A *parallel system* functions when one or more of the components function. Thus $\phi(\boldsymbol{x})$ assumes the value 0 when $x_1 = x_2 = \cdots = x_n = 0$, and 1 otherwise.

$$\phi(\boldsymbol{x}) = \begin{cases} 0 & \text{if } x_i = 0 \text{ for all } i = 1, 2, \ldots, n \\ 1 & \text{if there exists an } i \text{ such that } x_i = 1 \end{cases}$$

$$= \max\{x_1, x_2, \ldots, x_n\}$$

$$= 1 - \prod_{i=1}^{n} (1 - x_i)$$

As in the case of the series system, the three ways of defining $\phi(\boldsymbol{x})$ are equivalent. The block diagram of a parallel arrangement of n components is shown in Figure 3.2. A parallel arrangement of components is appropriate when all components must fail for the system to fail. A two-component parallel system, for instance, is the brake system on an automobile that contains two reservoirs for brake fluid. Arranging components in parallel is also known as *redundancy*. ∎

Series and parallel systems are special cases of k-out-of-n systems, which function if k or more of the n components function. A series system is an n-out-of-n system, and a parallel system is a 1-out-of-n system. A suspension bridge that needs only k of its n cables to support the bridge or an automobile engine that needs only k of its n cylinders to run are examples of k-out-of-n systems.

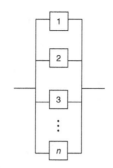

FIGURE 3.2 A parallel system.

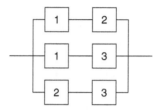

FIGURE 3.3 A 2-out-of-3 system.

Example 3.3

The structure function of a k-out-of-n system is

$$\phi(\boldsymbol{x}) = \begin{cases} 0 & \text{if } \sum_{i=1}^{n} x_i < k \\ 1 & \text{if } \sum_{i=1}^{n} x_i \geq k \end{cases}$$

The block diagram for a k-out-of-n system is difficult to draw in general, but for specific values of k and n it can be drawn by repeating components in the block diagram. The block diagram for a 2-out-of-3 system, for example, is shown in Figure 3.3. The block diagram indicates that if all three, or exactly two out of three components (in particular 1 and 2, 1 and 3, or 2 and 3) function, then the system functions. The structure function for a 2-out-of-3 system is

$$\phi(\boldsymbol{x}) = 1 - (1 - x_1 x_2)(1 - x_1 x_3)(1 - x_2 x_3)$$
$$= x_1 x_2 + x_1 x_3 + x_2 x_3 - x_1^2 x_2 x_3 - x_1 x_2^2 x_3 - x_1 x_2 x_3^2 + (x_1 x_2 x_3)^2 \qquad \blacksquare$$

Most real-world systems have a more complex arrangement of components than a k-out-of-n arrangement. The next example illustrates how to combine series and parallel arrangements to determine the appropriate structure function for a more complex system.

Example 3.4

An airplane has four propellers, two on each wing. The airplane will fly (function) if at least one propeller on each wing functions. In this case, the four propellers are denoted by components 1, 2, 3, and 4, with 1 and 2 being on the left wing and 3 and 4 on the right wing. For the moment, if the plane is considered to consist of two wings (not considering individual propellers), then the wings are arranged in series, as failure of the propulsion on either wing results in system failure. Each wing can be modeled as a two-component parallel subsystem of propellers, as only one propeller on each wing is required to function.

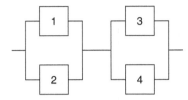

FIGURE 3.4   A four-propeller system.

The appropriate block diagram for the system is shown in Figure 3.4. The structure function is the product of the structure functions of the two parallel subsystems:

$$\phi(\boldsymbol{x}) = [1 - (1 - x_1)(1 - x_2)][1 - (1 - x_3)(1 - x_4)]$$

■

To avoid studying structure functions that are unreasonable, a subset of all possible systems of n components, namely *coherent systems*, has been defined. A system is coherent if $\phi(\boldsymbol{x})$ is nondecreasing in $\boldsymbol{x}$ [e.g., $\phi(x_1, \ldots, x_{i-1}, 0, x_{i+1}, \ldots, x_n) \leq \phi(x_1, \ldots, x_{i-1}, 1, x_{i+1}, \ldots, x_n)$ for all i] and there are no irrelevant components. The condition that $\phi(\boldsymbol{x})$ be nondecreasing in $\boldsymbol{x}$ implies that the system will not degrade if a component upgrades. A component is irrelevant if its state has no impact on the structure function. Many theorems related to coherent systems have been proven; one of the more useful is that redundancy at the component level is more effective than redundancy at the system level. This is an important consideration in *reliability design*, where a reliability engineer decides the components to choose (reliability allocation) at appropriate positions in the system (reliability optimization).

3.2.2   Reliability Functions

The discussion of structure functions so far has been completely deterministic in nature. We now introduce probability into the mix by first defining reliability. The paragraphs following the definition expand on the italicized words in the definition.

DEFINITION 3.3   The *reliability* of an *item* is the *probability* that it will *adequately perform* its specified *purpose* for a specified *period of time* under specified *environmental conditions*.

The definition implies that the object of interest is an *item*. The definition of the item depends on the purpose of the study. In some situations, we will consider an item to be an interacting arrangement of components; in other situations, the component level of detail in the model is not of interest.

Reliability is defined as a *probability*. Thus, the axioms of probability apply to reliability calculations. In particular, this means that all reliabilities must be between 0 and 1 inclusive, and that the results derived from the probability axioms must hold. For example, if two independent components have 1000-hour reliabilities of p_1 and p_2, and system failure occurs when either component fails (i.e., a two-component series system), then the 1000-hour system reliability is $p_1 p_2$.

Adequate performance for an item must be stated unambiguously. A *standard* is often used to determine what is considered adequate performance. A mechanical part may require tolerances that delineate adequate performance from inadequate performance. The performance of an item is related to the mathematical model used to represent the condition of the item. The simplest model for an item is a binary model, which was introduced earlier, in which the item is in either the functioning or failed state. This model is easily applied to a light bulb; it is more difficult to apply to items that gradually degrade over time, such as a machine tool. To apply a binary model to an item that degrades gradually, a threshold value must be determined to separate the functioning and failed states.

The definition of reliability also implies that the *purpose* or intended use of the item must be specified. Machine tool manufacturers, for example, often produce two grades of an item: one for professional use and another for consumer use.

The definition of reliability also indicates that *time* is involved in reliability, which implies five consequences. First, the units for time need to be specified (e.g., minutes, hours, years) by the modeler to perform any analysis. Second, many lifetime models use the random variable T (rather than X, which is common in probability theory) to represent the failure time of the item. Third, time need not be taken literally. The number of miles may represent time for an automobile tire; the number of cycles may represent time for a light switch. Fourth, a time duration associated with a reliability must be specified. The reliability of a component, for example, should not be stated as simply 0.98, as no time is specified. It is equally ambiguous for a component to have a 1000-hour life without indicating a reliability for that time. Instead, it should be stated that the 1000-hour reliability is 0.98. This requirement of stating a time along with a reliability applies to systems as well as components. Finally, determining what should be used to measure the lifetime of an item may not be obvious. Reliability analysts must consider whether continuous operation or on/off cycling is more effective for items such as motors or computers.

The last aspect of the definition of reliability is that *environmental conditions* must be specified. Conditions such as temperature, humidity, and turning speed all affect the lifetime of a machine tool. Likewise, the driving conditions for an automobile will influence its reliability. Included in environmental conditions is the preventive maintenance to be performed on the item.

We now return to the mathematical models for determining the reliability of a system. Two additional assumptions need to be made for the models developed here. First, the n components comprising a system must be *nonrepairable*. Once a component changes from the functioning to the failed state, it cannot return to the functioning state. This assumption was not necessary when structure functions were introduced, as a structure function simply maps the component states to the system state. The structure function can be applied to a system with nonrepairable or repairable components. The second assumption is that the components are independent. Thus, failure of one component does not influence the probability of failure of other components. This assumption is not appropriate if the components operate in a common environment where there may be common-cause failures. Although the independence assumption makes the mathematics for modeling a system simpler, the assumption should not be automatically applied.

Previously, x_i was defined to be the state of component i. Now X_i is a random variable with the same meaning.

DEFINITION 3.4   The random variable denoting the state of component i, denoted by X_i, is

$$X_i = \begin{cases} 0 & \text{if component } i \text{ has failed} \\ 1 & \text{if component } i \text{ is functioning} \end{cases}$$

for $i = 1, 2, \ldots, n$.

These n values can be written as a random system state vector $\boldsymbol{X}$. The probability that component i is functioning at a certain time is given by $p_i = P(X_i = 1)$, which is often called the *reliability* of the ith component, for $i = 1, 2, \ldots, n$. The P function denotes probability. These n values can be written as a reliability vector $\boldsymbol{p} = (p_1, p_2, \ldots, p_n)$.

The *system reliability*, denoted by r, is defined by

$$r = P[\phi(\boldsymbol{X}) = 1]$$

where r is a quantity that can be calculated from the vector $\boldsymbol{p}$, so $r = r(\boldsymbol{p})$. The function $r(\boldsymbol{p})$ is called the *reliability function*. In some of the examples in this section, the components have identical reliabilities (that is, $p_1 = p_2 = \cdots = p_n = p$), which is indicated by the notation $r(p)$.

Several techniques are used to calculate system reliability. We will illustrate two of the simplest techniques: definition and expectation. The first technique for finding the reliability of a coherent system of n independent components is to use the definition of system reliability directly, as illustrated in the example.

Example 3.5

The system reliability of a *series* system of n components is easily found using the definition of $r(\boldsymbol{p})$ and the independence assumption.

$$r(\boldsymbol{p}) = P[\phi(\boldsymbol{X}) = 1]$$

$$= P\left[\prod_{i=1}^{n} X_i = 1\right]$$

$$= \prod_{i=1}^{n} P[X_i = 1]$$

$$= \prod_{i=1}^{n} p_i$$

The product in this formula indicates that system reliability is always less than the reliability of the least reliable component. This "chain is only as strong as its weakest link" result indicates that improving the weakest component causes the largest increase in the reliability of a series system. ∎

In the special case when all components are identical, the reliability function reduces to $r(p) = p^n$, where $p_1 = p_2 = \cdots = p_n = p$. The plot in Figure 3.5 of component reliability versus system reliability for several values of n shows that highly reliable components are necessary to achieve reasonable system reliability, even for small values of n.

The second technique, *expectation*, is based on the fact that $P[\phi(\boldsymbol{X}) = 1]$ is equal to $E[\phi(\boldsymbol{X})]$, because $\phi(\boldsymbol{X})$ is a Bernoulli random variable. Consequently, the expected value of $\phi(\boldsymbol{X})$ is the system reliability $r(\boldsymbol{p})$, as illustrated in the next example.

Example 3.6

Since the components are assumed to be independent, the system reliability for a parallel system of n components using the expectation technique is

$$r(\boldsymbol{p}) = E[\phi(\boldsymbol{X})]$$

$$= E\left[1 - \prod_{i=1}^{n} (1 - X_i)\right]$$

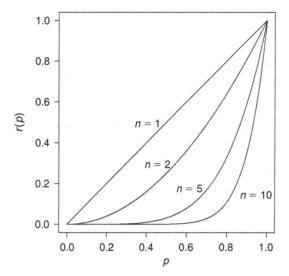

FIGURE 3.5   Reliability of a series system of n components.

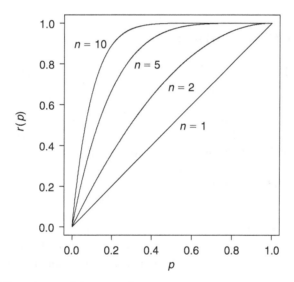

FIGURE 3.6   Reliability of a parallel system of n components.

$$= 1 - E\left[\prod_{i=1}^{n}(1 - X_i)\right]$$

$$= 1 - \prod_{i=1}^{n} E[1 - X_i]$$

$$= 1 - \prod_{i=1}^{n}(1 - p_i)$$

In the special case of identical components, this expression reduces to $r(p) = 1 - (1 - p)^n$. Figure 3.6 shows component reliability versus system reliability for a parallel system of n identical components. The law of diminishing returns is apparent from the graph when a fixed component reliability is considered. The marginal gain in reliability decreases dramatically as more components are added to the system.   ∎

There are two systems that appear to be similar to parallel systems on the surface, but they are not true parallel systems such as the one considered in the previous example. The first such system is a *standby system*. In a standby system, not all the components function simultaneously, and components are switched to standby components upon failure. Examples of standby systems include a spare tire for an automobile and having three power sources (utility company, backup generator, and batteries) for a hospital. In contrast, all components are functioning simultaneously in a true parallel system.

The second such system is a *shared-parallel system*. In a shared-parallel system, all components are online, but the component reliabilities change when one component fails. The lug nuts that attach a wheel to an automobile are an example of a five-component shared-parallel system. When one lug nut fails (i.e., loosens or falls off), the load on the remaining functioning lug nuts increases. Thus, the static reliability calculations presented in this section are not appropriate for a wheel attachment system. In contrast, the failure of a component in a true parallel system does not affect the reliabilities of any of the other components in the system.

3.3   Lifetime Distributions

Reliability has only been considered at one particular instant of time. Reliability is generalized to be a function of time in this section, and various lifetime distribution representations that are helpful in describing the evolution of the risks to which an item is subjected over time are introduced. In particular, four lifetime distribution representations are presented: the *survivor* function, the *probability density* function, the *hazard* function, and the *cumulative hazard* function. These four distribution representations apply to both continuous (e.g., a fuse) and discrete (e.g., software executed daily) lifetimes, although the focus here is on continuous lifetimes. These four representations are not the only ways to define the distribution of the continuous, nonnegative random variable T, referred to generically here as a "lifetime." Other methods include the moment generating function $E[e^{sT}]$, the characteristic function $E[e^{isT}]$, the Mellin transform $E[T^s]$, and the mean residual life function $E[T - t | T \geq t]$. The four representations used here have been chosen because of their intuitive appeal, their usefulness in problem solving, and their popularity in the literature.

3.3.1   Survivor Function

The first lifetime distribution representation is the *survivor function*, $S(t)$. The survivor function is a generalization of reliability. Whereas reliability is defined as the probability that an item is functioning at one particular time, the survivor function is the probability that an item is functioning at any time t:

$$S(t) = P[T \geq t]   t \geq 0$$

It is assumed that $S(t) = 1$ for all $t < 0$. A survivor function is also known as the reliability function [since $S(t)$ is the reliability at time t] and the complementary cumulative distribution function [since $S(t) = 1 - F(t)$ for continuous random variables, where $F(t) = P[T \leq t]$ is the cumulative distribution function]. All survivor functions must satisfy three conditions:

$$S(0) = 1   \lim_{t \to \infty} S(t) = 0   S(t) \text{ is nonincreasing}$$

There are two interpretations of the survivor function. First, $S(t)$ is the probability that an individual item is functioning at time t. This is important, as will be seen later, in determining the lifetime distribution of a system from the distribution of the lifetimes of its individual components. Second, if there is a large population of items with identically distributed lifetimes, $S(t)$ is the expected fraction of the population that is functioning at time t.

The survivor function is useful for comparing the survival patterns of several populations of items. If $S_1(t) \geq S_2(t)$, for all t values, for example, it can be concluded that the items in population 1 are superior to those in population 2 with regard to reliability.

3.3.2 Probability Density Function

The second lifetime distribution representation, the *probability density function*, is defined by $f(t) = -S'(t)$, where the derivative exists. It has the probabilistic interpretation

$$f(t)\Delta t = P[t \leq T \leq t + \Delta t]$$

for small values of Δt. Although the probability density function is not as effective as the survivor function in comparing the survival patterns of two populations, a graph of $f(t)$ indicates the likelihood of failure for a new item over the course of its lifetime. The probability of failure between times a and b is calculated by an integral:

$$P[a \leq T \leq b] = \int_a^b f(t)\mathrm{d}t$$

All probability density functions for lifetimes must satisfy two conditions:

$$\int_0^\infty f(t)\mathrm{d}t = 1 \quad f(t) \geq 0 \text{ for all } t \geq 0$$

It is assumed that $f(t) = 0$ for all $t < 0$.

3.3.3 Hazard Function

The *hazard function*, $h(t)$, is perhaps the most popular of the five representations for lifetime modeling due to its intuitive interpretation as the amount of *risk* associated with an item at time t. The hazard function goes by several aliases: in reliability it is also known as the hazard rate or failure rate; in actuarial science it is known as the force of mortality or force of decrement; in point process and extreme value theory it is known as the rate or intensity function; in vital statistics it is known as the age-specific death rate; and in economics its reciprocal is known as Mill's ratio.

The hazard function can be derived using conditional probability. First, consider the probability of failure between t and $t + \Delta t$:

$$P[t \leq T \leq t + \Delta t] = \int_t^{t+\Delta t} f(\tau)\mathrm{d}\tau = S(t) - S(t + \Delta t)$$

Conditioning on the event that the item is working at time t yields

$$P[t \leq T \leq t + \Delta t | T \geq t] = \frac{P[t \leq T \leq t + \Delta t]}{P[T \geq t]} = \frac{S(t) - S(t + \Delta t)}{S(t)}$$

If this conditional probability is averaged over the interval $[t, t + \Delta t]$ by dividing by Δt, an average rate of failure is obtained:

$$\frac{S(t) - S(t + \Delta t)}{S(t)\Delta t}$$

As $\Delta t \rightarrow 0$, this average failure rate becomes the instantaneous failure rate, which is the hazard function

$$h(t) = \lim_{\Delta t \to 0} \frac{S(t) - S(t + \Delta t)}{S(t)\Delta t}$$

$$= -\frac{S'(t)}{S(t)}$$

$$= \frac{f(t)}{S(t)} \quad t \geq 0$$

Thus, the hazard function is the ratio of the probability density function to the survivor function. Using the previous derivation, a probabilistic interpretation of the hazard function is

$$h(t)\Delta t = P[t \leq T \leq t + \Delta t | T > t]$$

for small values of Δt, which is a conditional version of the interpretation for the probability density function. All hazard functions must satisfy two conditions:

$$\int_0^\infty h(t)\mathrm{d}t = \infty \quad h(t) \geq 0 \text{ for all } t \geq 0$$

The *units* on a hazard function are typically given in failures per unit time, for example, $h(t) = 0.03$ failures per hour. Since the magnitude of hazard functions can often be quite small, they are often expressed in scientific notation, for example, $h(t) = 3.8$ failures per 10^6 hours, or the time units are chosen to keep hazard functions from getting too small, for example, $h(t) = 8.4$ failures per year.

The shape of the hazard function indicates how an item ages. The intuitive interpretation as the amount of *risk* an item is subjected to at time t indicates that when the hazard function is large the item is under greater risk, and when the hazard function is small the item is under less risk. The three hazard functions plotted in Figure 3.7 correspond to an increasing hazard function (labeled IFR for increasing failure rate), a decreasing hazard function (labeled DFR for decreasing failure rate), and a bathtub-shaped hazard function (labeled BT for bathtub-shaped failure rate).

The increasing hazard function is probably the most likely situation of the three. In this case, items are more likely to fail as time passes. In other words, items wear out or degrade with time. This is almost certainly the case with mechanical items that undergo wear or fatigue. The second situation, the decreasing hazard function, is less common. In this case, the item is less likely to fail as time passes. Items with this type of hazard function improve with time. Some metals, for example, work-harden through use and thus have increased strength as time passes. Another situation for which a decreasing hazard function might be appropriate for modeling is in working the bugs out of computer programs. Bugs are more likely to appear initially, but the likelihood of them appearing decreases as time passes.

The third situation, a bathtub-shaped hazard function, occurs when the hazard function decreases initially and then increases as items age. Items improve initially and then degrade

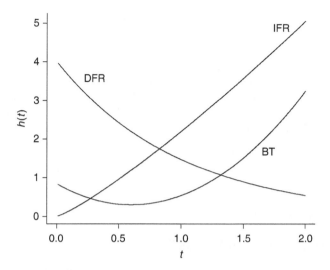

FIGURE 3.7 Hazard functions.

as time passes. One instance where the bathtub-shaped hazard function arises is in the life-times of manufactured items. Often manufacturing, design, or component defects cause early failures. The period in which these failures occur is sometimes called the *burn-in* period. If failure is particularly catastrophic, this part of the lifetime will often be consumed by the manufacturer in a controlled environment. The time value during which early failures have been eliminated may be valuable to a producer who is determining an appropriate warranty period. Once items pass through this early part of their lifetime, they have a fairly constant hazard function, and failures are equally likely to occur at any point in time. Finally, as items continue to age, the hazard function increases without limit, resulting in *wear-out* failures. Bathtub-shaped hazard functions are also used for modeling human lifetimes.

Care must be taken to differentiate between the hazard function for a population and the hazard function for an individual *item* under consideration. Consider the lifetimes of a certain model of a laptop computer as an illustration. Consider the following question: do two laptops operating in identical environments necessarily have the same hazard function? The answer is no. The laptops have their own individual hazard functions as they might have been manufactured at different facilities that may have included differing conditions (e.g., temperature, raw materials, parts suppliers). So, although a hazard function could be drawn for all laptop computers, it would be an aggregate hazard function representing the population, and individual laptops may be at increased or decreased risk.

3.3.4 Cumulative Hazard Function

The fourth lifetime distribution representation, the *cumulative hazard function $H(t)$*, can be defined by

$$H(t) = \int_0^t h(\tau)\mathrm{d}\tau \quad t \geq 0$$

The cumulative hazard function is also known as the integrated hazard function. All cumulative hazard functions must satisfy three conditions:

$$H(0) = 0 \quad \lim_{t \to \infty} H(t) = \infty \quad H(t) \text{ is nondecreasing}$$

The cumulative hazard function is valuable for variate generation in Monte Carlo simulation, implementing certain procedures in statistical inference, and defining certain distribution classes.

The four distribution representations presented here are equivalent in the sense that each completely specifies a lifetime distribution. Any one lifetime distribution representation implies the other four. Algebra and calculus can be used to find one lifetime distribution representation given that another is known. For example, if the survivor function is known, the cumulative hazard function can be determined by

$$H(t) = \int_0^t h(\tau)\mathrm{d}\tau = \int_0^t \frac{f(\tau)}{S(\tau)}\mathrm{d}\tau = -\log S(t)$$

where log is the natural logarithm (log base e).

3.3.5 Expected Values and Fractiles

Once a lifetime distribution representation for a particular item (which may be a component or an entire system) is known, it may be of interest to compute a moment or a fractile of the distribution. Moments and fractiles contain less information than a lifetime distribution representation, but they are often useful ways to summarize the distribution of a random lifetime. Examples of these performance measures include the mean time to failure, $E[T]$, the median, $t_{0.50}$, and the 99th fractile of a distribution, $t_{0.99}$.

A formula for the expectation of some function of the random variable T, say $u(T)$, is

$$E[u(T)] = \int_0^\infty u(t)f(t)\mathrm{d}t$$

The most common measure associated with a distribution is its *mean*, or first moment,

$$\mu = E[T] = \int_0^\infty tf(t)\mathrm{d}t = \int_0^\infty S(t)\mathrm{d}t$$

where the last equality is proved using integration by parts and is based on the assumption that $\lim_{t\to\infty} tS(t) = 0$. The mean is a measure of the central tendency or average value that a lifetime distribution assumes and is known as the center of gravity in physics. It is often abbreviated by MTTF (mean time to failure) for nonrepairable items. For repairable items that can be completely renewed by repair, it is often abbreviated by MTBF (mean time between failures). Another value associated with a distribution is its *variance*, or second moment about the mean,

$$\sigma^2 = V[T] = E[(T - \mu)^2] = E[T^2] - (E[T])^2$$

which is a measure of the dispersion of a lifetime distribution about its mean. The positive square root of the variance is known as the *standard deviation*, which has the same units as the random variable T.

Fractiles of a distribution are the times to which a specified proportion of the items survives. The definition of the pth fractile of a distribution, t_p, (often called the pth quantile or $100p$th percentile) satisfies

$$F(t_p) = P[T \leq t_p] = p$$

or, equivalently,

$$t_p = F^{-1}(p)$$

Example 3.7

The *exponential* distribution has survivor function

$$S(t) = e^{-\lambda t} \quad t \geq 0$$

where λ is a positive parameter known as the failure rate. Find the mean, variance, and the pth fractile of the distribution. The mean can be found by integrating the survivor function from 0 to infinity:

$$\mu = E[T] = \int_0^\infty S(t)\mathrm{d}t = \int_0^\infty e^{-\lambda t}\mathrm{d}t = \frac{1}{\lambda}$$

Since the probability density function is $f(t) = -S'(t) = \lambda e^{-\lambda t}$ for $t > 0$, the second moment about the origin is

$$E[T^2] = \int_0^\infty t^2 f(t)\mathrm{d}t = \int_0^\infty t^2 \lambda e^{-\lambda t}\mathrm{d}t = \frac{2}{\lambda^2}$$

using integration by parts twice. Therefore, the variance is

$$\sigma^2 = V[T] = E[T^2] - (E[T])^2 = \frac{2}{\lambda^2} - \frac{1}{\lambda^2} = \frac{1}{\lambda^2}$$

Finally, the pth fractile of the distribution, t_p, is found by solving

$$1 - e^{-\lambda t_p} = p$$

for t_p, yielding $t_p = -\frac{1}{\lambda}\log(1-p)$. ∎

3.3.6   System Lifetime Distributions

To this point, the discussion concerning the four lifetime representations $S(t), f(t), h(t)$, and $H(t)$ has assumed that the variable of interest is the *lifetime* of an *item*. For systems of components, both the individual components and the system have random lifetimes whose lifetime distributions can be defined by any of the four lifetime distributions. We now integrate reliability functions from Section 3.2 and the lifetime distribution representations from this section, which allows a modeler to find the distribution of the system lifetime, given the distributions of the component lifetimes. The component lifetime representations are denoted by $S_i(t), f_i(t), h_i(t)$, and $H_i(t)$, for $i = 1, 2, \ldots, n$, and the system lifetime representations are denoted by $S(t), f(t), h(t)$, and $H(t)$.

The survivor function is a time-dependent generalization of reliability. Whereas reliability always needs an associated time value (e.g., the 4000-hour reliability is 0.96), the survivor function is the reliability at any time t.

To find the reliability of a system at any time t, the component survivor functions should be used as arguments in the reliability function, that is,

$$S(t) = r(S_1(t), S_2(t), \ldots, S_n(t))$$

Once $S(t)$ is known, it is straightforward to determine any of the other four lifetime representations, moments, or fractiles, as illustrated in the following examples.

Example 3.8

Two independent components with survivor functions

$$S_1(t) = e^{-t} \quad \text{and} \quad S_2(t) = e^{-2t}$$

for $t \geq 0$, are arranged in series. Find the survivor and hazard functions for the system lifetime. Since the reliability function for a two-component series system is $r(\boldsymbol{p}) = p_1 p_2$, the system survivor function is

$$S(t) = S_1(t)\, S_2(t)$$

$$= e^{-t} e^{-2t}$$

$$= e^{-3t} \quad t \geq 0$$

which can be recognized as the survivor function for an exponential distribution with $\lambda = 3$. The hazard function for the system is

$$h(t) = -\frac{S'(t)}{S(t)} = 3 \quad t \geq 0$$

Thus, if two independent components with exponential times to failure are arranged in series, the time to system failure is also exponentially distributed with a failure rate that is the sum of the failure rates of the individual components. This result can be generalized to series systems with more than two components. If the lifetime of component i in a series system of n independent components has an exponential distribution with failure rate λ_i, then the system lifetime is exponentially distributed with failure rate $\sum_{i=1}^{n} \lambda_i$. ∎

The next example considers a parallel system of components having exponential lifetimes.

Example 3.9

Two independent components have hazard functions

$$h_1(t) = 1 \quad \text{and} \quad h_2(t) = 2$$

for $t \geq 0$. If the components are arranged in parallel, find the hazard function of the time to system failure and the mean time to system failure.

The survivor functions of the components are

$$S_1(t) = e^{-H_1(t)} = e^{-\int_0^t h_1(\tau)\mathrm{d}\tau} = e^{-t}$$

for $t \geq 0$. Likewise, $S_2(t) = e^{-2t}$ for $t \geq 0$. Since the reliability function for a two-component parallel system is $r(\mathbf{p}) = 1 - (1 - p_1)(1 - p_2)$, the system survivor function is

$$S(t) = 1 - (1 - S_1(t))(1 - S_2(t))$$

$$= 1 - (1 - e^{-t})(1 - e^{-2t})$$

$$= e^{-t} + e^{-2t} - e^{-3t} \quad t \geq 0$$

The hazard function is

$$h(t) = -\frac{S'(t)}{S(t)} = \frac{e^{-t} + 2e^{-2t} - 3e^{-3t}}{e^{-t} + e^{-2t} - e^{-3t}} \quad t \geq 0$$

To find the mean time to system failure, the system survivor function is integrated from 0 to infinity:

$$\mu = \int\limits_0^\infty S(t)\mathrm{d}t = \int\limits_0^\infty (\mathrm{e}^{-t} + \mathrm{e}^{-2t} - \mathrm{e}^{-3t})\mathrm{d}t = 1 + \frac{1}{2} - \frac{1}{3} = \frac{7}{6}$$

The mean time to failure of the stronger component is 1 and the mean time to failure of the weaker component is $1/2$. The addition of the weaker component in parallel with the stronger only increases the mean time to system failure by $1/6$. This is yet another illustration of the law of diminishing returns for parallel systems. ∎

3.4   Parametric Models

The survival patterns of a machine tool, a fuse, and an aircraft are vastly different. One would certainly not want to use the same failure time distribution with identical parameters to model these diverse lifetimes. This section introduces two distributions that are commonly used to model lifetimes. To adequately survey all the distributions currently in existence would require an entire textbook, so detailed discussion here is limited to the exponential and Weibull distributions.

3.4.1   Parameters

We begin by describing parameters, which are common to all lifetime distributions. The three most common types of parameters used in lifetime distributions are location, scale, and shape. Parameters in a lifetime distribution allow modeling of such diverse applications as light bulb failure time, patient postsurgery survival time, and the failure time of a muffler on an automobile by a single lifetime distribution (e.g., the Weibull distribution).

Location (or *shift*) parameters are used to shift the distribution to the left or right along the time axis. If c_1 and c_2 are two values of a location parameter for a lifetime distribution with survivor function $S(t; c)$, then there exists a real constant α such that $S(t; c_1) = S(\alpha + t; c_2)$. A familiar example of a location parameter is the mean of the normal distribution.

Scale parameters are used to expand or contract the time axis by a factor of α. If λ_1 and λ_2 are two values for a scale parameter for a lifetime distribution with survivor function $S(t; \lambda)$, then there exists a real constant α such that $S(\alpha t; \lambda_1) = S(t; \lambda_2)$. A familiar example of a scale parameter is λ in the exponential distribution. The probability density function always has the same shape, and the units on the time axis are determined by the value of λ.

Shape parameters are appropriately named because they affect the shape of the probability density function. Shape parameter values might also determine whether a distribution belongs to a particular distribution class such as IFR or DFR. A familiar example of a shape parameter is κ in the Weibull distribution.

In summary, location parameters *translate* survival distributions along the time axis, scale parameters *expand* or *contract* the time scale for survival distributions, and all other parameters are shape parameters.

3.4.2   Exponential Distribution

Just as the normal distribution plays an important role in classical statistics because of the central limit theorem, the exponential distribution plays an important role in reliability and lifetime modeling because it is the only continuous distribution with a constant hazard function. The exponential distribution has often been used to model the lifetime of electronic components and is appropriate when a used component that has not failed is statistically as good as a new component. This is a rather restrictive assumption. The exponential distribution is presented first because of its simplicity. The Weibull distribution, a more complex two-parameter distribution that can model a wider variety of situations, is presented subsequently. The exponential distribution has a single positive scale parameter λ, often called the *failure rate*, and the four lifetime distribution representations are

$$S(t) = e^{-\lambda t} \quad f(t) = \lambda e^{-\lambda t} \quad h(t) = \lambda \quad H(t) = \lambda t \text{ for } t \geq 0$$

There are several probabilistic properties of the exponential distribution that are useful in understanding how it is unique and when it should be applied. In the properties to be outlined below, the nonnegative lifetime T typically has the exponential distribution with parameter λ, which denotes the number of failures per unit time. The symbol $\sim$ means "is distributed as." Proofs of these results are given in most reliability textbooks.

THEOREM 3.1   (*Memoryless Property*) *If* $T \sim exponential(\lambda)$, *then*

$$P[T \geq t] = P[T \geq t + s | T \geq s] \quad t \geq 0; s \geq 0$$

As shown in Figure 3.8 for $\lambda = 1$ and $s = 0.5$, the memoryless property indicates that the conditional survivor function for the lifetime of an item that has survived to time s is identical to the survivor function for the lifetime of a brand new item. This used-as-good-as-new assumption is very strong. The exponential lifetime model should not be applied to mechanical components that undergo wear (e.g., bearings) or fatigue (e.g., structural supports) or electrical components that contain an element that burns away (e.g., filaments) or degrades with time (e.g., batteries). An electrical component for which the exponential lifetime assumption may be justified is a fuse. A fuse is designed to fail when there is a power surge that causes the fuse to burn out. Assuming that the fuse does not undergo any weakening or degradation over time and that power surges that cause failure occur with

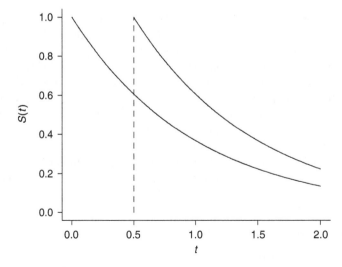

FIGURE 3.8   The memoryless property.

equal likelihood over time, the exponential lifetime assumption is appropriate, and a used fuse that has not failed is as good as a new one.

The exponential distribution should be applied judiciously as the memoryless property restricts its applicability. It is usually misapplied for the sake of simplicity as the statistical techniques for the exponential distribution are particularly tractable, or small sample sizes do not support more than a one-parameter distribution.

THEOREM 3.2 *The exponential distribution is the only continuous distribution with the memoryless property.*

This result indicates that the exponential distribution is the only continuous lifetime distribution for which the conditional lifetime distribution of a used item is identical to the original lifetime distribution. The only discrete distribution with the memoryless property is the geometric distribution.

THEOREM 3.3 *If $T \sim exponential(\lambda)$, then*

$$E[T^s] = \frac{\Gamma(s+1)}{\lambda^s} \quad s > -1$$

where the gamma function *is defined by*

$$\Gamma(\alpha) = \int_0^\infty x^{\alpha-1} e^{-x} dx$$

When s is a nonnegative integer, this expression reduces to $E[T^s] = s!/\lambda^s$. By setting $s = 1$ and 2, the mean and variance can be obtained:

$$E[T] = \frac{1}{\lambda} \quad V[T] = \frac{1}{\lambda^2}$$

THEOREM 3.4 *(Self-Reproducing) If $T_1, T_2, \ldots, T_n$ are independent, $T_i \sim$ exponential(λ_i), for $i = 1, 2, \ldots, n$, and $T = min\{T_1, T_2, \ldots, T_n\}$, then*

$$T \sim exponential \left(\sum_{i=1}^n \lambda_i \right)$$

This result indicates that the minimum of n exponential random lifetimes also has the exponential distribution, as alluded to in the previous section. This is important in two applications. First, if n independent exponential components, each with exponential times to failure, are arranged in series, the distribution of the system failure time is also exponential with a failure rate equal to the sum of the component failure rates. When the n components have the same failure rate λ, the system lifetime is exponential with failure rate $n\lambda$. Second, when there are several independent, exponentially distributed *causes* of failure competing for the lifetime of an item (e.g., failing by open or short circuit for an electronic item or death by various risks for a human being), the lifetime can be modeled as the minimum of the individual lifetimes associated with each cause of failure.

THEOREM 3.5   *If $T_1, T_2, \ldots, T_n$ are   independent   and   identically   distributed exponential(λ) random variables, then*

$$2\lambda \sum_{i=1}^{n} T_i \sim \chi^2(2n)$$

where $\chi^2(2n)$ denotes the chi-square distribution with $2n$ degrees of freedom.

This property is useful for determining a confidence interval for a λ based on a data set of n independent exponential lifetimes. For instance, with probability $1 - \alpha$,

$$\chi^2_{2n, 1-\alpha/2} < 2\lambda \sum_{i=1}^{n} T_i < \chi^2_{2n, \alpha/2}$$

where the left- and right-hand sides of this inequality are the $\alpha/2$ and $1 - \alpha/2$ fractiles of the chi-square distribution with $2n$ degrees of freedom, that is, the second subscript denotes *right*-hand tail areas. Rearranging this expression yields a $100(1 - \alpha)\%$ confidence interval for λ:

$$\frac{\chi^2_{2n, 1-\alpha/2}}{2\sum_{i=1}^{n} T_i} < \lambda < \frac{\chi^2_{2n, \alpha/2}}{2\sum_{i=1}^{n} T_i}$$

THEOREM 3.6   *If $T_1, T_2, \ldots, T_n$ are   independent   and   identically   distributed exponential(λ) random variables, $T_{(1)}, T_{(2)}, \ldots, T_{(n)}$ are the corresponding order statistics (the observations sorted in ascending order), $G_k = T_{(k)} - T_{(k-1)}$ for $k = 1, 2, \ldots, n$, and if $T_{(0)} = 0$, then*

- $P[G_k \geq t] = e^{-(n-k+1)\lambda t}$; $t \geq 0$; $k = 1, 2, \ldots, n$.
- $G_1, G_2, \ldots, G_n$ *are independent.*

This property is most easily interpreted in terms of a life test of n items with exponential(λ) lifetimes. Assume that the items placed on the life test are *not* replaced with new items when they fail. The ith item fails at time $t_{(i)}$, and $G_i = t_{(i)} - t_{(i-1)}$ is the time between the $(i-1)$st and ith failure, for $i = 1, 2, \ldots, n$, as indicated in Figure 3.9 for $n = 4$. The result states that these gaps (G_i's) between the failure times are independent and exponentially distributed. The proof of this theorem relies on the memoryless property and the self-reproducing property of the exponential distribution, which implies that when the ith failure occurs the time until the next failure is the minimum of $n - i$ independent exponential random variables.

THEOREM 3.7   *If $T_1, T_2, \ldots, T_n$ are   independent   and   identically   distributed exponential(λ) random variables and $T_{(r)}$ is the rth order statistic, then*

$$E[T_{(r)}] = \sum_{k=1}^{r} \frac{1}{(n-k+1)\lambda}$$

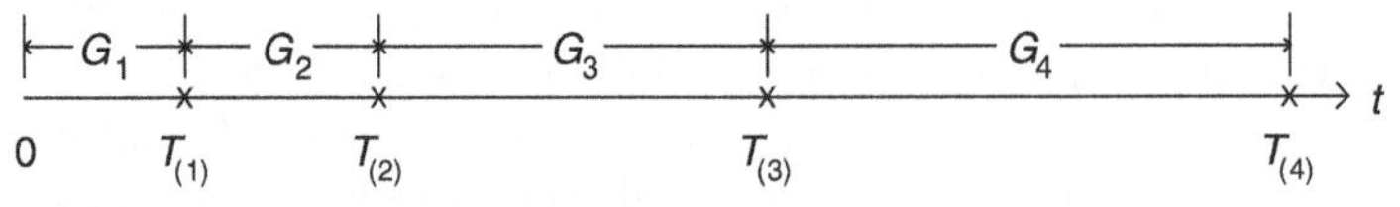

FIGURE 3.9   Order statistics and gap statistics.

and

$$V[T_{(r)}] = \sum_{k=1}^{r} \frac{1}{[(n-k+1)\lambda]^2}$$

for $r = 1, 2, \ldots, n$.

The expected value and variance of the rth-ordered failure are simple functions of n, λ, and r. The proof of this result is straightforward, as the gaps between order statistics are independent exponential random variables from the previous theorem, and the rth ordered failure on a life test of n items is the sum of the first r gaps. This result is useful in determining the expected time to complete a life test that is discontinued after r of the n items on test fail.

THEOREM 3.8 *If $T_1, T_2, \ldots$ are independent and identically distributed exponential(λ) random variables denoting the interevent times for a point process, then the number of events in the interval $[0, t]$ has the Poisson distribution with parameter λt.*

This property is related to the memoryless property and can be applied to a component that is subjected to shocks occurring randomly over time. It states that if the time between shocks is exponential(λ) then the number of shocks occurring by time t has the Poisson distribution with parameter λt. This result also applies to the failure time of a cold standby system of n identical exponential components in which nonoperating units do not fail and sensing and switching are perfect. The probability of fewer than n failures by time t (the system reliability) is

$$\sum_{k=0}^{n-1} \frac{(\lambda t)^k}{k!} e^{-\lambda t}$$

The exponential distribution, for which the item under study does not age in a probabilistic sense, is the simplest possible lifetime model. Another popular distribution that arises in many reliability applications is the Weibull distribution, which is presented next.

3.4.3   Weibull Distribution

The exponential distribution is limited in applicability because of the memoryless property. The assumption that a lifetime has a constant failure rate is often too restrictive or inappropriate. Mechanical items typically degrade over time and hence are more likely to follow a distribution with a strictly increasing hazard function. The Weibull distribution is a generalization of the exponential distribution that is appropriate for modeling lifetimes having constant, strictly increasing, and strictly decreasing hazard functions. The four lifetime distribution representations for the Weibull distribution are

$$S(t) = e^{-(\lambda t)^\kappa} \quad f(t) = \kappa \lambda^\kappa t^{\kappa-1} e^{-(\lambda t)^\kappa} \quad h(t) = \kappa \lambda^\kappa t^{\kappa-1} \quad H(t) = (\lambda t)^\kappa$$

for all $t \geq 0$, where $\lambda > 0$ and $\kappa > 0$ are the scale and shape parameters of the distribution. The hazard function approaches zero from infinity for $\kappa < 1$, is constant for $\kappa = 1$, the exponential case, and increases from zero when $\kappa > 1$. Hence, the Weibull distribution can attain hazard function shapes in both the IFR and DFR classes and includes the exponential distribution as a special case. One other special case occurs when $\kappa = 2$, commonly known as the Rayleigh distribution, which has a linear hazard function with slope $2\lambda^2$. When $3 < \kappa < 4$, the shape of the probability density function resembles that of a normal probability density function.

Using the expression

$$E[T^r] = \frac{r}{\kappa \lambda^r} \Gamma\left(\frac{r}{\kappa}\right)$$

for $r = 1, 2, \ldots$, the mean and variance for the Weibull distribution are

$$\mu = \frac{1}{\lambda \kappa} \Gamma\left(\frac{1}{\kappa}\right)$$

and

$$\sigma^2 = \frac{1}{\lambda^2} \left\{ \frac{2}{\kappa} \Gamma\left(\frac{2}{\kappa}\right) - \left[\frac{1}{\kappa} \Gamma\left(\frac{1}{\kappa}\right)\right]^2 \right\}$$

Example 3.10

The lifetime of a machine used continuously under known operating conditions has the Weibull distribution with $\lambda = 0.00027$ and $\kappa = 1.55$, where time is measured in hours. (Estimating the parameters for the Weibull distribution from a data set will be addressed subsequently, but the parameters are assumed to be known for this example.) What is the mean time to failure and the probability the machine will operate for 5000 hours?

The mean time to failure is

$$\mu = E[T] = \frac{1}{(0.00027)(1.55)} \Gamma\left(\frac{1}{1.55}\right) = 3331 \text{ hours}$$

The probability that the machine will operate for 5000 hours is

$$S(5000) = e^{-[(0.00027)(5000)]^{1.55}} = 0.203 \qquad \blacksquare$$

The Weibull distribution also has the self-reproducing property, although the conditions are slightly more restrictive than for the exponential distribution. If $T_1, T_2, \ldots, T_n$ are independent component lifetimes having the Weibull distribution with identical shape parameters, then the minimum of these values has the Weibull distribution. More specifically, if $T_i \sim \text{Weibull}(\lambda_i, \kappa)$ for $i = 1, 2, \ldots, n$, then

$$\min\{T_1, T_2, \ldots, T_n\} \sim \text{Weibull}\left(\left(\sum_{i=1}^{n} \lambda_i^\kappa\right)^{1/\kappa}, \kappa\right)$$

Although the exponential and Weibull distributions are popular lifetime models, they are limited in their modeling capability. For example, if it were determined that an item had a bathtub-shaped hazard function, neither of these two models would be appropriate. Dozens of other models have been developed over the years, such as the gamma, lognormal, inverse Gaussian, exponential power, and log logistic distributions. These distributions provide further modeling flexibility beyond the exponential and Weibull distributions.

3.5 Parameter Estimation in Survival Analysis

This section investigates fitting the two distributions presented in Section 3.4, the exponential and Weibull distributions, to a data set of failure times. Other distributions, such as the gamma distribution or the exponential power distribution, have analogous methods of parameter estimation. Two sample data sets are introduced and are used throughout this section.

The analysis in this section assumes that a random sample of n items from a population has been placed on a test and subjected to typical field operating conditions. The data values are assumed to be independent and identically distributed random lifetimes from a particular population distribution. As with all statistical inference, care must be taken to ensure that a random sample of lifetimes is collected. Consequently, random numbers should be used to determine which n items to place on test. Laboratory conditions should adequately mimic field conditions. Only representative items should be placed on test because items manufactured using a previous design may have a different failure pattern than those with the current design.

A data set for which all failure times are known is called a complete data set. Figure 3.10 illustrates a complete data set of $n = 5$ items placed on test, where the X's denote failure times. (The term "items placed on test" is used instead of "sample size" or "number of observations" because of potential confusion when right censoring is introduced.) The likelihood function for a complete data set of n items on test is given by

$$L(\boldsymbol{\theta}) = \prod_{i=1}^{n} f(t_i)$$

where $t_1, t_2, \ldots, t_n$ are the failure times and $\boldsymbol{\theta}$ is a vector of unknown parameters. (Although lowercase letters are used to denote the failure times here to be consistent with the notation for censoring times, the failure times are nonnegative random variables.)

Censoring occurs frequently in lifetime data because it is often impossible or impractical to observe the lifetimes of all the items on test. A censored observation occurs when only a bound is known on the time of failure. If a data set contains one or more censored observations, it is called a *censored data set*. The most frequent type of censoring is known as *right censoring*. In a right-censored data set, one or more items have only a lower bound known on the lifetime. The number of items placed on test is still denoted by n and the number of observed failures is denoted by r.

One special case of right censoring is considered here. Type II or *order statistic* censoring corresponds to terminating a study upon one of the ordered failures. Figure 3.11 shows the case of $n = 5$ items are placed on a test that is terminated when $r = 3$ failures are observed. The third and fourth items on test had their failure times right censored.

Writing the likelihood function for a censored data set requires some additional notation. As before, let $t_1, t_2, \ldots, t_n$ be lifetimes sampled randomly from a population. The corresponding right-censoring times are denoted by $c_1, c_2, \ldots, c_n$. The set U contains the indexes

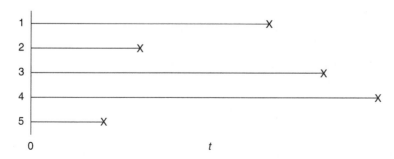

FIGURE 3.10 A complete data set with $n = 5$.

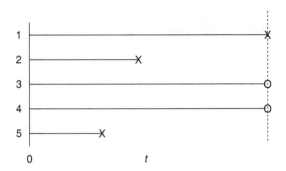

FIGURE 3.11   A right-censored data set with $n = 5$ and $r = 3$.

of the items that are observed to fail during the test (the uncensored observations):

$$U = \{i | t_i \leq c_i\}$$

The set C contains the indexes of the items whose failure time exceeds the corresponding censoring time (they are right censored):

$$C = \{i | t_i > c_i\}$$

Let $x_i = \min\{t_i, c_i\}$. The likelihood function is:

$$L(\boldsymbol{\theta}) = \prod_{i \in U} f(t_i) \prod_{i \in C} S(c_i) = \prod_{i \in U} f(x_i) \prod_{i \in C} S(x_i)$$

The reason that the survivor function is the appropriate term in the likelihood function for a right-censored observation is that $S(c_i)$ is the probability that item i survives to censoring time c_i. The log likelihood function is

$$\log L(\boldsymbol{\theta}) = \sum_{i \in U} \log f(x_i) + \sum_{i \in C} \log S(x_i)$$

As the density function is the product of the hazard function and the survivor function, the log likelihood function can be simplified to

$$\log L(\boldsymbol{\theta}) = \sum_{i \in U} \log h(x_i) + \sum_{i \in U} \log S(x_i) + \sum_{i \in C} \log S(x_i)$$

or

$$\log L(\boldsymbol{\theta}) = \sum_{i \in U} \log h(x_i) + \sum_{i=1}^{n} \log S(x_i)$$

Finally, as $H(t) = -\log S(t)$, the log likelihood can be written in terms of the hazard and cumulative hazard functions only as

$$\log L(\boldsymbol{\theta}) = \sum_{i \in U} \log h(x_i) - \sum_{i=1}^{n} H(x_i)$$

The choice of which of these three expressions for the log likelihood may be used for a particular distribution depends on the particular forms of $S(t), f(t), h(t)$, and $H(t)$.

3.5.1 Data Sets

Two lifetime data sets are presented here that are used to illustrate inferential techniques for survivor (reliability) data. The two types of lifetime data sets presented here are a complete data set, where all failure times are observed, and a Type II right-censored data set (order statistic right censoring).

Example 3.11

A complete data set of $n = 23$ ball bearing failure times to test the endurance of deep-groove ball bearings has been extensively studied (e.g., Meeker and Escobar, 1998, page 4). The ordered set of failure times measured in 10^6 revolutions is

| 17.88 | 28.92 | 33.00 | 41.52 | 42.12 | 45.60 | 48.48 | 51.84 | 51.96 |

| 54.12 | 55.56 | 67.80 | 68.64 | 68.64 | 68.88 | 84.12 | 93.12 | 98.64 |

| 105.12 | 105.84 | 127.92 | 128.04 | 173.40 |

Example 3.12

A Type II right-censored data set of $n = 15$ automotive a/c switches is given by Kapur and Lamberson (1977, pages 253–254). The test was terminated when the fifth failure occurred. The $r = 5$ ordered observed failure times measured in number of cycles are

| 1410 | 1872 | 3138 | 4218 | 6971 |

Although the choice of "time" as cycles in this case is discrete, the data will be analyzed as continuous data.

3.5.2 Exponential Distribution

The exponential distribution is popular due to its tractability for parameter estimation and inference. Using the failure rate λ to parameterize the distribution, recall that the survivor, density, hazard, and cumulative hazard functions are

$$S(t) = e^{-\lambda t} \quad f(t) = \lambda e^{-\lambda t} \quad h(t) = \lambda \quad H(t) = \lambda t \text{ for all } t \geq 0$$

We begin with the analysis of a complete data set consisting of failure times $t_1, t_2, \ldots, t_n$. Since all of the observations belong to the index set U, the log likelihood function derived earlier becomes

$$\log L(\lambda) = \sum_{i=1}^{n} \log h(x_i) - \sum_{i=1}^{n} H(x_i)$$

$$= \sum_{i=1}^{n} \log \lambda - \sum_{i=1}^{n} \lambda t_i$$

$$= n \log \lambda - \lambda \sum_{i=1}^{n} t_i$$

The maximum likelihood estimator for λ is found by maximizing the log likelihood function. To determine the maximum likelihood estimator for λ, the single element "score vector"

$$U(\lambda) = \frac{\partial \log L(\lambda)}{\partial \lambda} = \frac{n}{\lambda} - \sum_{i=1}^{n} t_i$$

often called the score statistic, is equated to zero, yielding

$$\hat{\lambda} = \frac{n}{\sum_{i=1}^{n} t_i}$$

Example 3.13

Consider the complete data set of $n = 23$ ball bearing failure times. For this particular data set, the total time on test is $\sum_{i=1}^{n} t_i = 1661.16$, yielding a maximum likelihood estimator

$$\hat{\lambda} = \frac{n}{\sum_{i=1}^{n} t_i} = \frac{23}{1661.16} = 0.0138$$

failure per 10^6 revolutions.

As the data set is complete, an exact 95% confidence interval for the failure rate of the distribution can be determined. Since $\chi^2_{46,0.975} = 29.16$ and $\chi^2_{46,0.025} = 66.62$, the confidence interval derived in Section 3.4:

$$\frac{\chi^2_{2n,1-\alpha/2}}{2\sum_{i=1}^{n} T_i} < \lambda < \frac{\chi^2_{2n,\alpha/2}}{2\sum_{i=1}^{n} T_i}$$

becomes

$$\frac{(0.0138)(29.16)}{46} < \lambda < \frac{(0.0138)(66.62)}{46}$$

or

$$0.00878 < \lambda < 0.0201$$

Note that, due to the use of the chi-square distribution for this confidence interval, the interval is not symmetric about the maximum likelihood estimator. For this and subsequent examples, care has been taken to perform intermediate calculations involving numeric quantities such as critical values or total time on test values to as much precision as possible; then final values are reported using only significant digits.

Figure 3.12 shows the empirical survivor function, which takes a downward step of $1/n = 1/23$ at each data point, along with the survivor function for the fitted exponential distribution. It is apparent from this figure that the exponential distribution is a very poor fit. This particular data set was chosen for this example to illustrate one of the shortcomings of using the exponential distribution to model any data set without assessing the adequacy of the fit. Extreme caution must be exercised when using the exponential distribution since, as indicated in Figure 3.12, the exponential distribution might be a poor fit. The appropriate distribution is probably in the IFR class, as the ball bearings are wearing out. As shown subsequently, the Weibull distribution is a much better approximation to this particular data set. As the exponential distribution can be fitted to any data set that has at least one observed failure, the adequacy of the model must always be assessed. The point and interval estimators associated with the exponential distribution are meaningful only if the data set is a random sample from an exponential population. ∎

The importance of model adequacy assessments, such as those indicated in the previous example, applies to all fitted distributions, not just the exponential distribution. Furthermore, if a modeler knows the failure physics (e.g., fatigue crack growth) underlying a process, then an appropriate model consistent with the failure physics should be chosen.

We now turn to the analysis of right-censored data sets drawn from exponential populations. The previous discussion concerning complete data sets is a special case of Type II

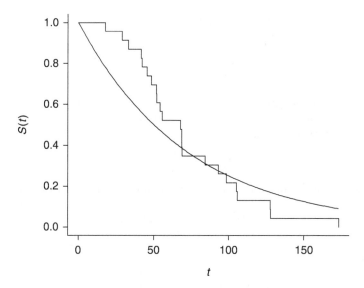

FIGURE 3.12   Empirical and exponential fitted survivor functions for the ball bearing data.

censoring when $r = n$. As before, assume that the failure times are $t_1, t_2, \ldots, t_n$, the test is terminated upon the rth ordered failure, the censoring times are $c_1 = c_2 = \cdots = c_n = t_{(r)}$ for all items, and $x_i = \min\{t_i, c_i\}$ for $i = 1, 2, \ldots, n$.

The log likelihood function is

$$\log L(\lambda) = \sum_{i \in U} \log h(x_i) - \sum_{i=1}^{n} H(x_i)$$

$$= \sum_{i \in U} \log \lambda - \sum_{i=1}^{n} \lambda x_i$$

$$= r \log \lambda - \lambda \sum_{i=1}^{n} x_i$$

Since there are r observed failures, the expression

$$\sum_{i=1}^{n} x_i$$

is often called the *total time on test* as it represents the total accumulated time that the n items accrue while on test. To determine the maximum likelihood estimator, the log likelihood function is differentiated with respect to λ,

$$U(\lambda) = \frac{\partial \log L(\lambda)}{\partial \lambda} = \frac{r}{\lambda} - \sum_{i=1}^{n} x_i$$

and is equated to zero, yielding the maximum likelihood estimator

$$\hat{\lambda} = \frac{r}{\sum_{i=1}^{n} x_i}$$

Exact confidence intervals and hypothesis tests concerning λ can also be derived in the Type II censoring case by using the result

$$2\lambda\sum_{i=1}^{n} x_i = \frac{2r\lambda}{\hat{\lambda}} \sim \chi^2(2r)$$

where $\chi^2(2r)$ is the chi-square distribution with $2r$ degrees of freedom. This result can be proved in an analogous fashion to the case of a complete data set. Using this fact, it can be stated with probability $1 - \alpha$ that

$$\chi^2_{2r,1-\alpha/2} < \frac{2r\lambda}{\hat{\lambda}} < \chi^2_{2r,\alpha/2}$$

Rearranging terms yields an exact $100(1-\alpha)\%$ confidence interval for the failure rate λ:

$$\frac{\hat{\lambda}\chi^2_{2r,1-\alpha/2}}{2r} < \lambda < \frac{\hat{\lambda}\chi^2_{2r,\alpha/2}}{2r}$$

Example 3.14

Consider the Type II right-censored data set of automotive switches, where $n = 15$ and there are $r = 5$ observed failures, which are

$$t_{(1)} = 1410, \ t_{(2)} = 1872, \ t_{(3)} = 3138, \ t_{(4)} = 4218, \ t_{(5)} = 6971$$

For this particular data set, the total time on test is $\sum_{i=1}^{n} x_i = 87{,}319$ cycles, yielding a maximum likelihood estimator

$$\hat{\lambda} = \frac{r}{\sum_{i=1}^{n} x_i} = \frac{5}{87{,}319} = 0.00005726$$

failure per cycle. Equivalently, the maximum likelihood estimator for the mean of the distribution is

$$\hat{\mu} = \frac{\sum_{i=1}^{n} x_i}{r} = \frac{87{,}319}{5} = 17{,}464$$

cycles to failure. As the data set is Type II right censored, an exact 95% confidence interval for the failure rate of the distribution can be determined. Using the chi-square critical values, $\chi^2_{10,0.975} = 3.247$ and $\chi^2_{10,0.025} = 20.49$, the formula for the confidence interval

$$\frac{\hat{\lambda}\chi^2_{2r,1-\alpha/2}}{2r} < \lambda < \frac{\hat{\lambda}\chi^2_{2r,\alpha/2}}{2r}$$

becomes

$$\frac{(0.00005726)(3.247)}{10} < \lambda < \frac{(0.00005726)(20.49)}{10}$$

or

$$0.00001859 < \lambda < 0.0001173$$

Taking reciprocals, this is equivalent to a 95% confidence interval for the mean number of cycles to failure of

$$8525 < \mu < 53{,}785$$

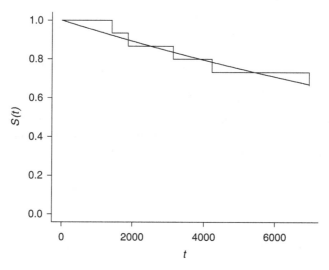

FIGURE 3.13   Empirical and exponential fitted survivor functions for the automotive a/c switch data.

Not surprisingly, with only five observed failures, this is a rather wide confidence interval for μ, and hence there is not as much precision as in the ball bearing example, where there were 23 observed failures. Assessing the adequacy of the fit is more difficult in the case of censoring, as it is impossible to determine what the lifetime distribution looks like after the last observed failure time, which is 6971 cycles in this case. Figure 3.13 shows the empirical survivor function and the associated fitted exponential survivor function. In this case, the exponential distribution appears to adequately model the lifetimes through 6971 cycles. ■

   Situations often arise when it is useful to compare the failure rates of two exponential populations based on data collected from each of the two populations. Examples include comparing the survival of one brand of integrated circuit versus another and comparing the survival of a single item at two levels of an environmental variable. Let the failure rate in the first population be λ_1 and the failure rate in the second population be λ_2.

   As in the previous subsections, x denotes the minimum of the lifetime t and the censoring time c. The two data sets are denoted by

$$x_{11}, x_{12}, \ldots, x_{1n_1}$$

the n_1 values from the first population, and

$$x_{21}, x_{22}, \ldots, x_{2n_2}$$

the n_2 values from the second population. Thus x_{ji} is observation i (failure or right-censoring time) from population j. Assume further that $r_1 > 0$ failures are observed in the first population and that $r_2 > 0$ failures are observed in the second population. For tractability, it is assumed that the tests performed on both populations use Type II censoring. This assumption allows exact confidence intervals for λ_1/λ_2 to be derived. Approximate methods exist when other types of censoring are used. In addition, these methods generalize to the case of comparing more than two populations.

Since $2\lambda_1\sum_{i=1}^{n_1} x_{1i}$ has the chi-square distribution with $2r_1$ degrees of freedom and $2\lambda_2\sum_{i=1}^{n_2} x_{2i}$ has the chi-square distribution with $2r_2$ degrees of freedom, the statistic

$$\frac{2\lambda_1\sum_{i=1}^{n_1} x_{1i}/(2r_1)}{2\lambda_2\sum_{i=1}^{n_2} x_{2i}/(2r_2)} = \frac{r_2\lambda_1\sum_{i=1}^{n_1} x_{1i}}{r_1\lambda_2\sum_{i=1}^{n_2} x_{2i}} = \frac{\lambda_1\hat{\lambda}_2}{\lambda_2\hat{\lambda}_1}$$

has the F distribution with $2r_1$ and $2r_2$ degrees of freedom. This is true because the ratio of two independent chi-square random variables divided by their respective degrees of freedom results in an F random variable. So with probability $1 - \alpha$

$$F_{2r_1,2r_2,1-\alpha/2} < \frac{\lambda_1\hat{\lambda}_2}{\lambda_2\hat{\lambda}_1} < F_{2r_1,2r_2,\alpha/2}$$

or

$$\frac{\hat{\lambda}_1}{\hat{\lambda}_2}F_{2r_1,2r_2,1-\alpha/2} < \frac{\lambda_1}{\lambda_2} < \frac{\hat{\lambda}_1}{\hat{\lambda}_2}F_{2r_1,2r_2,\alpha/2}$$

Two points are important to keep in mind with respect to this confidence interval. First, it is typically of interest to see whether this confidence interval contains 1, which indicates that there is no statistical evidence to conclude that the failure rates of the two populations are different. Second, if the null hypothesis

$$H_0 : \lambda_1 = \lambda_2$$

is to be tested directly, then there is cancellation in $\lambda_1\hat{\lambda}_2/\lambda_2\hat{\lambda}_1$ under H_0, so that the test statistic $\hat{\lambda}_2/\hat{\lambda}_1$ has the F distribution with $2r_1$ and $2r_2$ degrees of freedom.

3.5.3   Weibull Distribution

As mentioned earlier, the Weibull distribution is typically more appropriate for modeling the lifetimes of items with increasing and decreasing failure rates, such as mechanical items. We present the most general case of random censoring, rather than looking at each censoring mechanism individually.

As before, let $t_1, t_2, \ldots, t_n$ be the failure times, $c_1, c_2, \ldots, c_n$ be the censoring times, and $x_i = \min\{t_i, c_i\}$ for $i = 1, 2, \ldots, n$. Recall that the Weibull distribution has hazard and cumulative hazard functions

$$h(t) = \kappa\lambda(\lambda t)^{\kappa-1} \quad \text{and} \quad H(t) = (\lambda t)^{\kappa}$$

for $t \geq 0$. When there are r observed failures, the log likelihood function is

$$\log L(\lambda, \kappa) = \sum_{i\in U} \log h(x_i) - \sum_{i=1}^{n} H(x_i)$$

$$= \sum_{i\in U} (\log \kappa + \kappa\log\lambda + (\kappa - 1)\log x_i) - \sum_{i=1}^{n} (\lambda x_i)^{\kappa}$$

$$= r\log\kappa + \kappa r\log\lambda + (\kappa - 1)\sum_{i\in U} \log x_i - \lambda^{\kappa}\sum_{i=1}^{n} x_i^{\kappa}$$

and the 2×1 score vector has elements

$$U_1(\lambda, \kappa) = \frac{\partial \log L(\lambda, \kappa)}{\partial \lambda} = \frac{\kappa r}{\lambda} - \kappa\lambda^{\kappa-1}\sum_{i=1}^{n} x_i^{\kappa}$$

and

$$U_2(\lambda, \kappa) = \frac{\partial \log L(\lambda, \kappa)}{\partial \kappa} = \frac{r}{\kappa} + r \log \lambda + \sum_{i \in U} \log x_i - \sum_{i=1}^{n} (\lambda x_i)^{\kappa} \log(\lambda x_i)$$

When these equations are equated to zero, the simultaneous equations

$$\frac{\kappa r}{\lambda} - \kappa \lambda^{\kappa-1} \sum_{i=1}^{n} x_i^{\kappa} = 0$$

and

$$\frac{r}{\kappa} + r \log \lambda + \sum_{i \in U} \log x_i - \sum_{i=1}^{n} (\lambda x_i)^{\kappa} \log(\lambda x_i) = 0$$

have no closed-form solution for $\hat{\lambda}$ and $\hat{\kappa}$. One piece of good fortune, however, to avoid solving a 2×2 set of nonlinear equations, is that this first equation can be solved for λ in terms of κ as follows:

$$\lambda = \left(\frac{r}{\sum_{i=1}^{n} x_i^{\kappa}} \right)^{1/\kappa}$$

Using this expression for λ in terms of κ in the second element of the score vector yields a single, albeit more complicated, expression with κ as the only unknown. Applying some algebra, this equation reduces to

$$g(\kappa) = \frac{r}{\kappa} + \sum_{i \in U} \log x_i - \frac{r \sum_{i=1}^{n} x_i^{\kappa} \log x_i}{\sum_{i=1}^{n} x_i^{\kappa}} = 0$$

which must be solved iteratively using the Newton–Raphson technique or a fixed point method.

Example 3.15

It was seen in a previous example that the exponential distribution poorly approximated the ball bearing data set. The Weibull distribution is fit to the ball bearing failure times yielding maximum likelihood estimators $\hat{\lambda} = 0.0122$ and $\hat{\kappa} = 2.10$. Figure 3.14 shows the empirical survival function along with the exponential and Weibull fits to the data. It is clear that the Weibull distribution is far superior to the exponential for modeling the ball bearing failure times. This is due to the fact that the Weibull distribution is capable of modeling wear out for $\kappa > 1$.

In the case of the exponential distribution, we found a confidence *interval* for the parameter λ. In the case of the Weibull distribution, we desire a confidence *region* for the parameters λ and κ. Using the fact that the likelihood ratio statistic, $2[\log L(\hat{\lambda}, \hat{\kappa}) - \log L(\lambda, \kappa)]$, is asymptotically $\chi^2(2)$, a 95% confidence region for the parameters is all λ and κ satisfying

$$2[-113.691 - \log L(\lambda, k)] < 5.99$$

where $\log L(\hat{\lambda}, \hat{\kappa}) = -113.691$ and $\chi^2_{2,0.05} = 5.99$. The two degrees of freedom for the χ^2 distribution come from the fact that the Weibull distribution has two unknown parameters λ and κ. The 95% confidence region is shown in Figure 3.15, and, not surprisingly, the line $\kappa = 1$ is not interior to the region. This is further proof that the exponential distribution is not an appropriate model for this particular data set. The entire confidence region is in the $\kappa > 1$ region of the graph; this is statistically significant evidence provided by the data that the ball bearings are indeed wearing out. ∎

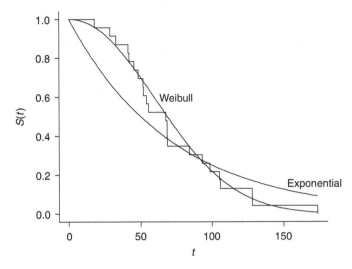

FIGURE 3.14 Empirical, exponential fitted, and Weibull fitted survivor functions for the ball bearing data.

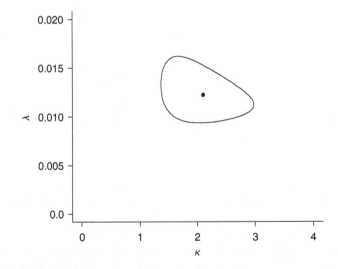

FIGURE 3.15 Confidence region for λ and κ for the ball bearing data.

3.6   Nonparametric Methods

In the previous sections, the focus was on developing parametric models for lifetimes and fitting them to a data set. The emphasis switches here to letting the data speak for itself, rather than approximating the lifetime distribution by one of the parametric models. There are several reasons to take this approach. First, it is not always possible to find a parametric model that adequately describes the lifetime distribution. This is particularly true of data arising from populations with nonmonotonic hazard functions. The most popular parametric models, such as the Weibull distribution, have monotonic hazard functions. A nonparametric analysis might provide more accurate estimates. Second, data sets are often so small that fitting a parametric model results in parameter estimators with confidence intervals that are so wide that the models are of little practical use. To cover all nonparametric methods used in reliability is not possible due to space constraints. We focus on just one such method here: the Kaplan–Meier product-limit survivor function estimate and the associated Greenwood's formula for assessing the precision of the estimate.

Let $y_1 < y_2 < \cdots < y_k$ be the k distinct failure times, and let d_j denote the number of observed failures at $y_j, j = 1, 2, \ldots, k$. Let $n_j = n(y_j)$ denote the number of items on test just before time $y_j, j = 1, 2, \ldots, k$, and it is customary to include any values that are censored at y_j in this count. Also, let $R(y_j)$ be the set of all indexes of items that are at risk just before time $y_j, j = 1, 2, \ldots, k$.

The search for a nonparametric survivor function estimator begins by assuming that the data were drawn from a discrete distribution with mass values at $y_1, y_2, \ldots, y_k$. For a discrete distribution, $h(y_j)$ is a conditional probability with interpretation $h(y_j) = P[T = y_j | T \geq y_j]$. For a discrete distribution, the survivor function can be written in terms of the hazard function at the mass values:

$$S(t) = \prod_{j \in R(t)'} [1 - h(y_j)] \quad t \geq 0$$

where $R(t)'$ is the complement of the risk set at time t. Thus a reasonable estimator for $S(t)$ is $\prod_{j \in R(t)'} [1 - \hat{h}(y_j)]$, which reduces the problem of estimating the survivor function to that of estimating the hazard function at each mass value. An appropriate element for the likelihood function at mass value y_j is

$$h(y_j)^{d_j} [1 - h(y_j)]^{n_j - d_j}$$

for $j = 1, 2, \ldots, k$. This expression is correct because d_j is the number of failures at y_j, $h(y_j)$ is the conditional probability of failure at y_j, $n_j - d_j$ is the number of items on test not failing at y_j, and $1 - h(y_j)$ is the probability of failing after time y_j conditioned on survival to time y_j. Thus the likelihood function for $h(y_1), h(y_2), \ldots, h(y_k)$ is

$$L(h(y_1), h(y_2), \ldots, h(y_k)) = \prod_{j=1}^{k} h(y_j)^{d_j} [1 - h(y_j)]^{n_j - d_j}$$

and the log likelihood function is

$$\log L(h(y_1), h(y_2), \ldots, h(y_k)) = \sum_{j=1}^{k} \{d_j \log h(y_j) + (n_j - d_j) \log [1 - h(y_j)]\}$$

The ith element of the score vector is

$$\frac{\partial \log L(h(y_1), h(y_2), \ldots, h(y_k))}{\partial h(y_i)} = \frac{d_j}{h(y_i)} - \frac{n_i - d_i}{1 - h(y_i)}$$

for $i = 1, 2, \ldots, k$. Equating this vector to zero and solving for $h(y_i)$ yields the maximum likelihood estimate:

$$\hat{h}(y_i) = \frac{d_i}{n_i}$$

This estimate for $\hat{h}(y_i)$ is sensible, since d_i of the n_i items on test at time y_i fail, so the ratio of d_i to n_i is an appropriate estimate of the conditional probability of failure at time y_i. This derivation may strike a familiar chord since, at each time y_i, estimating $h(y_i)$ with d_i divided by n_i is equivalent to estimating the probability of success, that is, failing at time y_i, for each of the n_i items on test. Thus, this derivation is equivalent to finding the maximum likelihood estimators for the probability of success for k binomial random variables.

Using this particular estimate for the hazard function at y_i, the survivor function estimate becomes

$$\hat{S}(t) = \prod_{j \in R(t)'} [1 - \hat{h}(y_j)]$$

$$= \prod_{j \in R(t)'} \left[1 - \frac{d_j}{n_j} \right]$$

commonly known as the Kaplan–Meier or product-limit estimate. One problem that arises with the product-limit estimate is that it is not defined past the last observed failure time. The usual way to handle this problem is to cut the estimator off at the last observed failure time y_k. The following example illustrates the product-limit estimate.

Example 3.16

An experiment is conducted to determine the effect of the drug 6-mercaptopurine (6-MP) on leukemia remission times (Lawless, 2003, page 5). A sample of $n = 21$ leukemia patients is treated with 6-MP, and the remission times are recorded. There are $r = 9$ individuals for whom the remission time is observed, and the remission times for the remaining 12 individuals are randomly censored on the right. There are $k = 7$ distinct observed failure times. Letting an asterisk denote a censored observation, the remission times (in weeks) are

$$6 \quad 6 \quad 6 \quad 6^* \quad 7 \quad 9^* \quad 10 \quad 10^* \quad 11^* \quad 13 \quad 16$$
$$17^* \quad 19^* \quad 20^* \quad 22 \quad 23 \quad 25^* \quad 32^* \quad 32^* \quad 34^* \quad 35^*$$

Find an estimate for $S(14)$.

Table 3.1 gives the values of y_j, d_j, n_j, and $1 - d_j/n_j$ for $j = 1, 2, \ldots, 7$. In particular, the product-limit survivor function estimate at $t = 14$ weeks is

$$\hat{S}(14) = \prod_{j \in R(14)'} \left[1 - \frac{d_j}{n_j} \right]$$

$$= \left[1 - \frac{3}{21} \right] \left[1 - \frac{1}{17} \right] \left[1 - \frac{1}{15} \right] \left[1 - \frac{1}{12} \right]$$

$$= \frac{176}{255}$$

$$\cong 0.69$$

TABLE 3.1   Product-Limit Calculations for the 6-MP Data

j	y_j	d_j	n_j	$1 - \dfrac{d_j}{n_j}$
1	6	3	21	$1 - \dfrac{3}{21}$
2	7	1	17	$1 - \dfrac{1}{17}$
3	10	1	15	$1 - \dfrac{1}{15}$
4	13	1	12	$1 - \dfrac{1}{12}$
5	16	1	11	$1 - \dfrac{1}{11}$
6	22	1	7	$1 - \dfrac{1}{7}$
7	23	1	6	$1 - \dfrac{1}{6}$

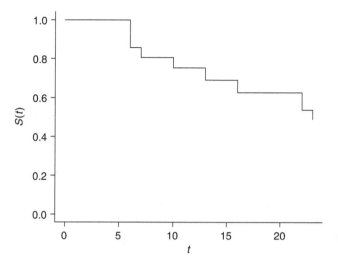

FIGURE 3.16 Product-limit survivor function estimate for the 6-MP data.

The product-limit survivor function estimate for all t values is plotted in Figure 3.16. Downward steps occur only at observed failure times. The effect of censored observations in the survivor function estimate is a larger downward step at the next subsequent failure time. If there are ties between the observations and a censoring time, as there is at time 6, our convention of including the censored values in the risk set means that there will be a larger downward step following this tied value. Note that the estimate is truncated at time 23, the last observed failure time. ∎

To find an estimate for the variance of the product-limit estimate is significantly more difficult than for the uncensored case. The Fisher and observed information matrices require a derivative of the score vector:

$$-\frac{\partial^2 \log L(h(y_1), h(y_2), \ldots, h(y_k))}{\partial h(y_i) \partial h(y_j)} = \frac{d_i}{h(y_i)^2} + \frac{n_i - d_i}{(1 - h(y_i))^2}$$

when $i = j$ and 0 otherwise, for $i = 1, 2, \ldots, k$, $j = 1, 2, \ldots, k$. Both the Fisher and observed information matrices are diagonal. Replacing $h(y_i)$ by its maximum likelihood estimate, the diagonal elements of the observed information matrix are

$$\left[-\frac{\partial^2 \log L(h(y_1), h(y_2), \ldots, h(y_k))}{\partial h(y_i)^2} \right]_{h(y_i) = d_i/n_i} = \frac{n_i^3}{d_i(n_i - d_i)}$$

for $i = 1, 2, \ldots, k$. Using this fact and some additional approximations, an estimate for the variance of the survivor function is

$$\hat{V}[\hat{S}(t)] = [\hat{S}(t)]^2 \sum_{j \in R(t)'} \frac{d_j}{n_j(n_j - d_j)}$$

commonly known as "Greenwood's formula." The formula can be used to find asymptotically valid confidence intervals for $S(t)$ by using the normal critical values as in the uncensored case:

$$\hat{S}(t) - z_{\alpha/2}\sqrt{\hat{V}[\hat{S}(t)]} < S(t) < \hat{S}(t) + z_{\alpha/2}\sqrt{\hat{V}[\hat{S}(t)]}$$

where $z_{\alpha/2}$ is the $1 - \alpha/2$ fractile of the standard normal distribution.

Example 3.17

For the 6-MP treatment group in the previous example, give a 95% confidence interval for the probability of survival to time 14.

The point estimator for the probability of survival to time 14 from the previous example is $\hat{S}(14) \cong 0.69$. Greenwood's formula is used to estimate the variance of the survivor function estimator at time 14:

$$\hat{V}[\hat{S}(14)] = [\hat{S}(14)]^2 \sum_{j \in R(14)'} \frac{d_j}{n_j(n_j - d_j)}$$

$$= (0.69)^2 \left[\frac{3}{21(21-3)} + \frac{1}{17(17-1)} + \frac{1}{15(15-1)} + \frac{1}{12(12-1)} \right]$$

$$\cong 0.011$$

Thus an estimate for the standard deviation of the survivor function estimate at $t = 14$ is $\sqrt{0.011} = 0.11$. A 95% confidence interval for $S(14)$ is

$$\hat{S}(14) - z_{0.025}\sqrt{\hat{V}[\hat{S}(14)]} < S(14) < \hat{S}(14) + z_{0.025}\sqrt{\hat{V}[\hat{S}(14)]}$$

$$0.69 - 1.96\sqrt{0.011} < S(14) < 0.69 + 1.96\sqrt{0.011}$$

$$0.48 < S(14) < 0.90$$

Figure 3.17 shows the 95% confidence bands for the survivor function for all t values. These have also been cut off at the last observed failure time, $t = 23$. The bounds are particularly wide as there are only $r = 9$ observed failure times. ∎

3.7   Assessing Model Adequacy

As there has been an emphasis on continuous lifetime distributions thus far in the chapter, the discussion here is limited to model adequacy tests for continuous distributions. The popular chi-square goodness-of-fit test can be applied to both continuous and discrete distributions, but suffers from the limitations of arbitrary interval widths and application only to large data sets. This section focuses on the Kolmogorov–Smirnov (KS) goodness-of-fit test for assessing model adequacy.

A notational difficulty arises in presenting the KS test. The survivor function $S(t)$ has been emphasized to this point in the chapter, but the cumulative distribution function, where $F(t) = P[T \leq t] = 1 - S(t)$ for continuous distributions, has traditionally been used to define the KS test statistic. To keep with this tradition, $F(t)$ is used in the definitions in this section.

The KS goodness-of-fit test is typically used to compare an empirical cumulative distribution function with a fitted or hypothesized parametric cumulative distribution function

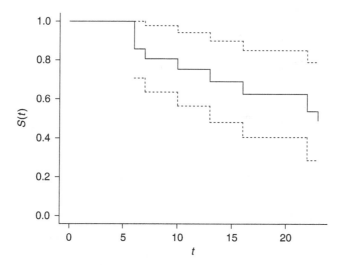

FIGURE 3.17 Confidence bands for the product-limit survivor function estimate for the 6-MP data.

for a continuous model. The KS test statistic is the maximum vertical difference between the empirical cumulative distribution function $\hat{F}(t)$ and a hypothesized or fitted cumulative distribution function $F_0(t)$. The null and alternative hypotheses for the test are

$$H_0 : F(t) = F_0(t)$$

$$H_1 : F(t) \neq F_0(t)$$

where $F(t)$ is the true underlying population cumulative distribution function. In other words, the null hypothesis is that data set of random lifetimes has been drawn from a population with cumulative distribution function $F_0(t)$. For a complete data set, the defining formula for the test statistic is

$$D_n = \sup_t |\hat{F}(t) - F_0(t)|$$

where *sup* is an abbreviation for *supremum*. This test statistic has intuitive appeal since larger values of D_n indicate a greater difference between $\hat{F}(t)$ and $F_0(t)$ and hence a poorer fit. In addition, D_n is independent of the parametric form of $F_0(t)$ when the cumulative distribution function is hypothesized. From a practical standpoint, computing the KS test statistic requires only a single loop through the n data values. This simplification occurs because $\hat{F}(t)$ is a nondecreasing step function and $F_0(t)$ is a nondecreasing continuous function, so the maximum difference must occur at a data value.

The usual computational formulas for computing D_n require a single pass through the data values. Let

$$D_n^+ = \max_{i=1,2,\ldots,n} \left(\frac{i}{n} - F_0(t_{(i)}) \right)$$

$$D_n^- = \max_{i=1,2,\ldots,n} \left(F_0(t_{(i)}) - \frac{i-1}{n} \right)$$

so that $D_n = \max\{D_n^+, D_n^-\}$. These computational formulas are typically easier to translate into computer code for implementation than the defining formula.

We consider only hypothesized (as opposed to fitted) cumulative distribution functions $F_0(t)$ here because the distribution of D_n is free of the hypothesized distribution specified.

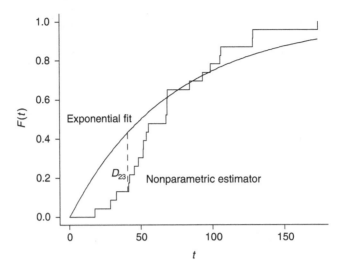

FIGURE 3.18   KS statistic for the ball bearing data set (exponential fit).

To illustrate the geometric aspects of the KS test statistic D_n, however, the fitted exponential distribution is compared to the empirical cumulative distribution function for the ball bearing data set. Figure 3.18 shows the empirical step cumulative distribution function $F(t)$ associated with the failure times of the $n = 23$ ball bearing failure times, along with the exponential fit $F_0(t)$. The maximum difference between these two cumulative distribution functions occurs just to the left of $t_{(4)} = 41.52$ and is $D_{23} = 0.301$ as indicated on the figure.

The test statistic for the KS test is nonparametric in the sense that it has the same distribution regardless of the distribution of the parent population under H_0 when all the parameters in the hypothesized distribution are known. The reason for this is that $F_0(t_{(1)}), F_0(t_{(2)}), \ldots, F_0(t_{(n)})$ have the same joint distribution as $U(0,1)$-order statistics under H_0 regardless of the functional form of F_0. These are often called *omnibus tests* as they are not tied to one particular distribution (e.g., the Weibull) and apply equally well to any hypothesized distribution $F_0(t)$. This also means that fractiles of the distribution of D_n depend on n only.

The rows in Table 3.2 denote the sample sizes and the columns denote several levels of significance. The values in the table are estimates of the $1 - \alpha$ fractiles of the distribution of D_n under H_0 in the all-parameters-known case (hypothesized, rather than fitted distribution) and have been determined by Monte Carlo simulation with one million replications. Not surprisingly, the fractiles are a decreasing function of n, as increased sample sizes will have lower sampling variability. Test statistics that exceed the appropriate critical value lead to rejecting H_0.

Example 3.18

Run the KS test (at $\alpha = 0.10$) to assess whether the ball bearing data set was drawn from a Weibull population with $\lambda = 0.01$ and $\kappa = 2$.

Note that the Weibull distribution in this example is a *hypothesized*, rather than *fitted* distribution, so the all-parameters-known case for determining critical values is appropriate. The goodness-of-fit test

$$H_0 : F(t) = 1 - e^{-(0.01t)^2}$$

$$H_1 : F(t) \neq 1 - e^{-(0.01t)^2}$$

TABLE 3.2 Selected Approximate KS Percentiles for Small Sample Sizes

n	$\alpha = 0.20$	$\alpha = 0.10$	$\alpha = 0.05$	$\alpha = 0.01$
1	0.900	0.950	0.975	0.995
2	0.683	0.776	0.842	0.930
3	0.565	0.636	0.708	0.829
4	0.493	0.565	0.624	0.733
5	0.447	0.509	0.563	0.668
6	0.410	0.468	0.519	0.617
7	0.381	0.436	0.483	0.576
8	0.358	0.409	0.454	0.542
9	0.339	0.388	0.430	0.513
10	0.323	0.369	0.409	0.489
11	0.308	0.352	0.391	0.468
12	0.296	0.338	0.376	0.449
13	0.285	0.325	0.361	0.433
14	0.275	0.314	0.349	0.418
15	0.266	0.304	0.338	0.404
16	0.258	0.295	0.327	0.392
17	0.250	0.286	0.318	0.381
18	0.243	0.278	0.309	0.370
19	0.237	0.271	0.302	0.361
20	0.232	0.265	0.294	0.352
21	0.226	0.259	0.287	0.345
22	0.221	0.253	0.281	0.337
23	0.217	0.248	0.275	0.330
24	0.212	0.242	0.269	0.323
25	0.208	0.237	0.264	0.317

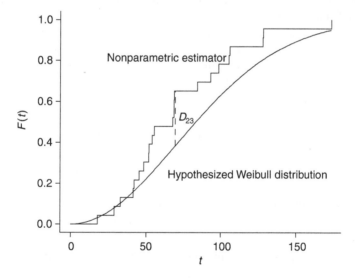

FIGURE 3.19 KS statistic for the ball bearing data set (Weibull fit).

does not involve any parameters estimated from data. The test statistic is $D_{23} = 0.274$. The empirical cumulative distribution function, the Weibull(0.01, 2) cumulative distribution function and the maximum difference between the two [which occurs at $t_{(15)} = 68.88$] are shown in Figure 3.19. At $\alpha = 0.10$, the critical value is 0.248, so H_0 is rejected. The test statistic is very close to the critical value for $\alpha = 0.05$, so the attained p-value for the test is approximately $p = 0.05$. ∎

The KS test can be extended in several directions. First, it can be adapted for the case the parameters are estimated from the data. Unfortunately, a separate table of critical values must be given for each fitted distribution. Second, the KS test can be adapted for right-censored data sets. Many researchers have devised approximate methods for determining the critical values for the KS test with random right censoring and parameters estimated from data. Finally, there are several variants of the KS test, such as the Anderson–Darling and Cramer–von Mises test, which improve on the power of the test.

3.8   Summary

The purpose of this chapter has been to introduce the mathematics associated with the design and assessment of systems with respect to their reliability. In specific, this chapter has:

- outlined basic techniques for describing the arrangement of components in a system by defining the structure function $\phi(\boldsymbol{x})$ that maps the states of the components to the state of the system;
- defined reliability as the probability that a nonrepairable item (component or system) is functioning at a specified time;
- introduced two techniques, definition and expectation, for determining the system reliability from component reliabilities;
- defined four functions, the survivor function $S(t)$, the probability density function $f(t)$, the hazard function $h(t)$, and the cumulative hazard function $H(t)$, which describe the distribution of a nonnegative random variable T, which denotes the lifetime of a component or system;
- reviewed formulas for calculating the mean, variance, and a fractile (percentile) of T;
- illustrated how to determine the system survivor function as a function of the component survivor functions;
- introduced two parametric lifetime distributions, the exponential and Weibull distributions, and outlined some of their properties;
- surveyed characteristics (e.g., right-censoring) of lifetime data sets;
- outlined point and interval estimation techniques for the exponential and distributions;
- derived a technique for comparing the failure rates of items with lifetimes drawn from two populations;
- derived and illustrated the nonparametric Kaplan–Meier product-limit estimate for the survivor function;
- introduced the Kolmogorov–Smirnov goodness-of-fit test for assessing model adequacy.

All of these topics are covered in more detail in the references. In addition, there are many topics that have not been covered at all, such as repairable systems, incorporating covariates into a survival model, competing risks, reliability growth, mixture models, failure modes and effects analysis, accelerated testing, fault trees, Markov models, and life testing. These topics and others are considered in the reliability literature, highlighted by the textbooks cited below. Software for reliability analysis has been written by several vendors and incorporated into existing statistical packages, such as SAS, S-Plus, and R.

References

1. Barlow, R. and Proschan, F. (1981), *Statistical Theory of Reliability and Life Testing Probability Models*, To Begin With, Silver Spring, MD.
2. Kalbfleisch, J.D. and Prentice, R.L. (2002), *The Statistical Analysis of Failure Time Data*, Second Edition, John Wiley & Sons, New York.
3. Kapur, K.C. and Lamberson, L.R. (1977), *Reliability in Engineering Design*, John Wiley & Sons, New York.
4. Lawless, J.F. (2003), *Statistical Models and Methods for Lifetime Data*, Second Edition, John Wiley & Sons, New York.
5. Meeker, W.Q. and Escobar, L.A. (1998), *Statistical Methods for Reliability Data*, John Wiley & Sons, New York.

4

Production Systems

Bobbie L. Foote
U.S. Military Academy

Katta G. Murty
University of Michigan

4.1 Production Planning Problem

The total problem of planning production consists of the following decisions: demand forecasting, the total floor space needed, the number and type of equipment, their aggregation into groups, the floor space needed for each group, the spatial relationship of each group relative to one another, the way the material and work pieces move within groups, the equipment and methods to move work between groups, the material to be ordered, the material to be produced in-house, the assembly layout and process, the quantities and timing of stock purchases, and the manner of storage of purchases and the items that are to be part of the information support system. There are no methods that determine the answers to these questions simultaneously. There are numerous models that consider each of these decisions as a subproblem to be optimized. When all the solutions are pieced together sometimes the whole will be suboptimal.

The production planning problem can also be looked at as a system of systems: forecasting, material handling, personnel, purchasing, quality assurance, production, assembly, marketing, design, finance, and other appropriate systems. At an advanced level one hopes that these systems integrate: that is, the design is such that it is easy to produce by using snap-in fasteners; materials easy to form; financial planning provides appropriate working

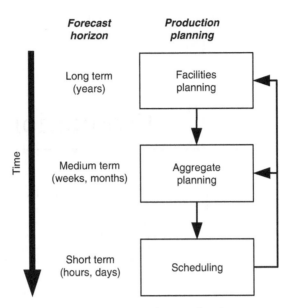

FIGURE 4.1   Planning and forecast horizons in production planning.

capital; purchases arrive on time and have appropriate quality. In this context, the interfaces of the systems become the appropriate design issue.

Forecasts form the basis of production planning. Strategic plans (e.g., where to locate factories, etc.) are based on long-term forecasts whereas aggregate production plans (allocating labor and capital resources for the next quarter's operations, say) are usually made a few months in advance, and production scheduling at the shop floor level may occur a few hours or days in advance of actual production. This also allows better information and a more detailed history to tailor production to demand or desired inventory levels. Figure 4.1 shows this graphically.

From the above comments, clearly forecasting is as much an art as a science. Given its importance, forecasting is introduced in Section 4.2. In production planning, demand forecasts are used for aggregate planning and more detailed materials requirement plans. These plans provide details on the components required and the relationships between the processing steps involved in production. This information is then used to design the facility (i.e., positioning of various machines and processing equipment within the facility) and schedule production. We assume that aggregate plans and more detailed material requirements plans are available and do not discuss these stages of production planning; the interested reader may consult Nahmias (1993) for details. Instead, we focus on the facility layout and scheduling issues in Sections 4.3 and 4.4, respectively.

4.2   Demand Forecasting

Forecasting is a key activity that influences many aspects of business planning and operation. In simple terms, a forecast is a prediction or estimate of the future. Forecasting is common at multiple levels, from macroeconomic forecasts (e.g., forecasts of oil prices) to microlevel estimates (of component requirements, say). We focus on forecasting in support of the production planning function. The *forecast horizon* refers to how far into the future we would like to make a prediction. Some common features of forecasts are as follows (Nahmias, 1993):

- *Forecasts are usually wrong*—Forecasts are *estimates* of a random variable (demand, price, etc.) and actual outcomes may be significantly different.
- *Aggregate forecasts are better*—Demand forecasts for an entire product family are more accurate (on a percentage basis) than those for each member of the product family.
- *Forecasting error increases with the forecast horizon*—Demand forecasts for the coming week are likely to be much more accurate than forecasts of next year's demand.
- *Environmental evidence*—The environment is very important. For example, if you are forecasting demand for spare parts for a military aircraft, then it is important to account for the type of operations in progress. For regular training, past history may be valid, but if the aircraft are being used in war, then peacetime history is likely to be invalid.

4.2.1 Commonly Used Techniques for Forecasting Demand

Models for controlling and replenishing of inventories in production have the aim of determining order quantities to minimize the sum of total overage costs (costs of excess inventory remaining at the end of the planning period) and underage costs (shortage costs, or costs of having less than the desired amount of stock at the end of the planning period). In production planning literature, the total overage (underage) cost is usually assumed to be proportional to the overage (shortage) amount or quantity, to make the analysis easier. Some companies find that a piecewise linear (PL) function provides a much closer representation of the true overage and underage costs. In these companies, there is a buffer with limited space in which excess inventory at the end of the planning period can be stored and retrieved later at a low cost (i.e., with minimum requirements of human-hours needed) per unit. Once this buffer is filled up, any remaining excess quantity has to be held at a location farther away that requires greater number of man hours for storing or retrieval/unit. A similar situation exists for underage cost as a function of the shortage amount. This clearly implies that the overage and underage costs are PL functions of the excess, shortage quantities. Determining optimum order quantities to minimize such unusual overage and underage cost functions is much harder with inventory control models using forecasting techniques in current literature. After reviewing the forecasting methods commonly used in production applications at present (Section 4.2.2), we will discuss a new nonparametric forecasting method (Section 4.2.3), which has the advantage of being able to accommodate such unusual overage and underage cost functions easily (Murty, 2006).

Almost all production management problems in practice are characterized by the uncertainty of demand during a future planning period; that is, this demand is a random variable. Usually, we do not have knowledge about its exact probability distribution, and the models for these problems have the objective of minimizing the sum of expected overage and underage costs. Successful production management systems depend heavily on good demand forecasts to provide data for inventory replenishment decisions. The output of forecasting is usually presented in the literature as the *forecasted demand quantity*; in reality it is an estimate of the expected demand during the planning period. Because of this, the purpose of forecasting is often misunderstood to be that of generating this single number, even though sometimes the standard deviation of demand is also estimated.

All commonly used methods for demand forecasting are parametric methods; they usually assume that demand is normally distributed, and they update its distribution by updating the parameters of the distribution, the mean μ, and the standard deviation σ. The most commonly used methods for updating the values of the parameters are the method of moving averages, and the exponential smoothing method.

The method of moving averages uses the average of n most recent observations on demand as the forecast for the expected demand for the next period. n is a parameter known as the order of the *moving average method* being used; typically it is between 3 to 6 or larger.

The other method, perhaps the most popular method in practice, is the exponential smoothing method introduced and popularized by Brown (1959). It takes $\hat{D}_{t+1}$, the forecast of expected demand during next period $t+1$, to be $\alpha x_t + (1-\alpha)\hat{D}_t$, where x_t is the observed demand during current period t, $\hat{D}_t$ is the forecasted expected demand for current period t, and $0 < \alpha \leq 1$ is a *smoothing constant*, which is the relative weight placed on the current observed demand. Typically, values of α between 0.1 and 0.4 are used, and normally the value of α is increased whenever the absolute value of the deviation between the forecast and observed demand exceeds a tolerance times the standard deviation. Smaller values of α (like 0.1) yield predicted values of expected demand that have a relatively smooth pattern, whereas higher values of α (like 0.4) lead to predicted values exhibiting significantly greater variation, but doing a better job of tracking the demand series. Thus using larger α makes forecasts more responsive to changes in the demand process, but will result in forecast errors with higher variance.

One disadvantage of both the method of moving averages and the exponential smoothing method is that when there is a definite trend in the demand process (either growing or falling), the forecasts obtained by them lag behind the trend. Variations of the exponential smoothing method to track trend linear in time in the demand process have been proposed (see Holt, 1957), but these have not proved very popular.

There are many more sophisticated methods for forecasting the expected values of random variables, for example, the Box–Jenkins ARIMA models (Box and Jenkins, 1970), but these methods are not popular for production applications, in which forecasts for many items are required.

4.2.2 Parametric Methods for Forecasting Demand Distributions

Using Normal Distribution with Updating of Expected Value and Standard Deviation in Each Period

As discussed in the previous section, all forecasting methods in the literature only provide an estimate of the expected demand during the planning period. The optimum order quantity to be computed depends of course on the entire probability distribution of demand, not just its expected value. So, almost everyone assumes that the distribution of demand is the normal distribution because of its convenience. One of the advantages that the normality assumption confers is that the distribution is fully characterized by only two parameters, the mean and the standard deviation, both of which can be very conveniently updated by the exponential smoothing or the moving average methods.

Let t be the current period, x_r the observed demand in period r for $r \leq t$, $\hat{D}_t$ the forecast (i.e., estimate) of expected demand in current period t (by either the exponential smoothing or the moving average methods, whichever is being used), and $\hat{D}_{t+1}, \hat{\sigma}_{t+1}$ the forecasts for expected demand, standard deviation of demand for the planning period which is the next period $t + 1$. Then these forecasts are:

Method	Forecast
Method of moving averages of order n	$\hat{D}_{t+1} = \frac{1}{n} \sum_{r=t-n+1}^{t} x_r$
Exponential smoothing method with smoothing constant α	$\hat{D}_{t+1} = \alpha x_t + (1 - \alpha)\hat{D}_t$
Method of moving averages of order n	$\hat{\sigma}_{t+1} = +\sqrt{(\sum_{r=t-n+1}^{t} (x_r - \hat{D}_{t+1})^2)/n}$

To get $\hat{\sigma}_{t+1}$ by the exponential smoothing method, it is convenient to use the mean absolute deviation (MAD), and use the formula: standard deviation $\sigma \approx (1.25)\text{MAD}$ when the distribution is the normal distribution. Let MAD_t denote the estimate of MAD for current period t. Then the forecasts obtained by the exponential smoothing method with smoothing parameter α for the next period $t + 1$ are:

$$\text{MAD}_{t+1} = \alpha|x_t - \hat{D}_t| + (1 - \alpha)\text{MAD}_t$$
$$\hat{\sigma}_{t+1} = (1.25)\text{MAD}_{t+1}$$

Usually $\alpha = 0.1$ is used to ensure stability of the estimates. And the normal distribution with mean $\hat{D}_{t+1}$ and standard deviation $\hat{\sigma}_{t+1}$ is taken as the forecast for the distribution of demand during the next period $t + 1$ for making any planning decisions under this procedure.

Using Normal Distribution with Updating of Expected Value and Standard Deviation Only when There Is Evidence of Change

In some applications, the distribution of demand is assumed to be the normal distribution, but estimates of its expected value and standard deviation are left unchanged until there is evidence that their values have changed. Foote (1995) discusses several statistical control tests on demand data being generated over time to decide when to re-estimate these parameters. Under this scheme, the method of moving averages is commonly used to estimate the expected value and the standard deviation from recent data whenever the control tests indicate that a change may have occurred.

Using Distributions Other Than Normal

In a few special applications in which the expected demand is low (i.e., the item is a slow-moving item), other distributions like the Poisson distribution are sometimes used, but by far the most popular distribution for making inventory management decisions is the normal distribution because of its convenience, and because using it has become a common practice historically.

For the normal distribution the mean is the mode (i.e., the value associated with the highest probability), and the distribution is symmetric around this value. If histograms of observed demand data of an item do not share these properties, it may indicate that the normal distribution is a poor approximation for the actual distribution of demand, in this case order quantities determined using the normality assumption may be far from being optimal.

These days, the industrial environment is very competitive with new products replacing the old periodically due to rapid advancements in technology. In this dynamic environment, the life cycles of components and end products are becoming shorter. Beginning with the introduction of the product, its life cycle starts with a growth period due to gradual market penetration of the product. This is followed by a stable period of steady demand. It is then followed by a final decline period of steadily declining demand, at the end of which the item disappears from the market. Also, the middle stable period seems to be getting shorter for many major components. Because of this constant rapid change, it is necessary to periodically update demand distributions based on recent data.

The distributions of demand for some components are far from being symmetric around the mean, and the skewness and shapes of their distributions also seem to be changing over time. Using a probability distribution like the normal defined by a mathematical formula, involving only a few parameters, it is not possible to capture changes taking place in the shapes of distributions of demand for such components. This is the disadvantage of existing forecasting methods based on an assumed probability distribution. Our conclusions can be erroneous if the true probability distribution of demand is very different from the assumed distribution.

Nonparametric methods use statistical learning, and base their conclusions on knowledge derived directly from data without any unwarranted assumptions. In the next section, we discuss a nonparametric method for forecasting the entire demand distribution (Murty, 2006) that uses the classical empirical probability distribution derived from the relative frequency histogram of time series data on demand. It has the advantage of being capable of updating all changes occurring in the probability distribution of demand, including those in the shape of this distribution.

Then, in the following section, we illustrate how optimal order quantities that optimize piecewise linear and other unusual cost functions discussed in the previous section can be easily computed using these empirical distributions.

4.2.3   A Nonparametric Method for Updating and Forecasting the Entire Demand Distribution

In production systems, the important random variables are daily or weekly (or whatever planning period is being used) demands of various items (raw materials, components, sub-assemblies, finished goods, spare parts, etc.) that companies either buy from suppliers, or sell to their customers. Observed values of these random variables in each period are generated automatically as a time series in the production process, and are usually available in the production databases of companies. In this section, we discuss a simple nonparametric method for updating changes in the probability distributions of these random variables using these data directly.

Empirical Distributions and Probability Density Functions

The concept of the probability distribution of a random variable evolved from the ancient practice of drawing histograms for the observed values of the random variable. The observed range of variation of the random variable is usually divided into a convenient number of value intervals (in practice about 10 to 25) of equal length, and the relative frequency of each interval is defined to be the proportion of observed values of the random variable that lie in that interval. The chart obtained by marking the value intervals on the horizontal axis, and erecting a rectangle on each interval with its height along the vertical axis equal to the relative frequency, is known as the relative frequency histogram of the random variable, or its discretized probability distribution. The relative frequency in each value interval I_i is the estimate of the probability p_i that the random variable lies in that interval; see Figure 4.2 for an example.

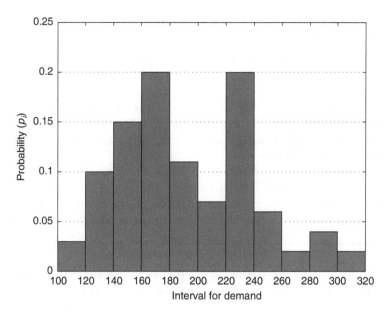

FIGURE 4.2 Relative frequency histogram for daily demand for a major component at a PC assembling plant in California.

Let $I_1, \ldots, I_n$ be the value intervals with $u_1, \ldots, u_n$ as their midpoints, and $p = (p_1, \ldots, p_n)$, the probability vector in the discretized probability distribution of the random variable. Let

$$\overline{\mu} = \sum_{i=1}^{n} u_i p_i, \quad \overline{\sigma} = \sqrt{\sum_{i=1}^{n} p_i (u_i - \overline{\mu})^2}$$

Then, $\overline{\mu}$, $\overline{\sigma}$ are estimates of the expected value μ and standard deviation σ of the random variable, respectively.

We will use the phrase *empirical distribution* to denote such a discretized probability distribution of a random variable, obtained either through drawing the histogram, or by updating a previously known discretized probability distribution based on recent data.

When mathematicians began studying random variables from the sixteenth century onwards, they found it convenient to represent the probability distribution of the random variable by the *probability density function*, which is the mathematical formula for the curve defined by the upper boundary of the relative frequency histogram in the limit as the length of the value interval is made to approach 0, and the number of observed values of the random variable goes to infinity. So the probability density function provides a mathematical formula for the height along the vertical axis of this curve as a function of the variable represented on the horizontal axis. Because it is a mathematically stated function, the probability density function lends itself much more nicely into mathematical derivations than the somewhat crude relative frequency histogram.

It is rare to see empirical distributions used in decision making models these days. Almost everyone uses mathematically defined density functions characterized by a small number of parameters (typically two or less) to represent probability distributions. In these decision making models, the only freedom we have in incorporating changes is to change the values of those parameters. This may be inadequate to capture all the dynamic changes occurring in the shapes of probability distributions from time to time.

Extending the Exponential Smoothing Method to Update the Empirical Probability Distribution of a Random Variable

We will now see that representing the probability distributions of random variables by their empirical distributions gives us unlimited freedom in making any type of change including changes in shape (Murty, 2002).

Let $I_1, \ldots, I_n$ be the value intervals, and $p_1, \ldots, p_n$ the probabilities associated with them in the present empirical distribution of a random variable. In updating this distribution, we have the freedom to change the values of all the p_i; this makes it possible to capture any change in the shape of the distribution.

Changes, if any, will reflect in recent observations on the random variable. The following table gives the present empirical distribution, histogram based on most recent observations on the random variable (e.g., the most recent k observations where k could be about 30), and x_i to denote the probabilities in the updated empirical distribution to be determined.

Value Interval	Probability vector in the		
	Present Empirical Distribution	Recent Histogram	Updated Empirical Distribution
I_1	p_1	f_1	x_1
$\vdots$	$\vdots$	$\vdots$	$\vdots$
I_n	p_n	f_n	x_n

$f = (f_1, \ldots, f_n)$ represents the estimate of the probability vector in the recent histogram, but it is based on too few observations. $p = (p_1, \ldots, p_n)$ is the probability vector in the empirical distribution at the previous updating. $x = (x_1, \ldots, x_n)$, the updated probability vector, should be obtained by incorporating the changing trend reflected in f into p. In the theory of statistics the most commonly used method for this incorporation is the weighted least squares method, which provides the following model (Murty, 2002) to compute x from p and f.

$$\text{Minimize} \quad (1 - \beta) \sum_{i=1}^{n} (p_i - x_i)^2 + \beta \sum_{i=1}^{n} (f_i - x_i)^2$$

$$\text{Subject to} \quad \sum_{i=1}^{n} x_i = 1 \tag{4.1}$$

$$x_i \geq 0, \quad i = 1, \ldots, n$$

where β is a weight between 0 and 1, similar to the smoothing constant α in the exponential smoothing method for updating the expected value (like α there, here β is the relative weight placed on the probability vector from the histogram composed from recent observations). x is taken as the optimum solution of this convex quadratic program. $\beta = 0.1$ to 0.4 works well; the reason for choosing the weight for the second term in the objective function to be small is because the vector f is based on only a small number of observations. As the quadratic model minimizes the weighted sum of squared forecast errors over all value intervals, when used periodically, it has the effect of tracking gradual changes in the probability distribution of the random variable.

The above quadratic program has a unique optimum solution given by the following explicit formula.

$$x = (1 - \beta)p + \beta f \tag{4.2}$$

So we take the updated empirical distribution to be the one with the probability vector given by Equation 4.2.

The formula for updating the probability vector in Equation 4.2 is exactly analogous to the formula for forecasting the expected value of a random variable using the latest observation, in exponential smoothing. Hence, the above formula can be thought of as the extension of the exponential smoothing method to update the probability vector in the empirical distribution of the random variable.

When there is a significant increase or decrease in the mean value of the random variable, new value intervals may have to be opened up at the left or right end. In this case, the probabilities associated with value intervals at the other end may become very close to 0, and these intervals may have to be dropped from further consideration at that time.

This procedure can be used to update the discretized demand distribution either at every ordering point, or periodically at every rth ordering point for some convenient r, using the most recent observations on demand.

4.2.4 An Application of the Forecasting Method for Computing Optimal Order Quantities

Given the empirical distribution of demand for the next period based on the forecasting method given in Section 4.2.3, the well-known newsvendor model (Murty, 2002) can be used to determine the optimal order quantity for that period that minimizes the sum of expected overage and underage costs very efficiently numerically. We will illustrate with a numerical example. Let the empirical distribution of demand (in units) for the next period be

$I_i = $ Interval for Demand	Probability p_i	$u_i = $ Mid-point of Interval i
100–120	0.03	110
120–140	0.10	130
140–160	0.15	150
160–180	0.20	170
180–200	0.11	190
200–220	0.07	210
220–240	0.20	230
240–260	0.06	250
260–280	0.02	270
280–300	0.04	290
300–320	0.02	310

The expected value of this distribution $\bar{\mu} = \sum_i u_i p_i = 192.6$ units, and its standard deviation $\bar{\sigma} = \sqrt{\sum_i (u_i - \bar{\mu})^2 p_i} = 47.4$ units.

Let us denote the ordering quantity for that period, to be determined, by Q, and let d denote the random variable that is the demand during that period. Then

$y = $ overage quantity in this period = amount remaining after the demand is completely fulfilled $= (Q - d)^+ = $ maximum$\{0, Q - d\}$

$z = $ underage quantity during this period = unfulfilled demand during this period $= (Q - d)^- = $ maximum$\{0, d - Q\}$.

Suppose the overage cost $f(y)$, is the following piecewise linear function of y:

Overage Amount $= y$	Overage Cost $f(y)$ in $\$$	Slope
$0 \leq y \leq 30$	$3y$	3
$30 \leq y$	$90 + 10(y - 30)$	10

Suppose the underage cost $g(z)$ in $\$$, is the fixed cost depending on the amount given below:

Underage Amount $= y$	Underage Cost $g(z)$ in $\$$
$0 \leq z \leq 10$	50
$10 < z$	150

To compute $E(Q) =$ the expected sum of overage and underage costs when the order quantity is Q, we assume that the demand value d is equally likely to be anywhere in the interval I_i with probability p_i. This implies, for example, that the probability that the demand d is in the interval 120–125 is $=$ (probability that d lies in the interval 120–140)$/4 = (0.10)/4 = 0.025$.

Let $Q = 185$. When the demand d lies in the interval 120–140, the overage amount varies from 65 to 45 and the overage cost varies from \$440 to \$240 linearly. So the contribution to the expected overage cost from this interval is $0.10(440 + 240)/2$.

Demand lies in the interval 140–160 with probability 0.15. In this interval, the overage cost is not linear, but it can be partitioned into two intervals 140–155 (with probability 0.1125), and 155–160 (with probability 0.0375) in each of which the overage cost is linear. In the interval $140 \leq d \leq 155$ the overage cost varies linearly from \$240 to 90; and in $155 \leq d \leq 160$ the overage cost varies linearly from \$90 to 75. So, the contribution to the expected overage cost from this interval is $\$(0.115(240 + 90)/2) + (0.0375(90 + 75)/2)$.

Proceeding this way, we see that $E(Q)$ for $Q = 185$ is: $\$(0.03(640 + 440)/2) + (0.10(440 + 240)/2) + [(0.115(240 + 90)/2) + (0.0375(90 + 75)/2)] + (0.20(75 + 15)/2) + [0.0275(15 + 0)/2] + 0.055(50) + 0.0275\ (150)] + (0.07 + 0.20 + 0.06 + 0.02 + 0.04 + 0.02)150 = \$140.87$.

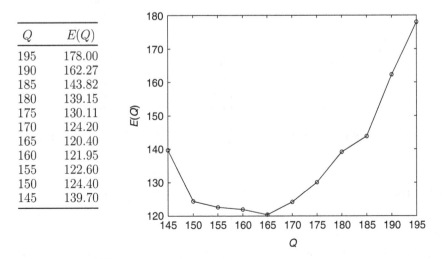

Q	$E(Q)$
195	178.00
190	162.27
185	143.82
180	139.15
175	130.11
170	124.20
165	120.40
160	121.95
155	122.60
150	124.40
145	139.70

FIGURE 4.3 Plot of $E(Q)$ for various values of Q.

In the same way, we computed the values of $E(Q)$ for different values of Q spaced 5 units apart, given by the side of Figure 4.3. The graph of Figure 4.3 is a plot of these values of $E(Q)$. Here, we computed $E(Q)$ at values of Q, which are multiples of 5 units, and it can be seen that $Q = 165$ is the optimum order quantity correct to the nearest multiple of 5. If the optimum is required to greater precision, the above calculation can be carried out for values of Q at integer (or closer) values between 150 and 170 and the best value of Q there chosen as the optimum order quantity.

The optimum value of Q can then be translated into the actual order quantity for the next period by subtracting the expected on-hand inventory at the end of the present period from it.

For each i, assuming that demand d is equally likely to be anywhere in the interval I_i with probability p_i makes the value of $E(Q)$ computed accurate for each Q. However, in many applications, people make the simpler assumption that p_i is the probability of demand being equal to u_i, the midpoint of the interval I_i. The values of $E(Q)$ obtained with this assumption will be approximate, particularly when the overage and underage costs are not linear (i.e., when they are piecewise linear etc.); but this assumption makes the computation of $E(Q)$ much simpler; that's why people use this simpler assumption.

4.2.5 How to Incorporate Seasonality in Demand into the Model

The discussion so far has dealt with the case when the values of demand in the various periods form a stationary time series. In some applications this series may be seasonal; that is, it has a pattern that repeats every N periods for some known value of N. The number of periods N, before the pattern begins to repeat, is known as the length of the season. To use seasonal models, the length of the season must be known.

For example, in the computer industry, the majority of the sales are arranged by sales agents who operate on quarterly sales goals. That's why the demand for components in the computer industry and demand for their own products tend to be seasonal with the quarter of the year as the season. The sales agents usually work much harder in the last month of the quarter to meet their quarterly goals; so demand for products in the computer industry tends to be higher in the third month of each quarter than in the beginning two months. As most of the companies are building to order nowadays, weekly production levels and demands for components inherit the same kind of seasonality.

At one company in this industry each quarter is divided into three homogeneous intervals. Weeks 1–4 of the quarter are slack periods; each of these weeks accounts a fraction of about 0.045 of the total demand in the quarter. Weeks 5–8 are medium periods; each of these weeks accounts for a fraction of about 0.074 of the total demand in the quarter. Weeks 9–13 are peak periods; each of these weeks accounts for a fraction of about 0.105 of the total demand in the quarter. This fraction of demand in each week of the season is called the seasonal factor of that week.

In the same way, in the paper industry, demand for products exhibits seasonality with each month of the year as the season. Demand for their products in the 2nd fortnight in each month tends to be much higher than in the 1st fortnight.

There are several ways of handling seasonality. One way is for each $i = 1$ to $N (=$ length of the season), consider demand data for the ith period in each season as a time series by itself, and make the decisions for this period in each season using this series based on methods discussed in earlier sections.

Another method that is more popular is based on the assumption that there exists a set of indices c_i, $i = 1$ to N called seasonal factors or seasonal indices (see Meybodi and Foote, 1995), where c_i represents the demand in the ith period of the season as a fraction of the demand during the whole season (as an example, see the seasonal factors given for the computer company described above). Once these seasonal factors are estimated, we divide each observation of demand in the original demand time series by the appropriate seasonal factor to obtain the de-seasonalized demand series. The time series of de-seasonalized demand amounts still contains all components of information of the original series except for seasonality. Forecasting is carried out using the methods discussed in the earlier sections, on the de-seasonalized demand series. Then estimates of the expected demand, standard deviation, and the optimal order quantities obtained for each period must be re-seasonalized by multiplying by the appropriate seasonal factor before being used.

4.3   Models for Production Layout Design

The basic engineering tasks involved in producing a product are

1. Developing a process plan,
2. Deciding whether to make or buy each component,
3. Deciding on the production doctrine: group technology, process oriented, or a mix,
4. Designing the layout, and
5. Creating the aisles and choosing the material handling system.

In this section, we will focus on tasks 3–5, assuming tasks 1 and 2 have been done. We will then have available the following data: a matrix that shows the flows between processes.

4.3.1   An Example Problem

Our first problem is how to judge between layouts. A common metric is to minimize $\sum d_i * f_i$. This metric uses the distance from the centers of the process area multiplied by the flow. The idea is that if this is minimized then the processes with the largest interactions will be closest together. This has an unrealistic side in that flow usually is door to door. We will look at two approaches. The first approach is the spanning tree concept and then an approximation for the above metric.

As a practical matter, we always want to look at an ideal layout. We would like for the material to be delivered and moved continuously through processes and be packaged in motion and moved right into the truck or railway car that carries the product to the store. Some examples already exist as logs of wood go to a facility that moves them directly through a few processes that cut them into 2×4's or 4×4's and capture the sawdust and create particle board, all of which with few stationary periods and go right into the trucks that move them to lumber yards. In this case, the product is simple; there are only a small number of processes that are easy to control by computers armed with expert system and neural net rules that control the settings of the processes. Consider Example 1, where the basic data on the first level is given in Table 4.1.

The matrix illustrates a factory with seven process areas and lists the number of containers that will flow between them. Let us first assume that each department is of the same unit size, say 100 sq. feet. We can draw a network to represent the flows between process areas as in Figure 4.4. We can act as if the flows of material are from center to center of departments by a conveyor. We can estimate a layout as in Figure 4.5. Note that we now look at the size of the flows and put P2 and P3 adjacent and the center for P4 is slightly further away. A similar reasoning locates the other departments. Note also the room left for expansion. This is not a mathematical principle but is based on experience. Just as traffic on a freeway always expands, production needs to expand if the business succeeds.

TABLE 4.1 Flow Values for Example 1

	P1	P2	P3	P4	P5	P6	P7
P1	0	20	30	15			
P2		0			20		
P3			0		28		
P4				0		16	
P5		5			0	28	
P6				4		0	48
P7							0

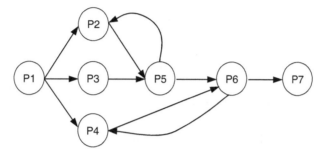

FIGURE 4.4 A process network for Example 1.

P1	P2	P6	P7
P3	P4	P5	Expansion area

FIGURE 4.5 Possible layout for Example 1.

The layout in Figure 4.5 has subjective virtues as department one is "door-to-door" close to the departments that it flows material to. An overhead conveyor down the middle has a straight line layout that is very convenient and is of minimum length. This is not necessarily an optimal layout but it provides a great starting point. If the departments are of different sizes, the layout problem is much more complex and natural adjacencies are sometimes infeasible.

Another heuristic approach to the layout adjacency problem is the use of the cut tree. If you look at Figure 4.4 and assume the arcs are conveyors, then you try to find the set of conveyors that, if inactive, will cut flow from 1 to 7. If the conveyors from 5 to 6 and 4 to 6 are inactive then flow from 1 to 7 cannot occur. This flow has value 44 and is the minimum of such cuts. A common sense interpretation is that P6 and P7 should be on one end and P1–P5 grouped on the other end. The flow between ends is the minimum it can be, which is 44. A cut can be made in the set P1–P5 and new smaller groups can be seen. Other cut trees can be computed and can provide a basis for several trials. The cut tree approach is discussed further in Section 4.3.5.

4.3.2 Optimal Plant Layouts with Practical Considerations

One of the difficulties in defining "optimal" plant layouts is choosing the metric that defines "optimal." Most of the metrics try to minimize the product of distance and flow. The problem is what is flowing. The weight and shape of what is moved varies tremendously. What is moved can be a box of parts, a chassis 7' by 4', a container of waste, or a bundle of electrical wire. At some point in time one needs to just find how many moves are made. Thirty boxes can cause a move or one waste container can cause a move. The movement

TABLE 4.2   Flow Values for Example 2

i/j	1	2	3	4	5
1		5	20		
2	5		20		
3	20	30			
4					100
5				100	

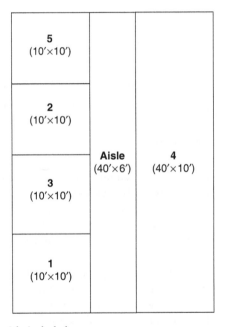

5	4
2	4
3	4
1	4

FIGURE 4.6   Conceptual layout without aisles.

FIGURE 4.7   Layout with aisle included.

can be along aisles or on overhead conveyors. If the move is along aisles then it comes out of doors. If it is moved by overhead conveyors, then the move is from a point inside the area that produces it and the area that needs it. If the move is on overhead conveyors, then the distance is Euclidean. If the move is along aisles, then distance is measured by the Manhattan metric, which is the sum of N-S and E-W moves. The problem is clearly NP complete, so that optimizing means optimizing in an ideal situation and then engineering the solution into a practical satisficing layout. An approach by Tretheway and Foote (1994) illustrates this approach.

Consider Example 2, where there are five departments with areas 100, 100, 100, 100, and 400. Let the flow values be as given in Table 4.2.

Let us assume a single aisle structure. Figures 4.6 and 4.7 illustrate an actual optimal layout for this aisle structure and set of areas. Each block is 100 sq. ft. Based on the flow and the aisle down the middle of the plant, this is optimal regardless of the material handling system. Notice that the aisle takes some area. This means that the overall structure must contain perhaps 1024 sq. feet if the aisle is 40′ by 6′, or that each department donates some area to the aisle.

4.3.3 The Tretheway Algorithm

Tretheway and Foote (1994) suggest a way to actually compute the two-dimensional coordinates of the location of departments in a layout. Their approach works for buildings that are rectangular, and variants of rectangular such as U, T, or L shaped structures. They adapt a concept by Drezner (1987) to develop a technique to actually draw the layout with aisles. The basic idea is to get a good adjacency layout on a line, then get a second layout on another line, use the cross product to obtain a two-dimensional relative position, then draw in aisles, account for loss of space to aisles, and create the drawing. "Rotations" of the aisles structure create alternative structures that can be evaluated in terms of practical criteria. An example is that a rotation which puts a test facility with a risk of explosions occurring during test in a corner would be the layout chosen if the aisle structure can support the material handling design.

NLP problem: Drezner's nonlinear program for n departments is given by

$$\min \sum_{ij} f_{ij} d_{ij} \quad i = 1, \ldots, n; \;\; j = 1, \ldots, m \tag{4.3}$$

where f_{ij}'s are flows (given data) and d_{ij} (decision variables) are distances between departments i and j. Utilizing a squared Euclidean distance metric, the nonlinear program is transformed to

$$\min \frac{\sum_{ij} f_{ij} \left[(x_i - x_j)^2 + (y_i - y_j)^2 \right]}{\sum_{i,j} (x_i - x_j)^2 + (y_i - y_j)^2} \tag{4.4}$$

The objective in one dimension becomes

$$\min \frac{\sum_{i,j} f_{ij} x_{ij}^2}{\sum_{i,j} x_{ij}^2} \tag{4.5}$$

Equation 4.5 can be optimized. The minimizing solution is the eigen vector associated with the second least eigen value of a matrix derived from the flow matrix.

To get a solution to the layout problem in a plane, we use the second and third least eigen values to create two eigen vectors $[x]$ and $[y]$, each with n values. The Cartesian product of these vectors creates a *scatter diagram* (a set of points in $\mathbb{R}^2$) that suggests an optimal spatial relationship in two dimensions for the n departments that optimizes the squared distance * flow relationship between the departments. Notice this relationship is not optimized based on practical flow measures. However, it may give us a set of alternatives that can be compared based on a practical measure. Further, this set of points can be expressed as a layout based on two practical considerations: the areas of the departments and the aisle structure selected. The Tretheway algorithm will illustrate how to do this. The algorithm can actually be done easily by hand, especially if we remember that the space requirements usually have some elasticity. The selection of an aisle structure first allows the departments to be painted in a simple way.

Example 3: Consider the input data for 12 departments shown in Table 4.3. For this input, consider an aisle structure with three vertical aisles. The scatter diagram with the

TABLE 4.3   Input Data for Example 3

	1	2	3	4	5	6	7	8	9	10	11	12
1	0	5	2	4	1	0	0	6	2	1	1	1
2	5	0	3	0	2	2	2	0	4	5	0	0
3	2	3	0	0	0	0	0	5	5	2	2	2
4	4	0	0	0	5	2	2	10	0	0	5	5
5	1	2	0	5	0	10	0	0	0	5	1	1
6	0	2	0	2	10	0	5	1	1	5	4	0
7	0	2	0	2	0	5	0	10	5	2	3	3
8	6	0	5	10	0	1	10	0	0	0	5	0
9	2	4	5	0	0	1	5	0	0	0	10	10
10	1	5	2	0	5	5	2	0	0	0	5	0
11	1	0	2	5	1	4	3	5	10	5	0	2
12	1	0	2	5	1	0	3	0	10	0	2	0

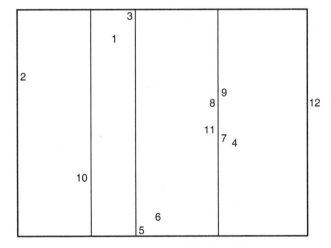

FIGURE 4.8   Example 3 scatter diagram for vertical aisles.

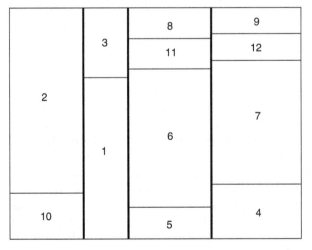

FIGURE 4.9   Example 3 final solution with vertical aisles.

overlaid *relative* position of the aisles is shown in Figure 4.8. Figure 4.9 shows the area proportioned final layout, with the heavy lines representing the aisles. For the same data input of Table 4.3, Figures 4.10 and 4.11 demonstrate the technique for one main aisle and two sub-aisles with the subaisles located between the aisle and the top wall. Note the different proximities for the two different aisle structure layouts.

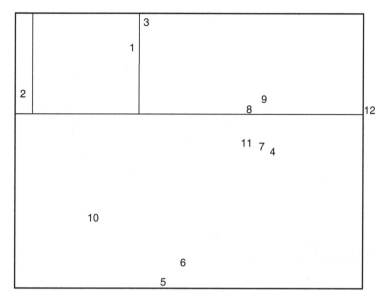

FIGURE 4.10 Example 3 scatter diagram with one main aisle and two sub-aisles.

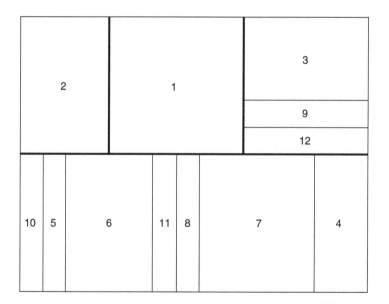

FIGURE 4.11 Example 3 final solution with one main aisle and two sub-aisles.

4.3.4 A Spanning Tree Network Model for Layout Design

Consider the flow network in Figure 4.12. The base example is from Ravindran et al. (1987).

The flows are $f_{12} = 3$, $f_{13} = 7$, $f_{14} = 4$, $f_{26} = 9$, $f_{36} = 6$, $f_{35} = 3$, $f_{46} = 3$, $f_{23} = 2$, and $f_{34} = 1$. These represent costs, not the flow of containers, if the associated departments are neighbors. Let us generate a spanning tree. This is a tree with no cycles such that all nodes can be reached by a path. According to Ravindran et al. (1987), we arbitrarily select node 1 to generate a tree such that the cost of each path segment is minimal. Thus, from 1 we go

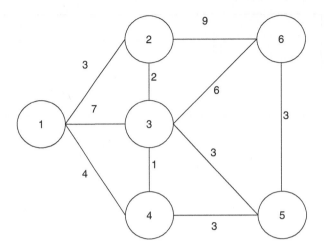

FIGURE 4.12   A material flow network.

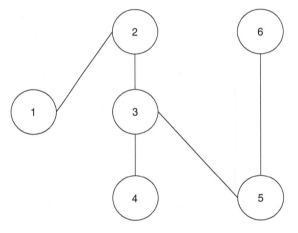

FIGURE 4.13   The minimum cost spanning tree.

5	6
3	4
2	1

FIGURE 4.14   A layout for equal area departments.

to 2 as 3 = min (3, 7, 4). We now add node 3 as its cost of 2 is less than any other cost from nodes 1 and 2. Now we look at arcs connecting from nodes 1, 2 and 3. Arc (3,4) is the best.

So we have [(1,2),(2,3),(3,4)]. Now (3,5) and (4,5) are tied, so we arbitrarily pick (3,5). (5,6) is left so we have [(1,2),(2,3),(3,4),(3,5),(5,6)]. This is the minimal spanning tree illustrated in Figure 4.13. We have five arcs which is one less than the number of nodes and hence the smallest number of paths. If all departments have the same area an optimal layout would be as given in Figure 4.14.

What if the departments do not have equal areas? Then an appropriate aisle structure could help. If you are lucky with areas and 1 has the biggest area and 2, 3, and 4 are not

1	2	6
1	3	5
1	4	5

FIGURE 4.15   A layout adjusted for areas.

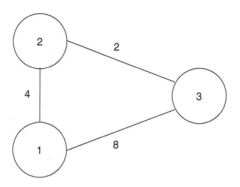

FIGURE 4.16   A network model of container flows.

too large and 5 and 6 are intermediate then we could have a layout as given in Figure 4.15. The aisle structure in Figure 4.15 could be two parallel aisles. This would work if the areas are roughly proportional to 1:300, 2,3,4:100, 5:200, and 6:100. Plant layout will always be NP complete relative to optimality in general with any metric, but in special cases apparent optimality can be achieved.

Optimizing (flow × distance) metrics must be constrained by unique circumstances. If department 3 in Figure 4.15 has some dangerous equipment that might explode, it has to go to a corner. This reduces the number of outer walls that can be blown out to two. It gets the area away from the center where all directions are in jeopardy. Buildings do not have to be in a rectangular shape. If space permits and modules are feasible, then the shape of the building can follow the spanning tree or the cut tree. U shapes can be solved by adding dummy departments to the Tretheway set and setting their artificial flows such that they will be in the middle at the end. We can also not have aisles and let the material handling system overhead be a minimum spanning tree with nodes at the appropriate points in departments or cells. See also Ravindran et al. (1989) for the use of shortest route network model to minimize overhead conveyor distances along with material movement control.

4.3.5   A Cut Tree Network Model for Layout Design

If you consider the numbers on arcs representing the flow of containers, then the cut tree provides a great basis for a layout that simply overlays a cut tree derived from the network flow model on a plant shape. Kim et al. (1995) give an algorithm for generating the cut tree and show an example layout for a nine area department cut tree. A cut tree divides the network model into a final spanning tree ($n - 1$ arcs) that assumes flows go along the arcs and that flows to other nodes pass through predecessor nodes. The following is a simple example.

Consider the example network with three departments D1, D2, and D3 given in Figure 4.16. We treat the arcs as conveyors. If we cut the arcs (2,1) and (2,3) we interrupt 6 units of flow to D1 and D3. If D1 is isolated, then a flow of 12 is cut. If D3 is isolated then a flow of 10 is cut. The minimum cut is 6. If we cut arc (2,3), the 2 units from D2 to D3 must move to D1 and flow from D1 to D3 increases to 10 as seen in Figure 4.17. If we move

FIGURE 4.17   The derived cut tree.

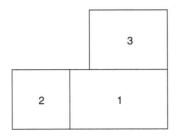

FIGURE 4.18   Layout of L shaped area using the cut tree solution.

material by conveyors, then we have the minimum flow on the conveyors and a layout of 2-1-3 pasted into a building shape.

In Figure 4.18 the building shape is deliberately L-shaped, as occurs often when areas inside a plant are re-engineered. The areas are different in size, but the material handling costs will be minimized. See Kim et al. (1995) and also Benson and Foote (1997) for other ways to use cut trees and door to door metrics to rate layouts.

4.3.6   Facility Shapes

The shapes of facilities can be designed with a few basic principles. For example, the shape with the minimum perimeter for a given area is a circle. This means that this simple shape should be explored for nursing wards (minimum walking per day and maximum visual surveillance of patients). This would also work for sleeping areas and jails of abused children. A circular shape for production cells gives the best visual and response for the operator. If walls are needed, they will have a minimum length. Of course wall construction costs also depend on other factors such as materials (brick is more costly for circular walls) and insulation (minimum perimeter means less insulation, and so on). Shopping mall designers want longer walks to expose more goods, so shapes and paths will meet this requirement. Long materials require long straight shapes. Towers work if work can trickle down by gravity.

4.4   Scheduling of Production and Service Systems

4.4.1   Definition of the Scheduling Decision

The basic function of scheduling is the decision to start a job on a given process and predict when it will finish. This decision is determined by the objective desired, the data known, and the constraints. The basic data required to make a mathematical decision are the process plan, the time of operation projected for each process, the quality criteria, the bill of materials (which may determine start dates if materials are not immediately available), the due date, if any, and the objectives that must be met, if any. A process plan is a list of the operations needed to complete the job and any sequence constraints. It is sometimes thought that there is always a strict precedence requirement, but that is not true. Many times one can paint or cut to shape in either order. A quality requirement may impose a precedence, but this is to be determined. The basic plan can be listed as cut, punch, trim, and smooth (emery wheel or other). This list is accompanied by instructions as to how to execute each process (usually in the form of a drawing or pictures) and standards of quality to be achieved; as well, there will be an estimate of the time to set up, perform the process, do quality checks, record required information for the information and tracking systems, and move. A decision is required when there are two or more jobs at a process and the decision has to be made as to which job to start next. Some jobs require only a single machine. The process could be welding, creating a photo from film, X-ray, creating an IC board, or performing surgery. The study of single machines is very important because this theory forms the basis of many heuristics for more complex manufacturing systems.

4.4.2 Single Machines

We are assuming here a job that only requires one process, and that the time to process is given as a deterministic number or a random value from a given distribution. The standard objectives are (1) to Min $\overline{F}$ the average flow time, (2) Min Max lateness L_{max}, (3) Min n_t, the number of tardy jobs, and combinations of earliness and lateness. The flow time F is the time to completion from the time the job is ready. If the job is ready at time equal to 0, flow time and completion time are the same. The function (finish time-due date) is the basic calculation. If the value is positive the job is late, if it is negative the job is early. The number of tardy jobs is the number of late jobs. Finish time is simply computed in the deterministic case as the time of start + processing time. There are surprisingly few objectives that have an optimal policy. This fact implies that for most systems we must rely on good heuristics. The following is a list of objectives and the policy that optimizes the objective. The list is not exclusive, as there are many types of problems with unusual assumptions.

Deterministic Processing Time

The objective is followed by the policy that achieves that objective:

1. Min $\overline{F}$: Schedule job with shortest processing time first (SPT). Break ties randomly or based on a second objective criterion. The basic idea is this: if the shortest job is worked first, that is the fastest a job can come out. Every job's completion time is the smallest possible (Figure 4.19).
2. Min Max lateness (L_{max}): Schedule the job with the earliest due first (EDD). Break ties randomly or based on a second objective criterion (Figure 4.20).
3. Min n_t: Execute Moore's algorithm. This algorithm is explained to illustrate a basic sequencing algorithm. Some notation is required (Figure 4.21).

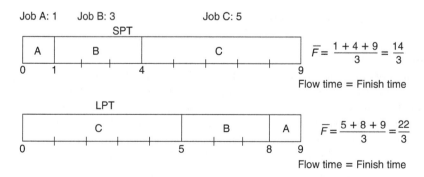

FIGURE 4.19 Optimality of SPT for min $\overline{F}$ (single machine Gantt charts comparing optimal with nonoptimal).

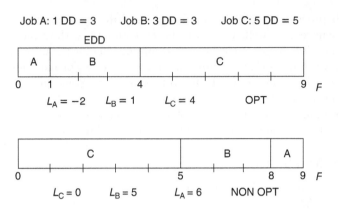

FIGURE 4.20   Optimality of EDD to Min Max lateness (single machine problem).

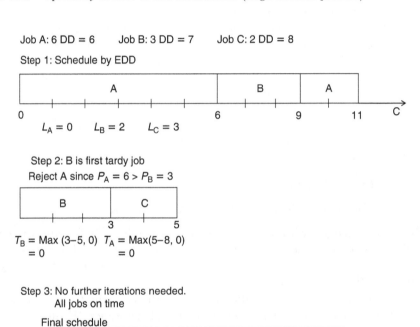

FIGURE 4.21   Use of Moore's algorithm.

Notation
Classification of problems: $n/m/A/B$, where n is the number of jobs, m the number of machines, A the flow pattern (blank when $m=1$, otherwise F for flow-shop, P for permutation job shop (jobs in same order on all machines), G the general job shop (where jobs can have different priorities at different process plan order), B the performance measure (objective), C the completion time of a job and equal to the ready time plus sum of waiting and processing times of jobs ahead of it, and D the due date of the job

L (lateness of job) $= C - D$

T (tardiness of job) $= \text{Max}(L,0)$

E (earliness of job) $= \text{Max}(-L,0)$

P_i (processing time of job i) $n_t = $ number of tardy jobs

Moore's algorithm solves the $n/1//n_t$ problem. The idea of the algorithm is simple. Schedule by EDD to min max tardiness. Then find the first tardy job and look at it and all the jobs in front of it. The problem job is the one with the largest processing time. It keeps other jobs from finishing; remove it. It could be a job that is not tardy. Now repeat the process. When no tardy jobs are found in the set of jobs that remain, schedule them in EDD order and place the rejected jobs at the end in any order.

Figure 4.21 illustrates Moore's algorithm and basic common sense ideas. For other single-machine problems that can be solved optimally, see French (1982) or any current text on scheduling (a fruitful path is to web-search Pinedo or Baker).

4.4.3   Flow Shops

More than one-third of production systems are flow shops. This means that all jobs have a processing plan that goes through the processes in the same order (some may take 0 time at a process). Only the two-process case has an optimal solution for min $F_{\max}$. A special case of the three-machine flow shop has an optimal solution for the same criteria. All other problems for m(number of processes) >2 do not have optimal solutions that are computable in a time proportional to some quadratic function of a parameter of the problem, such as n for any criteria. This will be addressed in more detail later.

Two-Process Flow Shops

It has been proved for flow shops that only schedules that schedule jobs through each process (machine) in the same sequence need be considered. This does not mean that deviations from this form will not sometimes get a criterion value that is optimal. When $m = 2$, an optimal solution is possible using Johnson's algorithm. The idea of the algorithm is to schedule jobs that have low processing times on the first machine and thus get them out of the way. Then schedule jobs later that have low processing times on the last machine, so that when they get on they are finished quickly and can make way for jobs coming up. In this way, the maximum flow time of all jobs is minimized (the time all jobs are finished).

Job	Time on Machine A	Time on Machine B
1	8	12
2	3	7
3	10	2
4	5	5
5	11	4

The algorithm is easy. Look at all the processing times and find the smallest. If it is on the first machine, schedule it as soon as possible, and if it is on the second machine, schedule it as late as possible. Then remove this job's data and repeat. The solution here has five positions. The following shows the sequence construction step by step. Schedule is represented by S.

$$S = [\,,\,,\,,3]$$
$$S = [2,\,,\,,\,,3]$$
$$S = [2,\,,\,,5,3]$$
$$S = [2,4,\,,5,3] \text{ or } [2,\,,4,5,3]$$

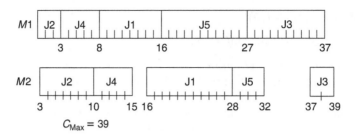

FIGURE 4.22 Gantt chart of an optimal schedule. For the example in two C_{max} problem schedule, using Johnson's algorithm.

Job 1 now goes in either sequence in the only position left: $S = [2,4,1,5,3]$ or $S = [2,1,4,5,3]$ (see Figure 4.22).

Heuristics for General Flow Shops for Some Selected Criteria

$$n/m/F/F_{max}$$

The Campbell–Dudek heuristic makes sense and gives good solutions. There are better heuristics if one is not solving by hand. The idea is common sense. Assume we have six machines. If we find a job that gets through the six machines quickly, schedule it first so it gets out of the way. If there is a job that gets through the first five quickly, schedule it as soon as you can. This follows up to a job that gets off the first machine quickly. Alternatively, if a job gets off the last machine quickly, schedule it last, or if it gets off the last two quickly, or the last three, and so on. This leads to solving five constructed two-machine flow shop problems (surrogates), finding the optimal sequences, and then picking the one with the best maximum flow. A sixth sequence can be tested by sequencing in order of the least total (on all machines) processing time. Here is a five-job, four-machine problem to illustrate the surrogate problems. There will be $m - 1$ or three of them.

Job	$m1$	$m2$	$m3$	$m4$	Total
1	5	7	4	11	27
2	2	3	6	7	18
3	6	10	1	3	20
4	7	4	2	4	17
5	1	1	I	2	5

$S = [5,4,2,3,1]$ when looking at total processing time.

Three Surrogate Problems
Surrogate one: times on first and last machines only (solution by Johnson's algorithm: [5,2,3,4,1])

Job	$m1'$	$m2''$
1	5	11
2	2	7
3	6	3
4	7	4
5	1	2

Surrogate two: times on first two and last two machines (solution by Johnson's algorithm: [5,2,1,4,3])

Job	$m1'$	$m2''$
1	12	15
2	5	13
3	16	4
4	11	6
5	2	3

Surrogate three: times on first three machines and last three machines (solution by Johnson's algorithm: [5,4,3,2,1])

Job	$m1'$	$m2''$
1	16	22
2	11	16
3	17	14
4	13	10
5	3	4

These four sequences are then Gantt-charted, and the one with the lowest F max is used. If we want to min average flow time, then these would also be good *solutions to try*. If we want to min max lateness, then try EDD, and also try these sequences. We could also put job 1 last and schedule by EDD as a commonsense approach. Later we will talk about computer heuristics for these types of problems.

The Concept of Fit in Flow Shops

Consider a case where job (i) precedes job (j), with a six-machine flow shop. Let their processing times be (i): [5 4 2 6 8 10] and (j): [3 2 5 8 9 15]. In the Gantt chart (see Figure 4.23) you will see that job (j) is always ready to work when job (i) is finished on a machine. No idle time is created (see Figure 4.23). If computer-processing time is available, then a matrix of job fit can be created to be used in heuristics.

Fit is computed with the following notation: t_{ij} is the processing time of job (i) on machine (j). Then let job (j) precede job (k):

$$\text{Fit} = \sum_{s=2}^{s=m} \max[t_{js} - t_{k,s-1}, 0]$$

Heuristics will try to make job (j) ahead of job (k) if fit is the largest positive found, that is, $[\ldots k, j, \ldots]$. By removing the max operator, one can get a measure of the overall fit or one can count the number of positive terms and use this to pick a pair that should be in sequence. If there are six machines, pick out all the pairs that have a count of 5, then 4, then 3, and so on, and try to make a common sense Gantt chart. Consider the following data: J1: [4 5 4 7], J2: [5 6 9 5], J3: [6 5 6 10], J4: [3 8 7 6], J5: [4 4 6 11]. The largest number of fits is 3.

$4 \rightarrow 3$ $3 \rightarrow 2$ $2 \rightarrow 1$ $1 \rightarrow 5$ have three positive terms in the fit formula. Hence, an optimal sequence is $S = [4\ 3\ 2\ 1\ 5]$ (Figure 4.24).

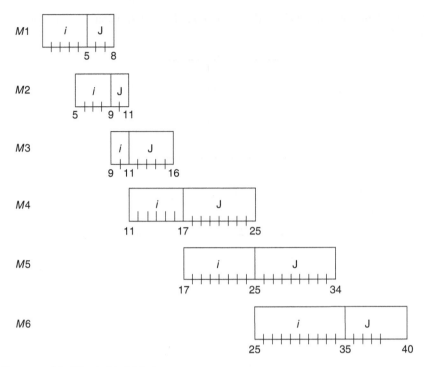

FIGURE 4.23   Job J fits behind job i.

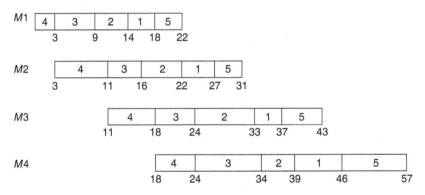

FIGURE 4.24   An optimal sequence since we have perfect fit.

4.4.4   Job Shops

Job shops are notoriously tough (see later discussion on NP-complete problems). In job shops, a job may have processing plans that have a variety of sequences. For large job shops some simulation studies have shown that job shops are dominated by their bottlenecks. A bottleneck is the machine that has the largest total work to do when all jobs are completed.

In practice, it has been effective to schedule the jobs on each machine by using the sequence determined by solving a one-machine problem with the bottleneck as the machine. For large job shops, jobs are constantly entering. Thus, the structure constantly changes. This means that at the beginning of the period, each machine is treated as a single machine, and single machine theory is used for scheduling. At the beginning of the next period, a new set of priorities is determined. The determination of period length is a matter still under

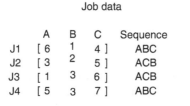

Job data

	A	B	C	Sequence
J1	[6	1	4]	ABC
J2	[3	2	5]	ACB
J3	[1	3	6]	ACB
J4	[5	3	7]	ABC

Work: A→15   B → 9   C → 22

A and C are top two bottlenecks

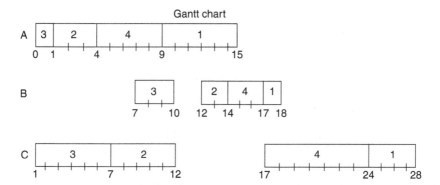

	A	C
1	6	4
2	3	5
3	1	6
4	5	7

Johnson's table

Johnson's solution → [3,2,4,1] = 5

Gantt chart

FIGURE 4.25   A heuristic approach to scheduling a job shop. Compare with other guesses and pick lowest C_{max}.

research. A period length of a multiple of total average job-processing time is a good start. A multiple such as 10 is a good middle ground to try. Perform some experiments and try going up if you have a lot of setup time and down if setup time is minimal.

Some other approaches are to take the top two bottlenecks and schedule the jobs as a two-machine flow shop, using only their times on the two bottlenecks. If you are using min average flow time, use flow shop heuristics designed for that purpose. If you are using min max flow time, use Johnson's algorithm. If a job does not go through the two bottlenecks, schedule it later, if you can. In the real world sometimes jobs can be worked faster than predicted. Most schedulers try to find such jobs. They then look for small processing time jobs and insert them. Schedulers also look for jobs that have remaining processes that are empty and pull them out if average flow time is an issue.

It is never good to operate at 100% capacity. Quality goes down under this kind of pressure. Studies have shown that around 90% is good. At times, random variation will put the shop at 100% even then. But a little gap is good for preventive maintenance and for training. Most maintenance will be off shift if unplanned, but off shift can be expensive. Figure 4.25 is an example using Johnson's algorithm on the two bottlenecks and then scheduling the three machines with that sequence.

Scheduling and Systems Analysis

If output from one subsystem goes to another, then there is a timing issue and hence a due date issue. Basic queuing issues come into play. If service centers are in a flow mode, then the jobs should be in SPT order. This gets them out of the early systems quickly and allows the other jobs get going. If work has a large variance, then low variance jobs go first. Simulation software analysis is a must for large systems. The biggest aid is total system knowledge. If a truck coming in has an accident and this is known by the warehouse, some healing action can be taken. Some borrowing from other trucks might happen. Work that will not be completed due to the accident can be put off. Top-level knowledge of demand from the primary data is a must. Inferring it from orders is a major mistake. This subsystem knowledge allows coordination.

Control

What does control mean for a system? If a policy is simulated, there will be a mean and standard deviation of the queue length. When a queue exceeds the mean plus two standard deviations something is likely wrong. So, what questions should you ask? Are the processing times different from predictions? Is a large amount of maintenance going on? Do we have critical material shortages? Have we instituted just-in-time scheduling when the variance in system parameters is too high? If your trucks go completely through urban areas, you may not be able to arrive in a just-in-time window.

It is now possible with computers to predict a range of conditions that are normal. Readers should refer to the literature for current information, journals, and books.

Solvability Problems

We now know of many areas that have no solution. The normal distribution function does not have a closed-form integral. It must be computed numerically. Most scheduling problems are such that to guarantee optimality, total enumeration or implied total enumeration (branch and bound) must be used. However, it has been shown over and over that heuristics can get optimal solutions or solutions that are only off by 1 or 2%. When data quality is considered, this will be satisfactory. In linear programming, any point in the vicinity of a basic solution will be near optimal and, in fact, due to round off errors in computing, the basis may even be better than the basis presented by the computation. Problems that require total enumeration to prove optimality are interesting, but in practice are solved and the solutions are good. Practically speaking, problems such as $n/m/F/n_t$ are what are called NP complete. They must have implied total enumeration to prove the answer is optimal. However, in practice, one can obtain good solutions. Moreover, for some problems, the optimal solution is obvious (for example, in flow shops where all jobs have unit processing times on all machines for the objective of min average flow time).

Bootstrapping

With the advent of modern computers and Monte Carlo simulation packages, it is possible to obtain statistical predictions of systems. By using heuristics and genetic algorithms to generate policies to simulate, a good statistical prediction can be made to set control standards with only three or four replications. Bootstrapping is the technique in which actual data are used for lead times, processing times, etc. These data can be arranged as a histogram from which values can be drawn randomly to determine how long a job will be processed, how long it will take to repair a machine, and so on. This uses the best evidence, the data, and has no chance of making an error in distribution assumption. Good data collection systems are a must. Distribution assumptions do have their virtues, in that we know repair times can occur that have not previously been observed. Running a Monte Carlo simulation both ways can throw much light on the matter.

Practical Issues

In both build-to-stock and build-to-order policies, the unexpected may happen. Machines break down, trucks get caught in traffic delays, materials are found to be defective, and workers do not show up. The plan and schedule must be repaired. A basic plan to illustrate repair is to assign workers to the bottleneck one at a time, then find the new bottleneck and assign the next worker there. Repairing, regaining feasibility, and extending the plan when new work arrives requires software if the system is large ("large" meaning that hand computation is simply not doable). An illustration of repair occurs when some jobs I are not ready in an $n/m/G$ or $F/C_{\max}$ problem at time 0. One then goes ahead and solves, assuming all i ready times are 0. If the solution requires a job to start before it is ready, repair must be done. These needs form the criteria for selecting commercial software from an available list. The software must be algorithmically correct and computationally efficient with good structures for search and information retrieval. Professional magazines such as *IE Solutions* provide lists of vendors to assess.

References

1. Baker, K.R. *Elements of Sequencing and Scheduling.* John Wiley & Sons, New York, revised and updated edition, 1995. URL http://www.dartmouth.edu/~kbaker/.

2. Baker, K. http://www.pearson.ch/pageid/34/artikel/28138PH/PrenticeHall/0130281387/ Scheduling Theory Algorithms.aspx, 1995 (also self published).

3. Benson, B. and B.L. Foote. DoorFAST: A constructive procedure to optimally layout a facility including aisles and door locations based on an aisle flow distance metric. *International Journal of Production Research*, 35(7):1825–1842, 1997.

4. Box, G.E.P. and G.M. Jenkins. *Time Series Analysis, Forecasting and Control.* Holden Day, San Francisco, 1970.

5. Brown, R.G. *Statistical Forecasting for Inventory Control.* McGraw-Hill, New York, 1959.

6. Drezner, Z. A heuristic procedure for the layout of a large number of facilities. *Management Science*, 33:907–915, 1987.

7. Foote, B.L. On the implementation of a control-based forecasting system for aircraft spare parts procurement. *IIE Transactions*, 27:210–216, 1995.

8. French, S. *Sequencing and Scheduling: An Introduction to the Mathematics of the Job-Shop.* John Wiley & Sons, New York, 1982.

9. Holt, C.C. Forecasting seasonals and trends by exponentially weighted moving averages. ONR Memo No. 52, 1957.

10. Kim, C-B, B.L. Foote, and P.S. Pulat. Cut-tree construction for facility layout. *Computers & Industrial Engineering*, 28(4):721–730, 1995.

11. Meybodi, M.Z. and B.L. Foote. Hierarchical production planning and scheduling with random demand and production failure. *Annals of Operations Research*, 59:259–280, 1995.

12. Montgomory, D.D. and L.A. Johnson. *Forecasting and Time Series Analysis.* McGraw-Hill, St. Louis, 1976.

13. Murty, K.G. Histogram, an ancient tool and the art of forecasting. Technical report, IOE Dept., University of Michigan, Ann Arbor, 2002.

14. Murty, K.G. Forcasting for supply chain management. Technical report, IOE dept., University of Michigan, Ann Arbor, 2006. Paper for a keynote talk at ICLS (International

Congress on Logistics and SCM Systems), Taiwan 2006, and appeared in the Proceedings of that Congress.

15. Nahmias, S. *Production and Operations Analysis*. Irwin, Boston, 2nd edition, 1993.

16. Pinedo, M. *Scheduling: Theory, Algorithms, and Systems*. Prentice Hall, Upper Saddle River, NJ, 2nd edition, 2001.

17. Pinedo, M.L. http://www.amazon.co.uk/exec/obidos/ASIN/0963974610/tenericbusine-21/026-2125298-8551636, Prentice Hall, Upper Saddle River, NJ, 2002.

18. Ravindran, A., D.T. Phillips, and J.J. Solberg. *Operations Research: Principles and Practice*. John Wiley & Sons, New York, 2nd edition, 1987.

19. Ravindran, A., B.L. Foote, A.B. Badiru, L.M. Leemis, and L. Williams. An application of simulation and network analysis to capacity planning and material handling systems at Tinker Air Force Base. *Interfaces*, 19(1):102–115, 1989.

20. Silver, E.A. and R. Peterson. *Decision Systems for Inventory Management and Production Planning*. John Wiley & Sons, New York, 2nd edition, 1979.

21. Tretheway, S.J. and B.L. Foote. Automatic computation and drawing of facility layouts with logical aisle structures. *International Journal of Production Research*, 32(7):1545–1555, 1994.

5

Energy Systems

C. Randy Hudson
Oak Ridge National Laboratory

Adedeji B. Badiru
Air Force Institute of Technology

5.1 Introduction

Energy is an all-encompassing commodity that touches the life of everyone. Managing energy effectively is of paramount importance in every organization and every nation. Energy issues have led to political disputes and wars. This chapter presents examples of the application of operations research (OR) to optimizing energy decisions. Kruger (2006) presents a comprehensive assessment of energy resources from different perspectives, which include historical accounts, fossil fuel, energy sustainability, consumption patterns, exponentially increasing demand for energy, environmental impact, depletion of energy reserves, renewable energy, nuclear energy, economic aspects, industrialization and energy consumption, and energy transportation systems. Badiru (1982) presents the application of linear programming (LP) technique of OR to energy management with a specific case example of deciding between alternate energy sources. Badiru and Pulat (1995) present an OR model for making investment decisions for energy futures. Hudson (2005) developed an OR-based spreadsheet model for energy cogeneration. All these and other previous applications of OR to energy systems make this chapter very important for both researchers and practitioners.

5.2   Definition of Energy

The word energy comes from the Greek word for "work." Indeed, energy is what makes everything *work*. Everything that we do is dependent on the availability of energy. Newton's law of conservation of momentum and energy states that energy cannot be created or destroyed. It can, however, be converted from one form to other. Recent energy-related events around the world have heightened the need to have full understanding of energy issues, from basic definitions to consumption and conservation practices. Tragic examples can be seen in fuel-scavenging practices that turn deadly in many energy impoverish parts of the world. In May 2006, more than 200 people died when a gasoline pipeline exploded in Nigeria while poor villagers were illegally tapping into the pipeline to obtain the much-needed energy source. This illustrates a major lack of understanding of the volatility of many energy sources.

Before we can develop a mathematical model of energy systems, we must understand the inherent characteristics of energy. There are two basic types of energy:

- Kinetic energy
- Potential energy

All other forms of energy are derived from the above two fundamental forms. Energy that is stored (i.e., not being used) is potential energy. *Kinetic energy* is found in anything that moves (e.g., waves, electrons, atoms, molecules, and physical objects). Electrical energy is the movement of electrical charges. Radiant energy is electromagnetic energy traveling in waves. Radiant energy includes light, X-rays, gamma rays, and radio waves. Solar energy is an example of radiant energy. Motion energy is the movement of objects and substances from one place to another. Wind is an example of motion energy. Thermal or heat energy is the vibration and movement of matter (atoms and molecules inside a substance). Sound is a form of energy that moves in waves through a material. Sound is produced when a force causes an object to vibrate. The perception of sound is the sensing (picking up) of the vibration of an object.

Potential energy represents stored energy as well as energy of position (e.g., energy due to fuel, food, and gravity). Chemical energy is energy derived from atoms and molecules contained in materials. Petroleum, natural gas, and propane are examples of chemical energy. Mechanical energy is the energy stored in a material by the application of force. Compressed springs and stretched rubber bands are examples of stored mechanical energy. Nuclear energy is stored in the nucleus of an atom. Gravitational energy is the energy of position and place. Water retained behind the wall of a dam is a demonstration of gravitational potential energy. Light is a form of energy that travels in waves. The light we see is referred to as visible light. However, there is also an invisible spectrum. Infrared or ultraviolet rays cannot be seen, but can be felt as heat. Getting a sunburn is an example of the effect of infrared. The difference between visible and invisible light is the length of the radiation wave, known as wavelengths. Radio waves have the longest rays while gamma rays have the shortest rays.

When we ordinarily talk about conserving energy, we are referring to adjusting the thermostat (for cooling or heating) to save energy. When scientists talk of conserving energy, they are referring to the law of physics, which states that energy cannot be created or destroyed. When energy is used (consumed), it does not cease to exist; it simply turns from one form to another. As examples, solar energy cells change radiant energy into electrical energy. As an automobile engine burns gasoline (a form of chemical energy), it is transformed from the chemical form to a mechanical form.

When energy is converted from one form to another, a useful portion of it is always lost because no conversion process is 100% efficient. It is the objective of energy analysts to minimize that loss, or convert the loss into another useful form. Therein lies the need to use OR techniques to mathematically model the interaction of variables in an energy system to achieve optimized combination of energy resources. The process of combined heat and power (CHP) to achieve these objectives can benefit from OR modeling of energy systems.

5.3   Harnessing Natural Energy

There is abundant energy in our world. It is just a matter of meeting technical requirements to convert it into useful and manageable forms. For example, every second, the sun converts 600 million tons of hydrogen into 596 million tons of helium through nuclear fusion. The balance of 4 million tons of hydrogen is converted into energy in accordance with Einstein's theory of relativity,

$$E = mc^2$$

This is a lot of energy that equates to 40,000 W per square inch on the visible surface of the sun. Can this be effectively harnessed for use on Earth? Although the Earth receives only one-half of a billionth of the sun's energy, this still offers sufficient potential for harnessing. Comprehensive technical, quantitative, and qualitative analysis will be required to achieve the harnessing goal. OR can play an importance role in that endeavor. The future of energy will involve several integrative decision scenarios involving technical and managerial issues such as:

- Micropower generation systems
- Negawatt systems
- Energy supply transitions
- Coordination of energy alternatives
- Global energy competition
- Green-power generation systems
- Integrative harnessing of sun, wind, and water energy sources
- Energy generation, transformation, transmission, distribution, storage, and consumption across global boundaries.

5.4   Mathematical Modeling of Energy Systems

The rapid industrialization of society, coupled with drastic population growth, has fueled increasingly complex consumption patterns that require optimization techniques to manage. It is essential to first develop mathematical representations of energy consumption patterns to apply OR methods of optimization. Some of the mathematical modeling options for energy profiles are linear growth model, exponential growth model, logistic curve growth model, regression model, and logarithmic functions. There is often a long history of energy data to develop appropriate mathematical models. The huge amount of data involved and the diverse decision options preclude the use of simple enumeration approaches. Thus, LP and other nonlinear OR techniques offer proven solution techniques. The example in the next section illustrates how the LP model is used to determine the optimal combination of energy sources to meet specific consumption demands.

5.5 Linear Programming Model of Energy Resource Combination

This example illustrates the use of LP for energy resource allocation (Badiru, 1982). Suppose an industrial establishment uses energy for heating, cooling, and power. The required amount of energy is presently being obtained from conventional electric power and natural gas. In recent years, there have been frequent shortages of gas, and there is a pressing need to reduce the consumption of conventional electric power. The director of the energy management department is considering a solar energy system as an alternate source of energy. The objective is to find an optimal mix of three different sources of energy to meet the plant's energy requirements. The three energy sources are

- Natural gas
- Commercial electric power grid
- Solar power

It is required that the energy mix yield the lowest possible total annual cost of energy for the plant. Suppose a forecasting analysis indicates that the minimum kilowatt-hour (kwh) needed per year for heating, cooling, and power are 1,800,000, 1,200,000, and 900,000 kwh, respectively. The solar energy system is expected to supply at least 1,075,000 kwh annually. The annual use of commercial electric grid must be at least 1,900,000 kwh due to a prevailing contractual agreement for energy supply. The annual consumption of the contracted supply of gas must be at least 950,000 kwh. The cubic foot unit for natural gas has been converted to kwh (1 cu. ft. of gas = 0.3024 kwh).

The respective rates of $6/kwh, $3/kwh, and $2/kwh are applicable to the three sources of energy. The minimum individual annual conservation credits desired are $600,000 from solar power, $800,000 from commercial electricity, and $375,000 from natural gas. The conservation credits are associated with the operating and maintenance costs. The energy cost per kilowatt-hour is $0.30 for commercial electricity, $0.05 for natural gas, and $0.40 for solar power. The initial cost of the solar energy system has been spread over its useful life of 10 years with appropriate cost adjustments to obtain the rate per kilowatt-hour. The sample data is summarized in Table 5.1. If we let x_{ij} be the kilowatt-hour used from source i for purpose j, then we would have the data organized as shown in Table 5.2. Note that the energy sources (solar, commercial electricity grid, and natural gas) are used to power devices to meet energy needs for cooling, heating, and power.

TABLE 5.1 Energy Resource Combination Data

Energy Source	Minimum Supply (1000's kwh)	Minimum Conservation Credit (1000's $)	Conservation Credit Rate ($/kwh)	Unit Cost ($/kwh)
Solar power	1075	600	6	0.40
Electricity grid	1900	800	3	0.30
Natural gas	950	375	2	0.05

TABLE 5.2 Tabulation of Data for LP Model

Energy Source	Heating	Type of Use Cooling	Power	Constraint
Solar power	x_{11}	x_{12}	x_{13}	$\geq 1075K$
Electric power	x_{21}	x_{22}	x_{23}	$\geq 1900K$
Natural gas	x_{31}	x_{32}	x_{33}	$\geq 950K$
Constraint	≥ 1800	≥ 1200	≥ 900	

TABLE 5.3 LP Solution to the Resource Combination Example (in kwh)

Energy Source	Heating	Type of Use Cooling	Power
Solar power	0	1075	0
Commercial Electricity	975	0	925
Natural gas	825	125	0

The optimization problem involves the minimization of the total cost function, Z. The mathematical formulation of the problem is presented below.

$$\text{Minimize:} \quad Z = 0.4 \sum_{j=1}^{3} x_{1j} + 0.3 \sum_{j=1}^{3} x_{2j} + 0.05 \sum_{j=1}^{3} x_{3j}$$

$$\text{Subject to:} \quad x_{11} + x_{21} + x_{31} \geq 1800$$

$$x_{12} + x_{22} + x_{32} \geq 1200$$

$$x_{13} + x_{23} + x_{33} \geq 900$$

$$6(x_{11} + x_{12} + x_{13}) \geq 600$$

$$3(x_{21} + x_{22} + x_{23}) \geq 800$$

$$2(x_{31} + x_{32} + x_{33}) \geq 375$$

$$x_{11} + x_{12} + x_{13} \geq 1075$$

$$x_{21} + x_{22} + x_{23} \geq 1900$$

$$x_{31} + x_{32} + x_{33} \geq 950$$

$$x_{ij} \geq 0, \qquad i,j = 1,2,3$$

Using the LINDO LP software, the solution presented in Table 5.3 was obtained. The table shows that solar power should not be used for heating or power; commercial electricity should not be used for cooling, and natural gas should not be used for power. In pragmatic terms, this LP solution may have to be modified before being implemented on the basis of the prevailing operating scenarios and the technical aspects of the facilities involved. The minimized value of the objective function is $1047.50 (in thousands).

5.6   Integer Programming Model for Energy Investment Options

This section presents an integer programming formulation as another type of OR modeling for energy decision making. Planning a portfolio of energy investments is essential in resource-limited operations. The capital rationing example presented here (Badiru and Pulat, 1995) involves the determination of the optimal combination of energy investments so as to maximize present worth of total return on investment. Suppose an energy analyst is given N energy investment options, $X_1, X_2, X_3, \ldots, X_N$, with the requirement to determine the level of investment in each option so that present worth of total investment return is maximized subject to a specified limit on available budget. The options are not mutually exclusive.

The investment in each option starts at a base level b_i $(i = 1, 2, \ldots, N)$ and increases by variable increments k_{ij} $(j = 1, 2, 3, \ldots, K_i)$, where K_i is the number of increments used for option i. Consequently, the level of investment in option X_i is defined as

$$x_i = b_i + \sum_{j=1}^{K_i} k_{ij}$$

where

$$x_i \geq 0 \quad \forall i$$

For most cases, the base investment will be 0. In those cases, we will have $b_i = 0$. In the modeling procedure used for this example, we have:

$$X_i = \begin{cases} 1 & \text{if the investment in option } i \text{ is greater than zero} \\ 0 & \text{otherwise} \end{cases}$$

and

$$Y_{ij} = \begin{cases} 1 & \text{if the increment of alternative } i \text{ is used} \\ 0 & \text{otherwise.} \end{cases}$$

The variable x_i is the actual level of investment in option i, while X_i is an indicator variable indicating whether or not option i is one of the options selected for investment. Similarly, k_{ij} is the actual magnitude of the jth increment, while Y_{ij} is an indicator variable that indicates whether or not the jth increment is used for option i. The maximum possible investment in each option is defined as M_i such that

$$b_i \leq x_i \leq M_i$$

There is a specified limit, B, on the total budget available to invest such that

$$\sum_i x_i \leq B$$

There is a known relationship between the level of investment, x_i, in each option and the expected return, $R(x_i)$. This relationship is referred to as the *utility function*, $f(.)$, for the option. The utility function may be developed through historical data, regression analysis, and forecasting models. For a given energy investment option, the utility function is used to determine the expected return, $R(x_i)$, for a specified level of investment in that option. That is,

$$R(x_i) = f(x_i)$$
$$= \sum_{j=1}^{K_i} r_{ij} Y_{ij}$$

where r_{ij} is the incremental return obtained when the investment in option i is increased by k_{ij}. If the incremental return decreases as the level of investment increases, the utility function will be concave. In that case, we will have the following relationship:

$$r_{ij} - r_{i,j+1} \geq 0$$

Thus,

$$Y_{ij} \geq Y_{i,j+1}$$

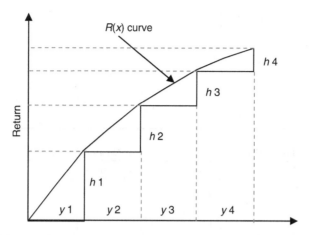

FIGURE 5.1   Utility curve for investment yield.

or

$$Y_{ij} - Y_{i,j+1} \geq 0$$

so that only the first n increments ($j = 1, 2, \ldots, n$) that produce the highest returns are used for project i. Figure 5.1 shows an example of a concave investment utility function.

If the incremental returns do not define a concave function, $f(x_i)$, then one has to introduce the inequality constraints presented above into the optimization model. Otherwise, the inequality constraints may be left out of the model, as the first inequality, $Y_{ij} \geq Y_{i,j+1}$, is always implicitly satisfied for concave functions. Our objective is to maximize the total investment return. That is,

$$\text{Maximize: } Z = \sum_i \sum_j r_{ij} Y_{ij}$$

Subject to the following constraints:

$$x_i = b_i + \sum_j k_{ij} Y_{ij} \quad \forall i$$

$$b_i \leq x_i \leq M_i \qquad \forall i$$

$$Y_{ij} \geq Y_{i,j+1} \qquad \forall i, j$$

$$\sum_i x_i \leq B$$

$$x_i \geq 0 \qquad \forall i$$

$$Y_{ij} = 0 \text{ or } 1 \qquad \forall i, j$$

Now suppose we are given four options (i.e., $N = 4$) and a budget limit of \$10 million. The respective investments and returns are shown in Tables 5.4 through 5.7.

All the values are in millions of dollars. For example, in Table 5.7, if an incremental investment of \$0.20 million from stage 2 to stage 3 is made in option 1, the expected incremental return from the project will be \$0.30 million. Thus, a total investment of \$1.20 million in option 1 will yield present worth of total return of \$1.90 million. The question addressed by the optimization model is to determine how many investment increments

TABLE 5.4   Investment Data for Energy Option 1

Stage (j)	Incremental Investment y_{1j}	Level of Investment x_1	Incremental Return r_{1j}	Total Return $R(x_1)$
0	—	0	—	0
1	0.80	0.80	1.40	1.40
2	0.20	1.00	0.20	1.60
3	0.20	1.20	0.30	1.90
4	0.20	1.40	0.10	2.00
5	0.20	1.60	0.10	2.10

TABLE 5.5   Investment Data for Energy Option 2

Stage (j)	Incremental Investment y_{2j}	Level of Investment x_2	Incremental Return r_{2j}	Total Return $R(x_2)$
0	—	0	—	0
1	3.20	3.20	6.00	6.00
2	0.20	3.40	0.30	6.30
3	0.20	3.60	0.30	6.60
4	0.20	3.80	0.20	6.80
5	0.20	4.00	0.10	6.90
6	0.20	4.20	0.05	6.95
7	0.20	4.40	0.05	7.00

TABLE 5.6   Investment Data for Energy Option 3

Stage (j)	Incremental Investment y_{3j}	Level of Investment x_3	Incremental Return r_{3j}	Total Return $R(x_3)$
0	0	—	—	0
1	2.00	2.00	4.90	4.90
2	0.20	2.20	0.30	5.20
3	0.20	2.40	0.40	5.60
4	0.20	2.60	0.30	5.90
5	0.20	2.80	0.20	6.10
6	0.20	3.00	0.10	6.20
7	0.20	3.20	0.10	6.30
8	0.20	3.40	0.10	6.40

TABLE 5.7   Investment Data for Energy Option 4

Stage (j)	Incremental Investment y_{4j}	Level of Investment x_4	Incremental Return r_{4j}	Total Return $R(x_4)$
0	—	0	—	0
1	1.95	1.95	3.00	3.00
2	0.20	2.15	0.50	3.50
3	0.20	2.35	0.20	3.70
4	0.20	2.55	0.10	3.80
5	0.20	2.75	0.05	3.85
6	0.20	2.95	0.15	4.00
7	0.20	3.15	0.00	4.00

should be used for each option. That is, when should we stop increasing the investments in a given option? Obviously, for a single option we would continue to invest as long as the incremental returns are larger than the incremental investments. However, for multiple investment options, investment interactions complicate the decision so that investment in one project cannot be independent of the other projects. The IP model of the capital rationing example was solved with LINDO software. The model is

Maximize: $Z = 1.4Y11 + .2Y12 + .3Y13 + .1Y14 + .1Y15 + 6Y21 + .3Y22 + .3Y23$
$+ .2Y24 + .1Y25 + .05Y26 + .05Y27 + 4.9Y31 + .3Y32 + .4Y33 + .3Y34$
$+ .2Y35 + .1Y36 + .1Y37 + .1Y38 + 3Y41 + .5Y42 + .2Y43 + .1Y44$
$+ .05Y45 + .15Y46$

Subject to:

$$.8Y11 + .2Y12 + .2Y13 + .2Y14 + .2Y15 - X1 = 0$$

$$3.2Y21 + .2Y22 + .2Y23 + .2Y24 + .2Y25 + .2Y26 + .2Y27 - X2 = 0$$

$$2.0Y31 + .2Y32 + .2Y33 + .2Y334 + .2Y35 + .2Y36 + .2Y37 + .2Y38 - X3 = 0$$

$$1.95Y41 + .2Y42 + .2Y43 + .2Y44 + .2Y45 + .2Y46 + .2Y47 - X4 = 0$$

$$X1 + X2 + X3 + X4 <= 10$$

$$Y12 - Y13 >= 0$$

$$Y13 - Y14 >= 0$$

$$Y14 - Y15 >= 0$$

$$Y22 - Y23 >= 0$$

$$\ldots\ldots$$

$$Y26 - Y27 >= 0$$

$$Y32 - Y33 >= 0$$

$$Y33 - Y34 >= 0$$

$$Y35 - Y36 >= 0$$

$$Y36 - Y37 >= 0$$

$$Y37 - Y38 >= 0$$

$$Y43 - Y44 >= 0$$

$$Y44 - Y45 >= 0$$

$$Y45 - Y46 >= 0$$

$$X_i >= 0 \quad \text{for } i = 1, 2, \ldots, 4$$

$$Y_{ij} = 0, 1 \quad \text{for all } i \text{ and } j$$

The solution indicates the following values for Y_{ij}.

5.6.1   Energy Option 1

Y11 = 1, Y12 = 1, Y13 = 1, Y14 = 0, Y15 = 0
Thus, the investment in option 1 is $X1 = \$1.20$ million. The corresponding return is \$1.90 million.

5.6.2   Option 2

Y21 = 1, Y22 = 1, Y23 = 1, Y24 = 1, Y25 = 0, Y26 = 0, Y27 = 0
Thus, the investment in option 2 is $X2 = \$3.80$ million. The corresponding return is \$6.80 million.

5.6.3   Option 3

Y31 = 1, Y32 = 1, Y33 = 1, Y34 = 1, Y35 = 0, Y36 = 0, Y37 = 0
Thus, the investment in option 3 is $X3 = \$2.60$ million. The corresponding return is \$5.90 million.

5.6.4   Option 4

Y41 = 1, Y42 = 1, Y43 = 1
Thus, the investment in option 4 is $X4 = \$2.35$ million. The corresponding return is \$3.70 million.

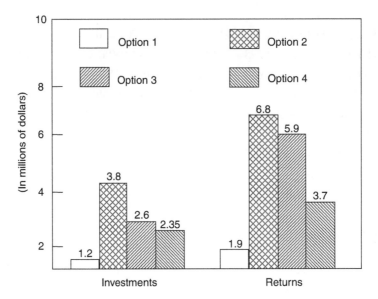

FIGURE 5.2   Comparison of investment levels.

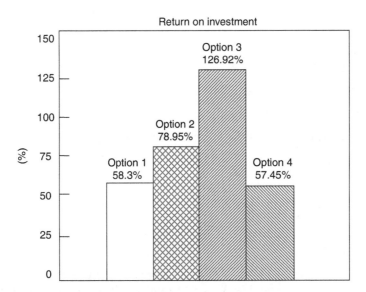

FIGURE 5.3   Histogram of returns on investments.

   The total investment in all four options is $9,950,000. Thus, the optimal solution indicates that not all of the $10,000,000 available should be invested. The expected present worth of return from the total investment is $18,300,000. This translates into 83.92% return on investment. Figure 5.2 presents histograms of the investments and the returns for the four options. The individual returns on investment from the options are shown graphically in Figure 5.3.

   The optimal solution indicates an unusually large return on total investment. In a practical setting, expectations may need to be scaled down to fit the realities of the investment environment. Not all optimization results will be directly applicable to real-world scenarios. Possible extensions of the above model of capital rationing include the incorporation of risk and time value of money into the solution procedure. Risk analysis would be relevant, particularly for cases where the levels of returns for the various levels of investment are not known with certainty. The incorporation of time value of money would be useful if the investment analysis is to be performed for a given planning horizon. For example, we might need to make investment decisions to cover the next 5 years rather than just the current time.

5.7 Simulation and Optimization of Distributed Energy Systems

Hudson (2005) has developed a spreadsheet-based adaptive nonlinear optimization model that utilizes the OR methods of simulation and nonlinear optimization to determine the optimal capacities of equipment in combined heat and power (CHP) applications, often called cogeneration. Evaluation of potential CHP applications requires an assessment of the operations and economics of a particular system in meeting the electric and thermal demands of a specific end-use facility. Given the electrical and thermal load behavior of a facility, the tariff structure for grid-supplied electricity, the price of primary fuel (e.g., natural gas), the operating strategy and characteristics of the CHP system, and an assumed set of installed CHP system capacities (e.g., installed capacity of prime mover and absorption chiller), one can determine the cost of such a system as compared to reliance solely on traditional, grid-supplied electricity and on-site boilers. Selecting the optimal capacities for a CHP system will help to produce optimal cost benefits and potentially avoid economic losses. The following material describes the methodology of the approach and provides an example of the results obtained.

5.8 Point-of-Use Energy Generation

Distributed energy is the provision of energy services at or near the point of use. It can take many forms, but a central element is the existence of a prime mover for generating electricity. Typical prime movers for current distributed energy applications are gas or light oil-fired turbines, fuel cells, or reciprocating engines fired with natural gas or diesel fuel. Such prime movers are only able to utilize roughly 30% of the input fuel energy in the production of electricity. The remaining energy can either be utilized as a thermal resource stream or must be exhausted to the atmosphere. When the waste heat is used to satisfy heating needs, the system is typically termed a cogeneration or combined heat and power system. Through the use of an absorption chiller, waste heat can also be utilized to provide useful cooling, in which case the system is considered a CHP application.

Generally, CHP systems are not the sole source of electricity and thermal resource for a facility. In most cases, these systems are merely alternatives to utility grid-supplied electricity, electric chillers, and electric or gas-fired on-site water heating. As a result, CHP systems are characteristic of the classic "make-or-buy" decision, and economic viability is relative to grid-based electricity and on-site boiler heating. An assessment of the economic viability of a particular CHP system requires an assumption regarding the installed equipment capacities of the system. As costs are a direct function of the installed capacities of these systems, the challenge is to determine the most economically optimal capacities of the equipment.

An important consideration in assessing the potential for CHP systems is recognition of the noncoincident behavior of the electric and thermal (i.e., heating and cooling) loads of a facility. That is, the load patterns for the three load streams are not perfectly correlated with each other through time. As a result, the peak of electrical demand will most likely not occur at the same point in time as either the heating or cooling demand peak. Absence of means to store electrical or thermal energy on-site, producing electricity with a distributed generator to track electrical demand (i.e., electric load following) may produce recovered thermal energy that cannot be used due to lack of thermal demand at that moment. Similarly, operating a CHP system in a thermal load following mode (i.e., tracking thermal demand), combined with limits on the sale of electricity into the grid, may also impact the degree to which the three demand streams can be satisfied by distributed energy.

5.9   Modeling of CHP Systems

To optimize a CHP system, the operational behavior of the system and the loads that it serves must be taken into account. One of the best ways to do this is by simulating the interaction of the system with its loads. With a properly detailed simulation, the variabilities of the load streams, the time-related prices of grid-based electricity, and the performance limitations of the CHP equipment can be recognized.

In addition to the use of the OR method of simulation, application of optimization methods is also required. An important distinction is the optimization of a system's *operation* as compared to the optimization of the system's installed *capacity*. A number of early works address the optimization of the operation of a given system. Baughman et al. (1989) developed a cogeneration simulation model in Microsoft Excel that sought optimal operation of industrial cogeneration systems over a 15-year planning horizon using the minimization of net present value of operating costs as the objective function. Consonni et al. (1989) developed an operations optimization simulator based on 36 separate sample day patterns to represent annual operations. Using a binary representation of equipment being either on or off, the model was a mixed integer linear program with an objective function of maximizing hourly profits from operation.

Regarding the optimization of installed system *capacity*, Yokoyama et al. (1991) introduced a coupled, "hierarchical" modeling concept, whereby the optimization of a system's installed capacity was an outer shell or layer serving to drive a separate inner operations optimization model based on mixed integer linear programming. Similar to Consonni, Yokoyama et al. used 36 sample day patterns to describe annual load behavior. Utilizing the hierarchical optimization process described by Yokoyama et al., Asano et al. (1992) considered the impact of time-of-use rates on optimal sizing and operations of cogeneration systems. Using 14 sample day patterns to represent the load behavior, Asano evaluated three commercial applications (hotel, hospital, and office building) and calculated optimal capacities ranging from 50% to 70% of peak electricity demand. Contemporaneously, a set of closed form equations for calculating the optimal generation capacity of an industrial cogeneration plant with stochastic input data was developed by Wong-Kcomt (1992). Wong-Kcomt's approach relied upon single unit prices for electricity (i.e., no separate demand charges) and assumed independent Gaussian distributions to describe aggregate thermal and electrical demand. The effects of hourly non-coincidence of loads were not addressed. Wong-Kcomt showed, however, that the solution space of the objective function (cost minimization) was convex in nature.

As an extension of the hierarchical model proposed by Yokoyama et al. in 1991, Gamou et al. (2002) investigated the impact that variation in end-use energy demands had on optimization results. Modeling demand (i.e., load) variation as a continuous random variable, probability distributions of electrical and thermal demands were developed. Dividing the problem into discrete elements, a piecewise LP approach was used to find the minimum cost objective function. It was observed that the capacity deemed as optimal using average data was, in fact, suboptimal when load variations were introduced. In characterizing the variability of electrical and thermal demands, the non-coincident behavior of the electrical and thermal loads led to the determination of a lower optimal capacity value when variability was recognized, relative to the value obtained when considering only average demands. A key finding from this work was that use of average demand data (e.g., sample day patterns) may not accurately determine the true optimal system capacity. Orlando (1996) states a similar conclusion in that "any averaging technique, even multiple load-duration curves, by definition, cannot fully model the interaction between thermal and electrical loads."

Within the last 10 years, cogeneration technology has evolved to include systems with smaller electric generation unit capacities in uses other than large, industrial applications. Termed "distributed energy" or "distributed generation," these systems are now being applied in commercial markets such as hospitals, hotels, schools, and retail stores. In addition, traditional cogeneration (i.e., the production of electricity and useful heat) has been expanded to include trigeneration, that is, the use of waste heat from electrical production to produce both useful heat and cooling. A number of works are focused on this recent development. Marantan (2002) developed procedures to evaluate a predefined list of candidate system capacities for an office building application, selecting the CHP system with the minimum net annual cost. Campanari et al. (2002) used a simulation model with 21 sample day patterns and a predefined list of operating scenarios to select the least cost operating strategy for a CHP system in commercial buildings. They did a somewhat reverse approach in investigating capacity-related optimization by varying the building size for a CHP system of fixed capacity. An important conclusion of their manual, trial and error optimization was that "due to the inherent large variability of heating, cooling, and electric demand typical of commercial buildings, the optimum size of a cogeneration plant is significantly lower than peak demand." A similar conclusion was found by Czachorski et al. (2002) while investigating the energy cost savings resulting from the use of CHP systems in hospitals, hotels, offices, retail, and educational facilities in the Chicago area. Through manual capacity enumeration, they found that, based on maximum annual energy cost savings, "the corresponding size of the power generator was between 60% and 80% of the maximum electric demand for CHP systems" in the applications considered. The study by Czachorski et al. also showed that annual energy cost savings exhibit a concave behavior with respect to generator capacity. While their work did not reflect life-cycle cost savings by including investment cost as a function of generator capacity, the inclusion of generation equipment capital cost should not eliminate the general concave behavior produced by the annual energy economics.

5.10   Economic Optimization Methods

Based on the modeling efforts described in the previous section, consideration is now given to the question of an appropriate method to apply in seeking an optimum of an economic objective function. A discussion of relevant optimization techniques cannot be made without some knowledge of the structure of the model in which the optimization will be conducted. Therefore, rather than providing a pedagogic recitation of the wide variety of optimization methods and algorithms that exist, this section will focus on the specific methods that are applicable to the problem at hand, which will then be further developed in the following section. There are, however, a number of good texts on optimization methods. Two examples are Bazaraa et al. (1993) and Gill et al. (1986).

As mentioned above, to determine an appropriate optimization method (i.e., to select the correct tool for the job), one must have some understanding of the system or model upon which the optimization will be applied. One approach to this selection is to consider the attributes of the system or model and proceed through somewhat of a classification process. A good initial step in the classification is to determine if the system or model is linear or nonlinear in either its objective function or its constraints. If the objective function and all constraints are linear, then linear optimization methods (e.g., linear programming) should be applied. If either the objective function or any constraint is nonlinear, then the nonlinear class of methods may be required. A further distinction is whether the independent variables are constrained. If the feasible region is defined by constraints, constrained optimization methods generally should be applied. In addition, if one or more of the independent variables can only take on integer values, specialized integer programming methods may be required.

With respect to the economic modeling of CHP systems, life-cycle cost, or alternatively, the life-cycle savings relative to some non-CHP alternative, as a function of installed equipment capacity, has been shown to be convex and concave, respectively (Wong-Kcomt, 1992; Czachorski et al., 2002). Therefore, using either life-cycle cost or savings as the objective function necessitates a nonlinear optimization approach. Beyond this, consideration must be given as to whether the current problem has independent variables that are constrained to certain sets of values (i.e., equality constraints) or somehow bounded (i.e., inequality constraints). In either case, constrained nonlinear optimization is generally performed by converting the problem in such a way that it can be solved using unconstrained methods (e.g., via Lagrangian multipliers or penalty methods) (Bazaraa et al., 1993). In this study, the independent variables are installed equipment capacities, which are assumed to be continuous and non-negative. Thus, the only constraints are simple bounds, defined as $x_i \geq 0$. Fletcher (1987) suggests a number of ways to handle such constraints, including variable transformation (e.g., $x = y^2$) and introduction of slack variables. With slack variables, a problem of the type Maximize $F(x)$ subject to $x_i \geq 0$ can be rewritten using slack variables as Maximize $F(x)$ subject to $x_i - w_i^2 = 0$, where w_i^2 is a squared slack variable. The solution can then follow through the development of the Karush–Kuhn–Tucker (KKT) conditions and the Lagrangian function. It has been shown that for linear constraints and a concave objective function, as in this study, the global optimum will be at a point satisfying the KKT conditions (Bazaraa et al., 1993; Winston, 1994).

Another method to extend equality-constraint methods to inequalities is through the use of a generalized reduced gradient (GRG) approach. The reduced gradient method seeks to reduce the number of degrees of freedom, and therefore, free variables, that a problem has by recognizing the constraints that are active (i.e., at their bounds) during each iteration. If a variable is at an active bound, it is excluded from calculations related to the determination of the incremental solution step. If no variables are at active constraints, the GRG method is very similar to the standard quasi-Newton method for unconstrained variables.

There are a number of methods available to perform unconstrained nonlinear optimization. A central distinction is whether the method relies on derivatives of the objective function. If derivatives are not available or are computationally difficult to obtain, nonderivative methods can be employed. Such methods are also needed when the objective function or gradient vector is not continuous. Methods that rely solely on function comparison are considered direct search methods (Gill et al., 1986). A common direct search method is the polytope or Nelder–Mead method in which prior functional evaluations are ordered such that the next iteration is a step in the direction away from the worst point in the current set of points. Another nonderivative method is the Hook and Jeeves method, which performs exploratory searches along each of the coordinate directions followed by pattern searches defined by the two most recent input vectors. The main disadvantage of these direct search methods is that they can be very slow to converge.

The two direct search methods mentioned above are considered sequential methods in that new trial inputs are the product of the previous result. Another class of the direct search method is the simultaneous direct search in which the trial points are defined a priori (Bazaraa et al., 1993). For variables in two dimensions, an example of this method would be a grid-pattern search, which is employed in this particular study.

If objective function derivatives are available, other, more efficient, methods can be brought to bear. One of the most fundamental procedures for optimizing a differentiable function is the method of steepest descent, also called the gradient method. In this method, the search direction is always the negative gradient, and the step size is calculated to minimize the objective function (assuming the function is convex). This is repeated until a stopping criterion, such as the gradient norm, is sufficiently small. However, it has been shown that following the direction of steepest descent does not necessarily produce rapid convergence, particularly near a stationary point, and that other derivative methods perform better (Bazaraa et al., 1993; Bartholomew-Biggs, 2005). For large problems (i.e., those with more than 100 decision variables), the conjugate gradient method is useful as it does not require storage of large matrices (Bazaraa et al., 1993). As this method is typically less efficient and less robust than other methods, and as the current problem concerns a small number of independent variables, the conjugate gradient method was not used for this application.

The remaining methods of interest are the Newton method and the related quasi-Newton method. The Newton method has been shown to be very efficient at unconstrained nonlinear optimization if the objective function has continuous first and second derivatives. If first and second derivatives are available, a Taylor-series expansion in the first three terms of the objective function yields a quadratic model of objective function that can subsequently be used to define a Newton direction for function minimization. As long as the Hessian is positive definite and the initial input values are in the neighborhood of the optimum, Newton's method converges to the optimum quadratically (Gill et al., 1986).

Due to the discrete form of the model in this study, analytical expressions for first and second derivatives are not available. In these situations, derivatives can be approximated using finite difference techniques. The lack of an exact expression for second derivatives means that curvature information, typically provided by calculating the Hessian matrix, is not directly available for use in a Newton method. The solution to this problem is to utilize the well-known quasi-Newton method, in which an approximation to the inverse Hessian is successively built-up during the iteration process. While typically expecting the first derivative to be analytically available in a quasi-Newton method, the additional lack of explicit first derivatives to form the gradient does not appear to be a fatal impediment. Van der Lee et al. (2001) successfully used this approach in studying the optimization of thermodynamic efficiency in power plant steam cycles. As Gill et al. (1986) state, "when properly implemented, finite-difference quasi-Newton methods are extremely efficient, and display the same robustness and rapid convergence as their counterparts with exact gradients."

With respect to the iterative update of the Hessian matrix, a number of Hessian update methods have been proposed over the years, including the Davidson–Fletcher–Powell (DFP) method, the Powell–Symmetric–Broyden (PSB) update, and the Broyden–Fletcher–Goldfarb–Shanno (BFGS) method. The literature indicates that the BFGS method is clearly accepted as the most effective update method currently available (Gill et al., 1986; Nocedal, 1992; Zhang and Xu, 2001; Bertsekas, 2004; Yongyou et al., 2004). Details of the BFGS algorithm will be provided in the following section.

A final element related to the quasi-Newton method is the use of line search methods when the full quasi-Newton step produces an objective function response that does not make satisfactory progress relative to the previous iteration, thus possibly indicating the passing of a local optimum. In that case, a "backtracking" process along the step direction is needed. As discussed by Dennis and Schnabel (1983), the backtracking approach should conform to the Armijo and Goldstein (AG) conditions to ensure satisfactory convergence. Dennis and Schnabel provide the classic quadratic fit using three previously calculated function values to solve for the optimum quasi-Newton step multiplier, followed by the cubic spline fit, should the new quadratic step not meet AG conditions.

It should be noted that the quadratic/cubic fit method is but one method to determine an appropriate step value. While less efficient in terms of computational requirements, line search methods that do not rely on the gradient, or in this case, an approximation to the gradient, can also be used. Sequential search methods such as the Fibonacci and related Golden section methods can be utilized to determine an acceptable step multiplier (Bazaraa et al., 1993).

5.11 Design of a Model for Optimization of CHP System Capacities

This section provides a detailed explanation of the simulation model as well as the approach used to determine an optimum set of equipment capacities for a CHP system. Similar to the approach used by Edirisinghe et al. (2000) and Yokoyama et al. (1991), the model consists of two nested sections: an outer, controlling optimization algorithm and an inner operation simulation routine. The overall flow of the optimization model is shown in Figure 5.4. Starting with an initial "guess" for the installed electrical generator and absorption chiller capacities, an hour-by-hour operation simulation is performed to develop a value of the objective function for the given generator and chiller capacities. Within the optimization algorithm, a stopping criterion is used to control the updating of the optimization routine and subsequent iterative looping back to the operation simulation with a new set of candidate installed capacities. The optimization algorithm seeks to maximize the net present value (NPV) savings produced by using the CHP system relative to a non-CHP scenario (where electricity is obtained solely from the grid and heating loads are met by an on-site boiler). The maximization of NPV savings (i.e., maximization of overall profitability) is an appropriate method for evaluating mutually exclusive alternatives (Sullivan et al., 2006).

In recognition of the problems identified earlier regarding the use of average or aggregated demand data (Gamou et al., 2002; Hudson and Badiru, 2004), this approach utilizes demand data expressed on an hourly basis, spanning a 1-year period. Use of hourly data has the advantage of explicitly capturing the seasonal and diurnal variations, as well as non-coincident behaviors, of electrical and thermal loads for a given application. In many cases, actual hourly demand data for an entire year may not be available for a specific site.

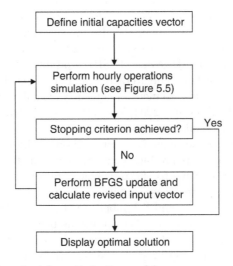

FIGURE 5.4   Overview flow chart for optimization model.

TABLE 5.8   Input Variables Used in Operation Simulation Model

Variable	Typical Units
Facility loads	
Hourly electrical demand (non-cooling related)	kW
Hourly heating demand	Btu/h
Hourly cooling demand	Btu/h
Electric utility prices	
Demand charge	$/kW-month
Energy charge	$/kWh
Standby charge	$/kW-month
On-site fuel price (LHV basis)	$/MMBtu
Equipment parameters	
Boiler efficiency (LHV)	Percent
Conventional chiller COP	Without units
Absorption chiller (AC) COP	Without units
Absorption chiller (AC) capacity	RT
AC minimum output level	Percent
AC system parasitic electrical load	kW/RT
Distributed generation (DG) capacity, net	kW
DG electric efficiency (LHV) at full output	Percent
DG minimum output level	Percent
DG power/heat ratio	Without units
Operating and maintenance (O&M) cost	$/kWh
Number of DG units	Units
DG capital cost	$/kW installed
AC capital cost	$/RT installed
General economic parameters	
Planning horizon	Years
Discount rate	Percent/year
Effective income tax rate	Percent

In these situations, building energy simulation programs, such as Building Energy Analyzer or BCHP Screening Tool, are available that can develop projected hourly loads for electricity, heating, and cooling on the basis of building application, size, location, and building design attributes (e.g., dimensions, insulation amounts, glazing treatments) (InterEnergy/GTI, 2005; Oak Ridge National Laboratory, 2005).

The data needed to simulate the operation of a CHP system are shown in Table 5.8. The input for the hourly facility electrical demand should include all facility electrical demand *except* for cooling-related demand. As cooling may be provided by an absorption chiller under CHP operation, electrical demand related to cooling is calculated explicitly within the simulation model. For the hourly heating and cooling demands, the input values are expressed on an end-use, as-consumed thermal basis.

The prices for utility-supplied electricity typically have a price component related to the amount of energy consumed (i.e., an energy charge) as well as a component proportional to the monthly peak rate of energy consumed (i.e., a demand charge). Some utilities will price their electricity at different rates to those who self-generate a portion of their electrical needs. In addition, some electric utilities charge a monthly standby fee for the availability of power that may be called upon should the distributed generation not be available. Utility tariff structures can also have unit prices that vary both seasonally and diurnally. Similar to electricity rates, the unit price for on-site fuel may be different for those who operate a CHP system.

The fuel assumed for on-site distributed generation and on-site water/steam heating in this study is natural gas, expressed on a $/million Btu (MMBtu) lower heating value (LHV) basis. The heating value of natural gas refers to the thermal energy content in the fuel, which can be expressed on a higher heating value (HHV) or lower heating value basis. The difference in the two heating values relates to the water formed as a product of combustion. The higher heating or gross value includes the latent heat of vaporization of the water vapor.

The lower heating or net value excludes the heat that would be released if the water vapor in the combustion products were condensed to a liquid. As DG/CHP systems try to limit exhaust vapor condensation due to corrosion effects, the usable heat from natural gas is typically the LHV. In the United States, natural gas is typically priced on a HHV basis, so care should be used in entering the proper value. For natural gas, the conversion between HHV and LHV is

$$\text{heat content}_{\text{HHV}} = \text{heat content}_{\text{LHV}} \times 1.11 \ (\text{Petchers, 2003})$$

The definitions for the equipment and economic parameters listed in Table 5.8 are as follows:

Boiler efficiency—The thermal efficiency of the assumed on-site source of thermal hot water/steam (e.g., boiler) for the baseline (non-CHP) scenario, expressed on an LHV basis.

Conventional chiller COP—The coefficient of performance (COP) for a conventional electricity-driven chiller. It is determined by dividing the useful cooling output by the electrical energy required to produce the cooling, adjusted to consistent units.

Absorption chiller COP—The coefficient of performance for the CHP system absorption chiller. It is determined by dividing the useful cooling output by the thermal energy required to produce the cooling, adjusted to consistent units. Parasitic electrical support loads (e.g., pump and fan loads) are addressed separately.

Absorption chiller capacity—The installed capacity of the absorption chiller in refrigeration tons (RT). This is an independent variable in the model.

AC minimum output level—The minimum percent operating level, relative to full output, for the absorption chiller. This is also known as the minimum turndown value.

AC system parasitic electrical load—The electrical load required to support the absorption chiller. The chiller load should include the chiller solution pump, the AC cooling water pump, and any cooling tower or induced draft fan loads related to the AC.

Distributed generation (DG) capacity—The installed capacity of the distributed electrical generator (i.e., prime mover), expressed in net kilowatts. This is an independent variable in the model.

DG electric efficiency (LHV) at full output—The electricity production efficiency of the DG prime mover at full output. This efficiency can be determined by dividing the electricity produced at full output by the fuel used on a LHV basis, adjusted to consistent units.

DG minimum output level—The minimum percent operating level, relative to full output, for the DG unit. Also known as the minimum economic turndown value.

DG power/heat ratio—The ratio of net electrical power produced to useful thermal energy available from waste heat, adjusted to consistent units.

O&M cost—The operating and maintenance cost of the total cooling, heating, and power system, expressed on a $/kWh of electricity generated basis.

Number of DG units—The number of prime mover units comprising the system. Currently, the model is limited to no more than two units, each identical in size and performance. The optimum capacity determined by the model is the total capacity of the CHP system, and for a two-unit system, that capacity is split equally between the units.

DG capital cost—The fully installed capital cost of the distributed generation system, expressed on a $/net kW basis.

AC capital cost—The fully installed capital cost of the absorption chiller system, expressed on a $/RT basis.

Planning horizon—The assumed economic operating life of the CHP system. The default value is 16 years to be consistent with U.S. tax depreciation schedules for a 15-year property. Currently, 16 years is the maximum allowed planning horizon in the model.

Discount rate—The rate used to discount cash flows with respect to the time-value of money.

Effective income tax rate—The income tax rate used in income tax-related calculations such as depreciation and expense deductions. The effective rate reflects any relevant state income tax and its deductibility from federal taxes.

The general flow of calculations within the operation simulation is shown in Figure 5.5. Once the electrical and thermal loads and general equipment/economic parameters are defined for each iteration of the optimization routine, a trial set of distributed generator and absorption chiller capacities are provided to the operations simulator. Two separate simulations must be performed. First, the hour-by-hour costs for satisfying the thermal and electric loads solely by a traditional utility grid/on-site boiler arrangement must be

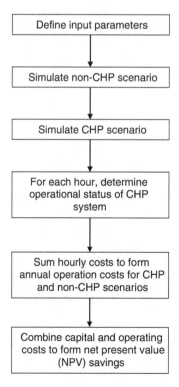

FIGURE 5.5 Operation simulation flow chart.

calculated. This is referred to as the non-CHP or grid-only scenario. In the non-CHP scenario, grid-supplied electricity is assumed to support all facility loads, including cooling but excluding heating. Heating is assumed to be provided by an on-site fossil-fired boiler. A second, separate calculation develops the hour-by-hour costs of meeting at least some part of the specified loads with a CHP system. The degree of contribution of the CHP system is determined hourly by the relative cost of using the CHP system (i.e., making) to satisfy the loads versus the traditional grid-purchase/on-site boiler operation (i.e., buying). If the operation of the CHP system is less expensive in a given hour than the grid/boiler approach, the CHP system will supply the loads, with the grid/boiler providing any supplemental energy needed to satisfy that hour's load. If the operation of the CHP system is more expensive, loads are satisfied in the traditional grid-purchase manner for that hour. As the price of grid-based electricity can change depending upon the time-of-day, this test is performed for each hour of the year. With typical time-of-day rates, the CHP system generally operates during the workday hours and is offline during the night.

Relative to the non-CHP scenario, developing the annual cost for a CHP-based system is substantially more complicated. There can be utility surcharges (e.g., standby fees) that are imposed as a result of operating self-generation equipment. In addition, the unit pricing for electricity may be different for customers using a CHP system than for those buying all their supply solely from the utility. The operational considerations related to the CHP system are of considerable influence as well. As an example, the fuel efficiency of electrical generation equipment is directly proportional to the relative output level. Typically, the highest efficiency (i.e., most electricity produced for the least fuel consumed) is at or near full-rated output. Depending upon the type of prime mover, electrical efficiencies at low part-load can be 65%–75% of full-load efficiency. As a result, there is a general lower limit on part-load operations. A typical minimum operating value is 50% of rated unit capacity. The limit becomes influential when the electrical demand is less than 50% of the rated unit capacity, requiring that electricity be purchased from the grid. Thus, there is an economic trade-off related to the size of the CHP generation capacity. A CHP system sized to meet peak electrical or thermal loads will incur higher utility standby charges and will have less ability to operate during periods of low demand. Conversely, a smaller sized system may be able to operate a larger fraction of time, but may result in a higher fraction of unmet load for the facility (resulting in higher utility purchases, typically at peak pricing). The economics are further influenced by the direct relationship of CHP electrical generation capacity and useful thermal energy available. Smaller electrical capacity means less useful thermal byproduct, which might then require more supplemental gas-boiler or electric chiller operation.

At the CHP/non-CHP scenario level, two sets of annual operating costs are then determined by summing the relevant hourly costs of meeting thermal and electric demands from either the grid and on-site boiler solely (i.e., the non-CHP scenario) or from CHP operations. A differential annual operating cost (or net annual savings, if the CHP scenario is less costly than the non-CHP scenario) is determined based on the annual cost difference between the non-CHP scenario and the CHP-available scenario. A net present value is then determined by calculating the present worth of the net annual savings over the number of years defined by the planning horizon at the defined discount rate and adding the installed capital costs of the CHP system, adjusted for income tax effects (e.g., depreciation).

As stated earlier, the planning horizon for this model can be up to 16 years. Unit prices for electricity and gas, as well as O&M unit costs, are not likely to remain constant over such a long period. Similarly, it is possible that electrical or thermal loads may change over such a period. As a result, the ability to reflect escalating unit prices and possible load changes is needed. As annual operations are calculated on an hourly basis, performing an explicit calculation for every hour within a 16-year duration would require 140,160 hourly calculations. It was felt that explicitly performing 140,160 hourly calculations would be intractable in the Excel platform. A solution to this dilemma was to express the variables that are subject to escalation as levelized values. A common method used in public utility economic analysis, a levelized value is determined by calculating the present value of the escalating annual stream and then applying an annual capital recovery factor to produce the levelized annual equivalent value (Park and Sharp-Bette, 1990). The levelized values are then used in the operation simulation calculations. Thus, in the material that follows, unless explicitly stated, values for electricity and gas unit prices, thermal and electric loads, unit O&M costs, and the resulting annual costs should be considered annual levelized values, spanning the duration of the planning horizon.

5.12 Capacity Optimization

As mentioned, the optimization goal is to maximize NPV cost savings by determining the optimum installed capacities for the electricity generation system and the absorption chiller. Given that only objective function values are directly available in this computational model (i.e., no analytical expressions for first or second derivatives), it is felt that, based on a review of current literature, the use of a quasi-Newton method with BFGS updates of the inverse Hessian is the most appropriate approach.

The quasi-Newton method is a variant of the Newton method and can be found in any good nonlinear optimization textbook (Gill et al., 1986; Fletcher, 1987; Bazaraa et al., 1993; Bertsekas, 2004; Bartholomew-Biggs, 2005). The Newton method relies on a three-term Taylor approximation of an assumed quadratic behavior of the objective function. As such, the quadratic model of the objective function, F, can be expressed as

$$F(x_k + p) \approx F_k + g_k^T p + \frac{1}{2} p^T G_k p$$

where g, p, and G are the gradient (Jacobian) in x, step direction, and Hessian in x, respectively. As we seek to find a stationary point of the function with respect to the step direction p, the objective function can be rewritten in p as

$$F(p) = g_k^T p + \frac{1}{2} p^T G_k p$$

A necessary condition for a stationary point is that the gradient vector vanishes at that point. Thus,

$$\nabla F(p) = g_k + G_k p = 0 \quad \text{or} \quad p = -G_k^{-1} g_k$$

If G is positive definite, then conditions are sufficient to state that p can be a minimum stationary point (Gill et al., 1986). In the case of maximization, G should be negative definite. The Newton method requires, however, that the Hessian of the objective function be known or determinable. In the current problem, the Hessian cannot be determined analytically. Thus, we rely on a sequential approximation to the Hessian as defined by the quasi-Newton method.

The typical quasi-Newton method assumes that the gradient of the objective function is available. The model used in this study has no analytic representation of either first or second derivatives. In this situation, a forward-difference approximation must be used to estimate the gradient vector. For the ith independent variable, x_i, the gradient is estimated by

$$g_i = \frac{1}{h}(F(x_i + h) - F(x_i))$$

where h is the finite-difference interval. For this study, a finite-difference interval of 10^{-4} was selected after evaluating choices ranging from 10^{-2} to 10^{-6}.

The general outline of the quasi-Newton method for maximization (Bartholomew-Biggs, 2005) is as follows:

- Choose some x_o as an initial estimate of the maximum of $F(x)$.
- Set the initial inverse Hessian, H_0, equal to the negative identity matrix (an arbitrary symmetric negative definite matrix).
- Repeat for $k = 0, 1, 2, \ldots$.
    - Determine $g_k = \nabla F(x_k)$ by forward-difference approximation.
    - Set the step length scalar, λ, equal to 1.
    - Calculate the full step direction $p_k = -H_k g_k$.
    - Evaluate whether the full step is appropriate by comparing $F(x_k + \lambda p_k)$ to $F(x_k)$. If $F(x_k + \lambda p_k) < F(x_k) + \rho \lambda g_k^T p_k$, solve for the step length λ that produces a univariate maximum $F(\lambda)$ for $0 \leq \lambda \leq 1$.
    - Set $x_{k+1} = x_k + \lambda p_k$, $y_k = g_{k+1} - g_k$, $d_k = x_{k+1} - x_k$.
    - Evaluate stopping criteria, and if not achieved,
    - Update the approximate inverse Hessian such that $H_{k+1} y_k = d_k$.
    - Increment k.

The stopping criteria used in this model are consistent with prior work by Edirisinghe et al. (2000) and Kao et al. (1997), in which the algorithm is terminated when either the change (i.e., improvement) in the objective function is less than a prescribed threshold amount or when the gradients of the objective function at a particular input vector are zero. The setting of the termination threshold value is a matter of engineering judgment. If a value is chosen that requires very small changes in the objective function before termination, the algorithm can cycle for a large number of iterations with very little overall improvement in the objective function. Conversely, a more relaxed threshold value can terminate the optimization algorithm prematurely, producing a suboptimal solution. A balance must therefore be struck between long execution times and less than total maximization of the objective function. As the objective function in this study is NPV cost savings over a multiyear period, one must select a value at which iterative improvements in NPV cost savings are considered negligible. There are two approaches used in setting this termination threshold. First, on an absolute basis, if the iterative improvement of the NPV cost savings is less than $50.00, it is considered reasonable to terminate the algorithm. In some cases, however, this absolute value can be a very small percentage of the overall savings, thus leading to long execution times with little relative gain. The second termination approach is based upon a relative measure on the objective function. If the change in NPV cost savings between iterations is greater than $50.00, but less than 0.00001 times the objective function value, then the algorithm terminates under the assumption that a change of less than 0.001% is insignificant.

In some situations, the objective function can exhibit multiple local optima of low magnitude relative to the average value within a neighborhood around the stationary point (i.e., low-level noise of the objective function). When this occurs, a means to help avoid getting "trapped" in a *near* optimum response space, particularly when the response surface is relatively flat, is to require two or three consecutive iterative achievements of the stopping criterion (Kim, D., personal communication, 2005). For this study, two consecutive achievements of the stopping criterion detailed in the previous paragraph were required to end the optimization process. In some cases with multiple local optima, the model may find a local optimum rather than the global optimum. A useful technique to improve the solution is to try different starting points for the optimization (Fylstra et al., 1998).

The updating of the matrix H, representing a sequential approximation of the inverse Hessian, is done using the BFGS method. As mentioned earlier, the BFGS update method is clearly considered to be the most efficient and robust approach available at this time. The BFGS formula for H_{k+1}, as presented by Zhang and Xu (2001) and Bartholomew-Biggs (2005), is:

$$H_{k+1} = H_k - \frac{H_k y_k d_k^T + d_k y_k^T H_k}{d_k^T y_k} + \left[1 + \frac{y_k^T H_k y_k}{d_k^T y_k}\right] \frac{d_k d_k^T}{d_k^T y_k}$$

There are a number of methods that can be employed in the backtracking search for the Newton step length λ that produces a maximum in the objective function. In this study, a quadratic and cubic spline fit was evaluated, but the method was not stable under some input conditions or required a large number of iterations before reaching the stopping criterion. This appears to be due to the lack of strict concavity of the objective function. As a result, the Golden sequential line search method was selected for its accuracy and stability. The Golden search was terminated when the interval of uncertainty (IOU) for the step length λ became less than 0.025. It should be noted that the step length can be unique to each variable rather than being a single scalar value. Such an approach was explored, but the additional computations did not seem to produce sufficiently improved results (i.e., faster optimization) to merit incorporating the approach in the final model.

To provide visual guidance regarding the surface behavior of the objective function within the overall solution space, a simultaneous uniform line search method was utilized as well. Using a 21×7 (electric generator capacity × absorption chiller capacity) grid, grid step sizes were selected to evaluate the complete range of possible CHP equipment capacities (i.e., $0 \leq \text{size} \leq \text{max load}$) for both the distributed generator and the absorption chiller. For each of the 147 cells, the capacity combination was used as an explicit input to the simulation model to determine the corresponding NPV cost savings. A contour plot of the NPV cost savings was produced to graphically display the overall solution space.

As mentioned earlier, there are simple lower bound constraints that require the capacities of the distributed generator and absorption chiller to be greater than or equal to zero. In an unconstrained method, it is possible that the direction vector could propose a solution that would violate the lower bound. This model checks for this condition, and if present, sets the capacity value to zero. As an added element to improving the efficiency of the algorithm, if the capacity of the distributed generation is set to zero, the capacity of the absorption chiller is also set to zero, as DG capacity is the energy source to operate the absorption chiller. This approach does not violate the quasi-Newton method as the effect of zeroing the capacity when a negative capacity is suggested is equivalent to reducing the Newton step size for that iteration. In this situation, the new x_{k+1} point is set to zero, and gradients are calculated at the new input vector for use in the quasi-Newton algorithm. Should the economic conditions of the problem be such that the maximum objective function (given the lower bound constraints) is truly at zero capacity for the distributed generator, the next iteration will yield the same adjusted input vector (owing to a direction vector pointing into the negative capacity space) and same NPV cost savings, which will appropriately terminate the optimization on the basis of similar NPV results, as discussed above.

5.13   Implementation of the Computer Model

To provide useful transparency of the calculations, the methods were implemented using Microsoft Excel. Excel spreadsheets allow others to view the computational formulae, which enhances understanding and confidence in the modeling approach. In addition, Microsoft Excel is a ubiquitous platform found on most personal computer (PC) systems. The model in this study, named the CHP Capacity Optimizer, was developed using Microsoft Office Excel 2003 on a PC running the Microsoft Windows XP Professional operating system (version 2002). The model makes use of Excel's Visual Basic for Applications (VBA) macro language to control movement to various sheets within the overall spreadsheet file and to initiate the optimization procedure. The Excel model and User's Manual are available from the author or can be downloaded from the Internet (search "CHP capacity optimizer").

5.13.1   Example Calculation

As an example of the use of the optimization tool, the potential use of CHP at a hospital in Boston, Massachusetts, will be evaluated. The hospital consists of five stories with a total floor area of 500,000 square feet. The maximum electrical load is 2275 kW; the maximum heating load is 17 million Btu/h; and the maximum cooling load is 808 refrigeration tons (RT). Hourly electrical and thermal demands for the facility were obtained using the building simulator program, Building Energy Analyzer (InterEnergy/GTI, 2005). Grid-based electricity prices were based on the Boston Edison T-2 time-of-use tariff. The price of natural gas in the initial year of operation was assumed to be $11.00/million Btu. Escalation assumptions, expressed in percent change from the previous year, for this example are provided in Table 5.9.

Other data needed to calculate the optimum capacity relate to equipment cost and performance and general modeling behavior (e.g., discount rate, planning horizon). The data assumed for the hospital in Boston are shown in Table 5.10.

TABLE 5.9   Sample Escalation Assumptions

Year	Fuel Price (%)	Elec Price (%)	O&M Cost (%)	Heat Load (%)	Cool Load (%)	Elec Load (%)
2	−0.5	0.5	0.5	0.0	0.0	0.0
3	0.0	1.0	0.5	0.0	0.0	0.0
4	0.0	1.0	0.5	0.0	0.0	0.0
5	0.0	1.0	0.5	0.0	0.0	0.0
6	0.0	1.0	0.5	0.0	0.0	0.0
7	0.5	0.5	0.5	0.0	0.0	0.0
8	0.5	0.5	1.0	0.0	0.0	0.0
9	0.5	0.5	1.0	0.0	0.0	0.0
10	0.5	0.5	1.0	0.0	0.0	0.0
11	0.5	0.5	1.0	0.0	0.0	0.0
12	1.0	1.0	1.0	0.0	0.0	0.0
13	1.0	1.0	1.0	0.0	0.0	0.0
14	1.0	1.0	2.0	0.0	0.0	0.0
15	1.0	1.0	2.0	0.0	0.0	0.0
16	1.0	1.0	2.0	0.0	0.0	0.0

TABLE 5.10   General Data for the
Boston Hospital Case

On-site boiler efficiency	82.0%
Conventional chiller COP	3.54
DG electric efficiency (full output)	29.0%
DG unit minimum output	50%
Absorption chiller COP	0.70
Absorption chiller min. output	25%
Abs chiller sys elec req (kW/RT)	0.20
CHP O&M cost ($/kWh)	0.011
DG power/heat ratio	0.65
Number of DG units	1
Type of prime mover	Recip
Discount rate	8.0%
Effective income tax rate	38.0%
DG capital cost ($/net kW installed)	1500
AC capital cost ($RT installed)	1000
Planning horizon (years)	16

Demands	Electricity	Heating	Cooling
Annual	12,406,742 kWh	37,074 MMBtu	1,617,306 RT-hr
Maximum	2275 kW	17.0 MMBtu/hr	808 RT
Minimum	934 kW	0.51 MMBtu/hr	0 RT

Installed DG capacity:	1130.1 kW (net)
Installed AC capacity:	210.5 RT
Installed capital cost:	$1,905,607

Hours of DG operation	6,717 hours/year
DG generated electricity	7,422,145 kWh/year
DG supplied heating	27,839 MMBtu/year
AC supplied cooling	535,793 RT-hr/year

Annual costs (before tax)	With CHP	No CHP
CHP system	$1,056,847	$0
Utility elec	$661,305	$1,785,547
Non-CHP fuel	$125,137	$502,367
Total	$1,843,290	$2,287,913

Annual operating savings (after tax):	$275,666
NPV savings:	$954,175

Optimum DG capacity:	1130.1 kW
Optimum AC capacity:	210.5 RT
NPV savings:	$954,175

FIGURE 5.6   Optimization results for a Boston hospital.

The numeric results of the optimization are shown in Figure 5.6. At the top of the figure is a summary of the electric and thermal loads, as estimated by the building simulation program mentioned above. The optimal capacities for a reciprocating engine prime mover and an absorption chiller are 1130 kW and 210.5 RT, respectively. As shown, the CHP system operates for 6717 h each year, producing 60% of the total electricity and over 75% of the total heating required by the facility. Owing to the relative economics of gas and electricity, it is preferable that waste heat first go to satisfying heating demands before contributing to cooling demands. As a result, only one-third of the total cooling demand is provided by the CHP system. The levelized total annual operating cost savings from the CHP system is $275,666/year. The resulting NPV cost savings, including capital investment, over the 16-year planning horizon is $954,175.

A summary of the operating frequency by hour of the day is provided in Figure 5.7. It can be observed from the figure that the frequency of the CHP system operation is influenced

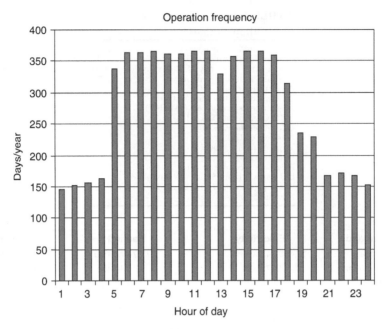

FIGURE 5.7   Operation frequency by time of day.

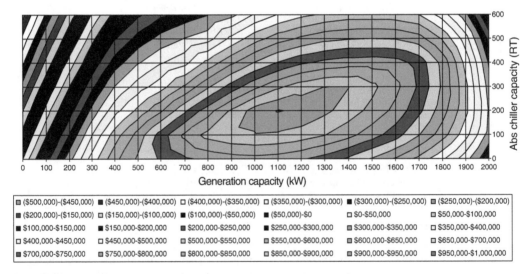

FIGURE 5.8   Optimization contour plot for a Boston hospital.

heavily by the time-of-use electricity rates, which are higher during the normal workday hours. Higher grid-based electricity rates increase the likelihood that the CHP system will be the less expensive alternative, and therefore the one to be selected, during those hours.

A contour plot of the objective function surface for the hospital in Boston is shown in Figure 5.8. Each iso-savings line represents a $50,000 increment. As shown, the surface behavior near the optimum is relatively flat. From a practical standpoint, this is a serendipitous result. Equipment capacities are offered in discrete sizes, and having a flat objective function surface in the neighborhood of the optimum gives a degree of flexibility in matching the calculated optimum set of capacities to near-values consistent with manufactured equipment sizes. As an example, vendors currently offer a reciprocating engine

TABLE 5.11   Summary Results

City	Parameter	Hospital	Hotel	Retail Store	Supermarket
Boston	Peak elec load (kW)	2275	1185	337	260
	Peak chiller load (RT)	808	492	147	40
	Optimal DG capacity (kW)	1130.1	417.2	118.3	117.7
	Optimal AC capacity (RT)	210.5	88.3	35.9	6.9
	NPV savings	$954,175	$341,877	$57,945	$67,611
	DG capacity % of peak load	50	35	35	45
	AC capacity % of peak load	26	18	24	17
San Francisco	Peak elec load (kW)	1949	1004	433	248
	Peak chiller load (RT)	452	334	131	17
	Optimal DG capacity (kW)	513.8	257.2	63.9	54.6
	Optimal AC capacity (RT)	30.9	36.2	19.1	5.0
	NPV savings	$431,123	$212,726	$15,954	$42,666
	DG capacity % of peak load	26	26	15	22
	AC capacity % of peak load	7	11	15	29

prime mover at 1100 kW and an absorption chiller at 210 RT. Manually substituting these capacities into the model produces a NPV cost savings of $951,861, which is a negligible difference relative to the value determined for the optimum capacity. Not all cases may have such a close match, but the flat gradient of the objective function near the optimum provides a reasonably wide range for matching actual equipment.

5.14   Other Scenarios

Due to the variation in fuel and electricity prices, site weather, and electrical and thermal loads of a specific facility, each potential CHP application will have a unique optimal solution that maximizes economic benefit. To illustrate, similar optimization runs were made for hotel, retail store, and supermarket applications in Boston as well as in San Francisco. It is instructive to consider the results of these eight cases together, looking in particular at the percent of peak load that the DG and AC optimum capacities represent. As shown in Table 5.11, the percent of peak load represented by the optimum capacities varies tremendously both by application and location. Thus, to maximize the economic benefit of a CHP system, each application should be individually evaluated by the methods described in this chapter.

References

1. Asano, H., S. Sagai, E. Imamura, K. Ito, and R. Yokoyama (1992). "Impacts of Time-of-Use Rates on the Optimal Sizing and Operation of Cogeneration Systems." *Power Systems, IEEE Transactions* **7**(4): 1444.

2. Badiru, A. B. (1982). "An Illustrative Example of the Application of Linear Programming to Energy Management," presented at the *4th Annual Conference on Computers & Industrial Engineering*, Orlando, Florida, March 1982.

3. Badiru, A. B. and P. S. Pulat (1995). *Comprehensive Project Management: Integrating Optimization Models, Management Principles, and Computers*. Prentice-Hall, Englewood Cliffs, NJ.

4. Bartholomew-Biggs, M. (2005). *Nonlinear Optimization with Financial Applications*. Kluwer, Boston.

5. Baughman, M. L., N. A. Eisner, and P. S. Merrill (1989). "Optimizing Combined Cogeneration and Thermal Storage Systems: An Engineering Economics Approach." *Power Systems, IEEE Transactions* **4**(3): 974.

6. Bazaraa, M. S., H. D. Sherali, and C. M. Shetty (1993). *Nonlinear Programming: Theory and Algorithms*. Wiley, New York.

7. Bertsekas, D. P. (2004). *Nonlinear Programming*. Athena Scientific, Belmont, MA.

8. Campanari, S., L. Boncompagni, and E. Macchi (2002). *GT-2002-30417 Microturbines and Trigeneration: Optimization Strategies and Multiple Engine Configuration Effects*, v. 4A & 4B, ASME Turbo Expo 2002, Amsterdam, the Netherlands. ASME, 2002.

9. Consonni, S., G. Lozza, and E. Macchi (1989). *Optimization of Cogeneration Systems Operation. Part B. Solution Algorithm and Examples of Optimum Operating Strategies*. ASME, New York.

10. Czachorski, M., W. Ryan, and J. Kelly (2002). *Building Load Profiles and Optimal CHP Systems*. American Society of Heating, Refrigerating and Air-Conditioning Engineers, Honolulu, HI.

11. Dennis, J. E. and R. B. Schnabel (1983). *Numerical Methods for Unconstrained Optimization and Nonlinear Equations*. Prentice-Hall, Englewood Cliffs, NJ.

12. Edirisinghe, N. C. P., E. I. Patterson, and N. Saadouli (2000). "Capacity Planning Model for a Multipurpose Water Reservoir with Target-Priority Operation." *Annals of Operations Research* **100**: 273–303.

13. Fletcher, R. (1987). *Practical Methods of Optimization*. Wiley, Chichester, UK.

14. Fylstra, D., L. Lasdon, J. Watson, and A. Waren (1998). "Design and Use of the Microsoft Excel Solver." *Informs Interfaces* **28**(5): 29–55.

15. Gamou, S., R. Yokoyama, and K. Ito (2002). "Optimal Unit Sizing of Cogeneration Systems in Consideration of Uncertain Energy Demands as Continuous Random Variables." *Energy Conversion and Management* **43**(9–12): 1349.

16. Gill, P. E., W. Murray, and M. H. Wright (1986). *Practical Optimization*. Elsevier Academic Press.

17. Hudson, Carl R. (2005). "Adaptive Nonlinear Optimization Methodology for Installed Capacity Decisions in Distributed Energy/Cooling Heat and Power Applications." http://etd.utk.edu/2005/HudsonCarlRandolph.pdf, Knoxville, The University of Tennessee. PhD: 133.

18. Hudson, C. R. and A. B. Badiru (2004). "Use of Time-Aggregated Data in Economic Screening Analyses of Combined Heat and Power Systems." *Cogeneration and Distributed Generation Journal* **19**(3): 7.

19. InterEnergy/GTI (2005). Building Energy Analyzer. http://www.interenergysoftware.com/BEA/BEAPROAbout.htm. Chicago, IL.

20. Kao, C., W. T. Song, and S.-P. Chen (1997). "A Modified Quasi-Newton Method for Optimization in Simulation." *International Transactions in Operational Research* **4**(3): 223.

21. Kruger, P. (2006). *Alternative Energy Resources: The Quest for Sustainable Energy*. John Wiley & Sons, New York.

22. Marantan, A. P. (2002). "Optimization of Integrated Microturbine and Absorption Chiller Systems in CHP for Buildings Applications." *DAI* **64**: 112.

23. Nocedal, J. (1992). "Theory of Algorithms for Unconstrained Optimization." *Acta Numerica*, 1992: 199–242.

24. Oak Ridge National Laboratory (2005). *BCHP Screening Tool*. Oak Ridge, TN.

25. Orlando, J. A. (1996). *Cogeneration Design Guide*. American Society of Heating, Refrigerating and Air-Conditioning Engineers, Atlanta, GA.

26. Park, C. S. and G. P. Sharp-Bette (1990). *Advanced Engineering Economics*. Wiley, New York.

27. Petchers, N. (2003). *Combined Heating, Cooling & Power Handbook: Technologies & Applications: An Integrated Approach to Energy Resource Optimization*. Fairmont Press, Lilburn, GA.

28. Sullivan, W. G., E. M. Wicks, and J. T. Luxhoj (2006). *Engineering Economy.* Pearson/Prentice Hall, Upper Saddle River, NJ.

29. Van der Lee, P. E. A., T. Terlaky, and T. Woudstra (2001). "A New Approach to Optimizing Energy Systems." *Computer Methods in Applied Mechanics and Engineering* **190** (40–41): 5297.

30. Winston, W. L. (1994). *Operations Research: Applications and Algorithms.* Belmont, CA: Duxbury Press.

31. Wong-Kcomt, J. B. (1992). "A Robust and Responsive Methodology for Economically Based Design of Industrial Cogeneration Systems." *DAI* **54**: 218.

32. Yokoyama, R., K. Ito, and Y. Matsumoto (1991). *Optimal Sizing of a Gas Turbine Cogeneration Plant in Consideration of Its Operational Strategy.* Budapest, Hungary, published by ASME, New York.

33. Yongyou, H., S. Hongye, and C. Jian (2004). "A New Algorithm for Unconstrained Optimization Problem with the Form of Sum of Squares Minimization." 2004 IEEE International Conference on Systems, Man and Cybernetics, IEEE.

34. Zhang, J. and C. Xu (2001). "Properties and Numerical Performance of Quasi-Newton Methods with Modified Quasi-Newton Equations." *Journal of Computational and Applied Mathematics* **137**(2): 269.

6

Airline Optimization

Jane L. Snowdon and
Giuseppe Paleologo
IBM T. J. Watson Research Center

6.1 Introduction

The main objective of this chapter is to describe and review airline optimization from an applied point of view. The airline industry was one of the first to apply operations research methods to commercial optimization problems. The combination of advancements in computer hardware and software technologies with clever mathematical algorithms and heuristics has dramatically transformed the ability of operations researchers to solve large-scale, sophisticated airline optimization problems over the past 60 years. As an illustration, consider United Airlines' eight fleet problem in Figure 6.1, which took nearly 9 h to solve in 1975 on an IBM System 370 running Mathematical Programming System Extended (MPSX), and only 18 s to solve in 2001 on an IBM Thinkpad running IBM's Optimization Subroutine Library (OSL) V3. Better models and algorithms supported by faster hardware achieved more than four orders of magnitude improvement in solution time for this problem instance over the course of 26 years.

In the subsequent sections, the state-of-the-art in the application of optimization to the airline industry is given for solving traditional airline planning problems and is arranged by functional area. The remainder of Section 6.1 presents an overview of the major historical events in the airline industry environment, gives a description of the high-level business processes for airline planning and operations, provides a representative list of major airline optimization vendors, and illustrates select cases where airlines have adopted and embraced optimization to realize productivity and efficiency gains and cost reductions. Section 6.2 covers schedule planning including the four steps of schedule design, fleet assignment, aircraft routing, and crew scheduling. Section 6.3 describes the four steps of revenue

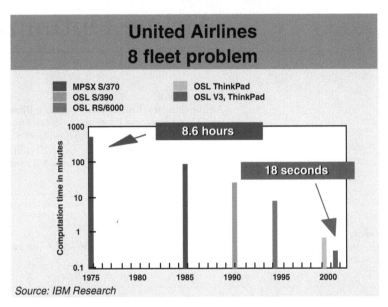

FIGURE 6.1 United Airlines' eight fleet problem.

management, which consist of forecasting, overbooking, seat inventory control, and pricing. Section 6.4 reviews aircraft load planning. Finally, Section 6.5 provides directions for future research.

6.1.1 Airline Industry Environment

The Chicago Convention was signed in December 1944 by Franklin Roosevelt and his peers to establish the basic rules for civil aviation. Over 60 years later, their vision of international civil aviation as a means for fostering friendship and understanding among nations and peoples worldwide has been realized. Air travel has not only become a common, affordable means of transportation for developed countries, but it has also brought enormous benefits to developing countries with potential for social progress and economic prosperity through trade and tourism.

According to the Air Transport Action Group (ATAG), aviation transported, in 2004, nearly 1.9 billion passengers and 38 million tons of freight worth $1.75 billion, which is over 40% of the world trade of goods by value [1]. Giovanni Bisignani, director general and CEO of the International Air Transport Association (IATA), predicted that 2.2 billion passengers will travel in 2006 during his speech at the State of the Industry Annual General Meeting and World Air Transport Summit. International passenger traffic increased by 15.6% in 2004, driven primarily by a strong rebound in Asia from SARS-affected levels in 2003, and by substantial increases in route capacity in the Middle East. However, international passenger traffic is forecast to grow at an annual average rate of 5.6% from 2005 to 2009, but may decline due to high oil prices [2]. Likewise, the 15.8% growth in international freight volumes in 2004 is expected to slow with an estimated forecast of a 6.3% annual average growth rate between 2005 and 2009 [2].

Aviation generated nearly $3 trillion, equivalent to 8% of the world's gross domestic product (GDP), in 2004. The air transport industry employs 5 million people directly and an additional 24 million indirectly or induced through spending by industry employees globally. The world's 900 airlines have a total fleet of nearly 22,000 aircraft [3], which serves an estimated 1670 airports [4] using a route network of several million kilometers managed by approximately 160 air navigation service providers [5]. Airbus forecasts that over 17,000 new planes will be needed by 2024 [6] and Boeing forecasts that over 27,000 new planes will be needed by 2025 [7] to meet growing demand for air transport, to replace older fleet, and to introduce more fuel-efficient planes.

Between 1944, when 9 million passengers traveled on the world's airlines, and 1949, when the first jet airliner flew, the modern era of airline optimization was born. George Dantzig designed the simplex method in 1947 for solving linear programming formulations of U.S. Air Force deployment and logistical planning problems. Applications to solve large-scale, real-world problems, such as those found in military operations, enabled optimization to blossom in the 1950s with the advent of dynamic programming (Bellman), network flows (Ford and Fulkerson), nonlinear programming (Kuhn and Tucker), integer programming (Gomory), decomposition (Dantzig and Wolfe), and the first commercial linear programming code (Orchard-Hays) as described in Nemhauser [8].

The Airline Group of the International Federation of Operational Research Societies (AGIFORS) was formed in 1961 and evolved from informal discussions between six airline operational research practitioners from Trans Canada, Air France, Sabena, BEA, and Swissair. Today AGIFORS is a professional society comprised of more than 1200 members representing over 200 airlines, aircraft manufacturers, and aviation associations dedicated to the advancement and application of operational research within the airline industry. AGIFORS conducts four active study groups in the areas of cargo, crew management, strategic and scheduling planning, and reservations and yield management, and holds an annual symposium.

In the late 1960s and early 1970s, United Airlines, American Airlines, British Airways, and Air France, recognizing the competitive advantage that decision technologies could provide, formed operations research groups. As described by Barnhart, Belobaba, and Odoni, these groups grew rapidly, developing decision support tools for a variety of airline applications, and in some cases offering their services to other airlines [9]. Most major airlines worldwide have launched similar groups.

A dramatic change in the nature of airline operations occurred with the Airline Deregulation Act of 1978 led by the United States. This Act is widely credited for having stimulated competition in the airline industry. For example, the year 1986, which was marked by economic growth and stable oil prices, saw the poor financial results of major airlines. This was just a symptom of the increased competitive pressures faced by the carriers. Effective 1997, the European Union's final stage of deregulation allows an airline from one member state to fly passengers within another member's domestic market. Since 1978, other important forces such as the rising cost of fuel, the entrance of low-cost carriers into the market, the abundance of promotional and discount fares, and consolidations have significantly shaped the competitive landscape of the airline industry. Hundreds of airlines have entered into alliances and partnerships, ranging from marketing agreements and code-shares to franchises and equity transfers, resulting in globalization of the industry. According to IATA, the three largest airline alliances combined today fly 58% of all passengers traveling each year, including Star Alliance (425 million passengers/year, 23.6% market share), Skyteam Alliance (372.9 million passengers/year, 20.7% market share), and oneworld (242.6 million passengers/year, 13.5% market share). Alliances between cargo airlines are also taking place—for example, WOW Alliance and SkyTeam Cargo.

6.1.2 Airline Planning and Operations Business Processes

Business processes for airline planning and operations may be described by four high-level processes as shown in Figure 6.2. First, resource planning consists of making strategic decisions with respect to aircraft acquisition, crew manpower planning, and airport resources such as slots. Resource planning is typically performed over a 1–2-year time horizon prior to the day of operations. Due to the competitive and proprietary nature of the data, little is published about strategic planning. Second, market planning and control consists of schedule design, fleet assignment, and yield management decisions. Market planning and control is typically performed over a 6 months' time horizon. Third, operations planning consists of making operational decisions with respect to aircraft routing and maintenance, crew resources, and airport operations. Operations planning is typically performed over a 1-month time horizon. Fourth, operations control consists of making tactical decisions with respect to aircraft rerouting, crew trip recovery, airport operations, and customer service. Operations control is typically performed over a 2-week time horizon and includes the day of operations.

6.1.3 Airline Optimization Solutions

Any discussion of airline optimization would not be complete without mentioning some of the major vendors in this space. While not exhaustive, this list is representative. First, global distribution systems (GDS) are operated by airlines to store and retrieve information and conduct transactions related to travel, such as making reservations and generating tickets. Four GDS dominate the marketplace: Amadeus (founded in 1987 by Air France, Iberia, Lufthansa, and SAS and now owned by InterActive Corporation, which also owns Expedia), Galileo International (founded in 1993 by 11 major North American and European airlines and acquired in 2001 by Cendant Corporation, which also owns Orbitz), Sabre (founded in 1960, it also owns Travelocity), and Worldspan (founded in 1990 and currently owned by affiliates of Delta Air Lines, Northwest Airlines, and American Airlines). Sabre also offers solutions for pricing and revenue management, flight and crew scheduling, operations control, among others.

Second, revenue management, also referred to as yield management, is the process of comprehending, predicting, and reacting to consumer behavior to maximize revenue. PROS Revenue Management is a market leader in revenue optimization. Their product suite includes forecasting demand, optimizing the allocation of inventory, providing dynamic

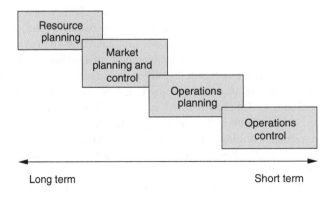

FIGURE 6.2 Airline planning and operations business processes.

packaging, and offering dynamic pricing. Recently JDA Software Group acquired Manugistics, another leading provider of solutions for demand management, pricing, and revenue management.

Third, the airline industry has seen a wave of consolidation of schedule planning providers. Jeppesen, a wholly owned subsidiary of Boeing Commercial Aviation Services, acquired Carmen Systems and its portfolio of manpower planning, crew pairing, crew rostering, crew tracking, crew control, fleet control, passenger recovery, and integrated operations control modules in 2006. Kronos purchased AD OPT Technologies' suite of customized crew planning, management, and optimization solutions for the airline industry in 2004. Navitaire, a wholly owned Accenture business that hosts technology solutions for reservations, direct distribution, operations recovery, decision-support and passenger revenue accounting, acquired Caleb Technologies (operations recovery decision-support and resource planning systems) in 2004 and Forte Solutions (operations and crew management control) in 2005.

Finally, a number of general purpose modeling and optimization tools are available including Dash Optimization's Xpress-MP, the IBM Optimization Solutions and Library, and ILOG's CPLEX.

6.1.4 Benefits of Airline Optimization

The practice of airline optimization has yielded significant business benefits. For example, the fractional aircraft operations system designed and developed by AD OPT Technologies, Bombardier, and GERAD to simultaneously maximize the use of aircraft, crew, and facilities generated an initial savings of $54 million with projected annual savings of $27 million as reported by Hicks et al. [10]. UPS claims savings of more than $87 million over 3 years for a system co-developed with MIT to optimize the design of service networks for delivering express packages by simultaneously determining aircraft routes, fleet assignments, and package routings as given in Armacost et al. [11]. The CrewSolver system developed by CALEB Technologies for recovering crew schedules in real-time when disruptions occur has saved Continental an estimated $40 million for major disruptions alone, as presented in Yu et al. [12]. Air New Zealand and Auckland University teamed to develop a suite of systems to optimize the airline's crew schedules, which has resulted in an annual savings of $10 million as explained in Butchers et al. [13]. Subrahmanian et al. [14] provide an account of solving a large-scale fleet assignment problem with projected savings to Delta Air Lines of $300 million over 3 years. According to Robert Crandell, the former president and chairman of American Airlines, yield management was the single most important technical development in transportation management since airline deregulation in 1979 and is estimated to have generated $1.4 billion in incremental revenue in 3 years, with projected annual savings of more than $500 million for the airline as quoted in Smith et al. [15]. These are but a few of many examples demonstrating the tremendous business impact and potential of airline optimization.

6.2 Schedule Planning

The problem of airline scheduling is among the largest and most complex of any industry. Airline scheduling includes determining the concurrent flows of passengers, cargo, aircraft, and flight crews through a space-time network, performing aircraft maintenance, and deploying ground-based resources such as gates. The problem is further complicated by air traffic control restrictions, noise curfews, company/union regulations, market structure (density, volume, and elasticity of demand), fare classes, and safety and security concerns. The overarching objective in airline schedule planning is to maximize profitability.

Etschmaier and Mathaisel [16], Rushmeier and Kontogiorgis [17], Barnhart and Talluri [18], Clarke and Smith [19], Klabjan [20], and Barnhart and Cohn [21] supply excellent overviews of the airline schedule development process. Airline schedule planning is traditionally broken down into four steps in such a way as to make each step computationally viable. No single optimization model has been formulated or solved to address the complex schedule planning task in its entirety, although attempts have been made to solve combinations of two or more of the steps in a single model. The steps in sequential order are:

1. Schedule Design—defining the markets to serve and how frequently, and determining the origin, destination, and departure times of flights.
2. Fleet Assignment—designating the aircraft size for each flight.
3. Aircraft Routing—selecting the sequence of flights to be flown by each aircraft during the course of every day and satisfying maintenance requirements.
4. Crew Scheduling—assigning crews to flights.

Next, each of these steps will be described in more detail.

6.2.1 Schedule Design

The scheduling process starts with the schedule design. The schedule design constructs the markets and how frequently to serve them, and determines the origin, destination, and departure times of flights. In general, schedules are dictated largely by differences in seasonal demand and to a smaller extent by month-to-month differences. Profitability and feasibility are two important criteria that affect how a flight schedule is designed. Schedule profitability is determined by the degree to which the origin–destination city pairs attract revenue from passengers and cargo compared to the operational costs of transporting the people and goods. Schedule feasibility is determined by the ability of the airline to assign sufficient resources, such as airplanes, crew, and airport facilities, to the flights.

Construction of the schedule typically begins 1 year in advance of its operation and 6 months in advance of its publication. The schedule construction process encompasses three steps:

1. Establishing the service plan (a list of routes and frequencies where factors such as demand forecasts, competitive information, and marketing initiatives are considered).
2. Developing an initial feasible schedule (establish flight departure and arrival times while satisfying maintenance planning and resource constraints, such as the number of aircraft).
3. Iteratively reviewing and modifying the schedule based upon evaluations of new routes, aircraft, or frequencies, and connection variations.

Decision support systems based upon operations research have been slow to develop for schedule design, primarily because of the inherent complexity and problem size, the lack of sufficient and timely data from the airline revenue accounting process, the method for collecting and forecasting market demand, and the difficulty of predicting competing airlines' responses to market conditions. Published literature on airline schedule design is sparse. One of the earliest descriptions of the schedule construction process can be found in Rushmeier et al. [22]. Erdmann et al. [23] address a special case of the schedule generation problem for charter airlines, model the problem as a capacitated network design problem using a path-based mixed integer programming formulation, and solve it using a branch-and-cut approach. Armacost et al. [24,25] apply a composite variable formulation approach by removing aircraft- and package-flow decisions as explicit decisions, and solve a set-covering instance to the next-day-air network-design problem for United Parcel Service (UPS).

6.2.2 Fleet Assignment

Given a flight schedule and a set of aircraft, the fleet assignment problem determines which type of aircraft, each having a different capacity, should fly each flight segment to maximize profitability, while complying with a large number of operational constraints. The objective of maximizing profitability can be achieved both by increasing expected revenue, for example, assigning aircraft with larger seating capacity to flight legs with high passenger demand, and by decreasing expected costs, for example, fuel, personnel, and maintenance. Operational constraints include the availability of maintenance at arrival and departure stations (i.e., airports), gate availability, and aircraft noise, among others. Assigning a smaller aircraft than needed on a flight results in spilled (i.e., lost) customers due to insufficient capacity. Assigning a larger aircraft than needed on a flight results in spoiled (i.e., unsold) seats and possibly higher operational costs.

Literature on the fleet assignment problem spans nearly 20 years. For interested readers, Sherali et al. [26] provide a comprehensive tutorial on airline fleet assignment concepts, models, and algorithms. They suggest future research directions including the consideration of path-based demands, network and recapture effects, and exploring the interactions between initial fleeting and re-fleeting. Additional review articles on fleet assignment include Gopalan and Talluri [27], Yu and Yang [28], Barnhart et al. [9], Clarke and Smith [19], and Klabjan [20].

Several of the earliest published accounts are by Abara [29], who describes a mixed integer programming implementation for American Airlines; Daskin and Panayotopoulos [30], who present an integer program that assigns aircraft to routes in a single hub and spoke network; and Subramanian et al. [14], who develop a cold-start solution (i.e., a valid initial assignment does not exist) approach for Delta Airlines that solves the fleet assignment problem initially as a linear program using an interior point method, fixes certain variables, and then solves the resulting problem as a mixed integer program. Talluri [31] proposes a warm-start solution (i.e., a valid initial assignment does exist) that performs swaps based on the number of overnighting aircraft for instances when it becomes necessary to change the assignment on a particular flight leg to another specified aircraft type. Jarrah et al. [32] propose a re-fleeting approach for the incremental modification of planned fleet assignments. Hane et al. [33] employ an interior point algorithm and dual steepest edge simplex, cost perturbation, model aggregation, and branching on set-partitioning constraints with prioritized branching order. Rushmeier and Kontogiorgis [17] focus on connect time rules and incorporate profit implications of connections to assign USAir's fleets resulting in an annual benefit of at least $15 million. Their approach uses a combination of dual simplex and a fixing heuristic to solve a linear programming relaxation of the problem to obtain an initial solution, which in turn is fed into a depth-first branch-and-bound process. Berge and Hopperstad [34] propose a model called Demand Driven Dispatch for dynamically assigning aircraft to flights to leverage the increased accuracy of the flight's demand forecast as the actual flight departure time approaches.

Attempts have been made to integrate the fleet assignment model with other airline processes such as schedule design, maintenance routing, and crew scheduling. Because of the interdependencies of these processes, the optimal solution for processes considered separately may not yield a solution that is optimal for the combined processes.

For example, integrated fleet assignment and schedule design models have the potential to increase revenues through improved flight connection opportunities. Desaulniers et al. [35] introduce time windows on flight departures for the fleet assignment problem and solve the multicommodity network by branch-and-bound and column generation, where the column generator is a specialized time-constrained shortest path problem. Rexing et al. [36] discretize time windows and create copies of each flight in the underlying graph to represent different departure time possibilities and then solve using a column generator that is a shortest path problem on an acyclic graph. Lohatepanont and Barnhart [37] present an integrated schedule design and fleet assignment solution approach that determines incremental changes to existing flight schedules to maximize incremental profits. The integrated schedule design and fleet assignment model of Yan and Tseng [38] includes path-based demand considerations and uses Lagrangian relaxation, where the Lagrangian multipliers are revised using a subgradient optimization method.

Integrated fleet assignment and maintenance, routing, and crew considerations have the potential for considerable cost savings and productivity improvements. Clarke et al. [39] capture maintenance and crew constraints to generalize the approach of Hane et al. [33] and solve the resulting formulation using a dual steepest-edge simplex method with a customized branch-and-bound strategy. Barnhart et al. [40] explicitly model maintenance issues using ground arcs and solve the integrated fleet assignment, maintenance, and routing problem using a branch-and-price approach where the column generator is a resource-constrained shortest path problem over the maintenance connection network. Rosenberger et al. [41] develop a fleet assignment model with hub isolation and short rotation cycles (i.e., a sequence of legs assigned to each aircraft) so that flight cancellations or delays will have a lower risk of impacting subsequent stations or hubs. Belanger et al. [42] present both a mixed-integer linear programming model and a heuristic solution approach for the weekly fleet assignment problem for Air Canada in the case where homogeneity of aircraft type is desired over legs sharing the same flight number, which enables easier ground service planning.

Models to integrate fleet assignment with passenger flows and fare classes adopt an origin–destination-based approach compared to the more traditional flight-leg approach. Examples of fleet assignment formulations that incorporate passenger considerations include Farkas [43], Kniker [44], Jacobs et al. [45], and Barnhart et al. [46].

The basic fleet assignment model (FAM) can be described as a multicommodity flow problem with side constraints defined on a time-space network and solved as an integer program using branch-and-bound. The time-space network has a circular time line representing a 24-hour period, or daily schedule, for each aircraft fleet at each city. Along a given time line, a node represents an event: either a flight departure or a flight arrival. Each departure (arrival) from the city splits an edge and adds a node to the time line at the departure (arrival + ground servicing) time. A decision variable connects the two nodes created at the arrival and departure cities and represents the assignment of that fleet to that flight.

Mathematically, the fleet assignment model may be stated as follows, which is an adaptation of Hane et al. [33] without required through-flights (i.e., one-stops). The objective is to minimize the cost of assigning aircraft types to flight legs as given in Equation 6.1. To be feasible, the fleet assignment must be done in such a way that each flight in the schedule is assigned exactly one aircraft type (i.e., cover constraints as given in Equation 6.2), the itineraries of all aircraft are circulations through the network of flights that can be repeated cyclically over multiple scheduling horizons (i.e., balance constraints as given in Equation 6.3), and the total number of aircraft assigned cannot exceed the number available in the fleet (i.e., plane count constraints as given in Equation 6.4). Additional inequality constraints may be incorporated to address such issues as through-flight assignments, maintenance, crew, slot allocation, and other issues.

$$\min \quad \sum_{j\in J}\sum_{i\in I} c_{ij}X_{ij} \tag{6.1}$$

$$\text{subject to:} \quad \sum_i X_{ij} = 1, \qquad\qquad \forall j \in J \tag{6.2}$$

$$\sum_d X_{idot} + Y_{iot-t} - \sum_d X_{iodt} - Y_{iott+} = 0, \quad \forall\{iot\} \in N \tag{6.3}$$

$$\sum_{j\in O(i)} X_{ij} + \sum_{o\in C} Y_{iot_nt_1} \le S(i), \qquad\qquad \forall i \in I \tag{6.4}$$

$$Y_{iott+} \ge 0, \qquad\qquad \forall\{iott^+\} \in N \tag{6.5}$$

$$X_{ij} \in \{0,1\}, \qquad\qquad \forall i \in I, \ \forall j \in J \tag{6.6}$$

The mathematical formulation requires the following notation with parameters:

C = set of stations (cities) serviced by the schedule,
I = set of available fleets,
$S(i)$ = number of aircraft in each fleet for $i \in I$,
J = set of flights in the schedule,
$O(i)$ = set of flight arcs, for $i \in I$, that contains an arbitrary early morning time (i.e., 4 AM, overnight),
N = set of nodes in the network, which are enumerated by the ordered triple $\{iot\}$ consisting of fleet $i \in I$, station $o \in C$, and t = takeoff time or landing time at o.
t^- = time preceding t,
t^+ = time following t,
$\{i_{ot_n}\}$ = last node in a time line, or equivalently, the node that precedes 4 AM,
$\{i_{ot_1}\}$ = successor node to the last node in a time line, and decision variables:
$X_{iodt} = X_{ij} = 1$ if fleet i is assigned to the flight leg from o to d departing at time t, and 0 otherwise;
Y_{iott+} = number of aircraft of fleet $i \in I$ on the ground at station $o \in C$ from time t to t^+.

6.2.3 Aircraft Routing

The Federal Aviation Administration (FAA) mandates some safety requirements for aircraft. Those are of four types:

- A-checks are routine visual inspections of major systems, performed every 65 block-hours or less. A-checks take 3–10 h, and are usually performed at night at the gate.
- B-checks are detailed visual inspections, and are performed every 3 months.
- C-checks are performed every 12–18 months depending on aircraft type and operational circumstances, and consist of in-depth inspections of many systems. C-checks require disassembly of parts of the aircraft, and must be performed in specially equipped spaces.
- D-checks are performed every 4–5 years, perform extensive disassembly and structural, chemical, and functional analyses of each subsystem, and can take more than 2 months.

Of these checks, A and B are performed on the typical planning horizon of fleet planning and crew scheduling, and must therefore be incorporated into the problem.

Aircraft routing determines the allocation of candidate flight segments to a specific aircraft tail number within a given fleet-type while satisfying all operational constraints, including maintenance. Levin [47] is the first author to analyze maintenance scheduling. One simplified formulation of the maintenance problem, following Barnhart et al. [40], is the following. We define a string as a sequence of connected flights (i.e., from airport a to airport b, then from b to c, until a final airport) performed by an individual aircraft. Moreover, a string satisfies the following requirements: the number of block-hours satisfies maintenance requirements, and the first and last nodes in the sequence of airports are maintenance stations. A string k has an associated cost c_k. Let S be the set of all augmented strings. For every node v, $I(v)$ and $O(v)$ denote the set of incoming and outgoing links, respectively. The problem has two types of decision variables: x_s is set to 1 if string s is selected as a route for the associated fleet, and 0 otherwise. y_m is the number of aircraft being serviced at service station m. The problem then becomes:

$$\min \qquad \sum_{s \in S} c_s x_s \qquad\qquad\qquad\qquad\qquad\qquad (6.7)$$

$$\text{subject to:} \quad \sum_{i \in S} x_s = 1 \qquad\qquad\qquad\qquad \text{for all flights } i \qquad (6.8)$$

$$\sum_{\substack{j \in O(v) \\ j \in s}} x_s - \sum_{\substack{j \in I(v) \\ j \in s}} x_s + y_{O(v)} - y_{I(v)} = 0 \quad \text{for all stations } v \qquad (6.9)$$

where $y \geq 0$, $x = 0/1$.

The problem can be formulated as a multicommodity flow problem (e.g., Cordeau et al. [48]), as a set partitioning problem (Feo and Bard [49]; Daskin and Papadopoulos [30]), and employing eulerian tours (Clarke et al. [50], Talluri [31], and Gopalan and Talluri [51]). Recent work includes that of Gabteni and Gronkvist [52], who provide a hybrid column generation and constraint programming optimization solution approach, Li and Wang [53], who present a path-based integrated fleet assignment and aircraft routing heuristic, Mercier et al. [54], who solve the integrated aircraft routing and crew scheduling model using Benders decomposition, and Cohn and Barnhart [55], who propose an extended crew pairing model that integrates crew scheduling and maintenance routing decisions.

6.2.4   Crew Scheduling

The area of crew scheduling involves the optimal allocation of crews to flights and can be partitioned into three phases: crew pairing, crew assignment based upon crew rostering or preferential bidding, and recovery from irregular operations. Each of these three phases will be described in more detail in this section. The literature on crew scheduling is plentiful over its nearly 40-year history with descriptions in comprehensive survey papers by Etschmaier and Mathaisel [16], Clarke and Smith [19], Klabjan [20], Barnhart and Cohn [21], Arabeyre et al. [56], Bornemann [57], Barutt and Hull [58], Desaulniers et al. [59], Barnhart et al. [60], and Gopalakrishnan and Johnson [61].

Crew Pairing

In crew pairing optimization, the objective is to find a minimum cost set of legal duties and pairings that covers every flight leg. An airline generally starts with the schedule of flight segments and their corresponding fleet assignments and decomposes the problem for different crew types (pilots, flight attendants) and for different fleet types. The flight legs are joined together to form duties, which are essentially work shifts. These duties are then combined into trips/pairings that range from 1 to 5 days in length and which start and end at a crew's home base. There are many FAA and union regulations governing the legality of duties and pairings. Some typical ones may include: a limit on the number of hours of flying time in a duty, a limit on the total length of a duty (possibly stricter for a nighttime duty), a limit on the number of duties in a pairing, and upper and lower bounds on the rest time between two duties. The generation of good crew pairings is complicated by the fact that a crew's pay is determined by several guarantees that are based on the structure of the duties in the pairing. Some common rules governing crew compensation may include: a crew is paid overtime for more than 5 h of flying in a duty, a crew is guaranteed a minimum level of pay per duty period, a crew is guaranteed that a certain percentage of the duty time will count as flying time, and a crew receives a bonus for overnight flying. Some companies also include additional constraints and objective function penalties to avoid marginally legal pairings, tight connections, and excessive ground transportation.

The area of crew pairing optimization is complex and challenging for two reasons. First, it is not practical to generate the complete set of feasible pairings, which can number in the billions for major carriers, for a problem. Second, the cost of a pairing usually incorporates linear components for hotel and per-diem charges and several nonlinear components based upon actual flying time, total elapsed work time, and total time away from the home base.

The crew pairing problem is normally solved in three stages for North American carriers, as described in Anbil et al. [62]. First, a daily problem is solved where all flights operate every day. Next, a weekly problem is solved, which reuses as much of the daily problem as possible, and also handles weekly exceptions. Finally, a dated problem is solved, which reuses as much of the weekly problem as possible, and also produces a monthly schedule that handles holidays and weekly transitions.

Several differences exist for European carriers. First, European crews typically receive fixed salaries, so the objective is simply to minimize the number of crews needed. Second, European airlines do not tend to run on a daily, repeating schedule or have hub-and-spoke networks. Third, European carriers must comply with government regulations and work with somewhat variable union rules, while for North American carriers, crew pairing is primarily determined by FAA regulations.

The crew pairing problem can be modeled as a set partitioning problem where the rows represent flights to be covered and the columns represent the candidate crew pairings as follows:

$$\min \qquad \sum_{j \in P} c_j x_j \qquad\qquad (6.10)$$

$$\text{subject to:} \quad \sum_{j \in P} a_{ij} x_j = 1, \quad \forall i \in F \qquad\qquad (6.11)$$

$$x_j \in \{0, 1\} \qquad\qquad (6.12)$$

The mathematical formulation requires the following notation with parameters:

$P =$ set of all feasible pairings,
$c_j =$ cost of pairing j,
$a_{ij} = 1$ if pairing j includes (covers) flight i, and 0 otherwise,
$F =$ set of all flights that must be covered in the period of time under consideration,

and decision variable:

$x_j = 1$ if the pairing j is used, and 0 otherwise.

The objective is to minimize the total cost as shown in Equation 6.10. Equation 6.11 ensures that each flight leg is covered once and only once. In practice, side constraints are often added to reflect the use of crew resources; for example, upper and lower bounds on the number of available crew at a particular base, and balancing pairings across crew bases with respect to cost, the number of days of pairings, or the number of duties.

A variety of solution methodologies have been suggested for the crew pairing problem and the majority are based upon pairing generation and pairing selection strategies. Pairings are frequently generated using either an enumerative or shortest path approximation approach on a graph network that can be represented either as a flight segment network or a duty network. Pairing generation approaches and selected illustrations include:

- *Row approach*: Start with a feasible (or artificial) initial solution. Choose a subset of columns and exhaustively generate all possible pairings from the flight segments covered by these columns, and solve the resulting set partitioning problem. If the solution to this sub-problem provides an improvement, replace the appropriate pairings in the current solution. Repeat until no further improvements are found or until a limit on the execution time is reached. Selected illustrations include Gershkoff [63], Rubin [64], Anbil et al. [65,66], and Graves et al. [67].

- *Column approach*: Consider all the rows simultaneously, explicitly generate all possible legal pairings by selecting a relatively small set of pairings that have small reduced cost from a relatively large set of legal pairings, and solve the subproblem until optimality. Continue until the solution to the sub-problem no longer provides an improvement. A selected illustration includes Hu and Johnson [68].

- *Network approach*: The network approach can utilize one of two strategies: the column approach where the columns are generated using the flight segment or duty network, or column generation based on shortest paths. Selected illustrations include Minoux [69], Baker et al. [70], Lavoie et al. [71], Barnhart et al. [72], Desaulniers et al. [73], Stojković et al. [74], and Galia and Hjorring [75].

Strategies for selecting pairings and solving the crew pairing optimization problem include TPACS [64], TRIP [65,66], SPRINT (Anbil et al. [62]), volume algorithm [76,77], branch-and-cut [78], branch-and-price [79–82], and branch-and-cut-and-price [62]. Others have proposed strategies using genetic algorithms [58,83–86].

Crew Assignment

In crew assignment, pairings are assembled into monthly work schedules and assigned to individual crew members. Depending upon the airline's approach, either crew rostering or preferential bidding is used for crew assignment. In crew rostering, a common process outside of North America, the objective is to assign pairings to individual crew members based upon the individual's preferences while minimizing overall costs. In preferential bidding, a common process used within North America, individual crew members bid on a set of cost-minimized schedules according to their relative preferences. The airline assigns specific schedules to the crew members based upon their bids and seniority.

The crew rostering and preferential bidding problems can be formulated as set partitioning problems. The problem decomposes based upon the aircraft equipment type and the certification of the crew member (e.g., Captain, First Officer, Flight Attendant). The crew assignment problem is complex, with FAA, union and company rules, and is compounded by additional rules involving several rosters, including language restrictions and married crew members preferring to fly together. The airline strives to produce equitable crew assignments in terms of equal flying time and number of days off.

The crew assignment problem can be modeled simply as follows:

$$\min \quad c_r^k x_r^k \tag{6.13}$$

$$\text{subject to:} \quad \sum_{\substack{k \in K \\ i \in r}} x_r^k \geq n_i, \quad \forall i \tag{6.14}$$

$$\sum_r x_r^k = 1, \quad \forall k \tag{6.15}$$

$$x \in \{0, 1\} \tag{6.16}$$

The mathematical formulation requires the following notation with parameters:

K = set of crew members,
c_r^k = cost of assigning roster r to crew member k,
n_i = number of crew members that are required for task i,

and decision variable:

$x_r^k = 1$ if roster r is assigned to crew member k, and 0 otherwise.

The objective is to minimize the total cost as shown in Equation 6.13. Equation 6.14 ensures that each task is covered. Equation 6.15 assigns a roster to every crew member.

Recent work in the area of crew assignment includes Kohl and Karisch [87], Cappanera and Gallo [88], Caprara et al. [89], Christou et al. [90], Dawid et al. [91], Day and Ryan [92], Gamache and Soumis [93], Gamache et al. [94,95], Shebalov and Klabjan [96], and Sohoni et al. [97]. Recent solution approaches to the crew assignment problem focus on branch-and-price where sub-problems are solved based upon constrained shortest paths, and on numerous heuristic approaches, for instance, simulated annealing as described in Lučić and Teodorvić [98] and Campbell et al. [99].

Recovery from Irregular Operations

In the face of disruptions such as bad weather, unscheduled aircraft maintenance, air traffic congestion, security problems, flight cancellations or delays, passenger delays, and sick or illegal crews on the day of operations, the three aspects of aircraft, crew, and passengers are impacted, often causing the airline to incur significant costs. Clarke [100] states that a major U.S. airline can incur more than $400 million annually in lost revenue, crew overtime pay, and passenger hospitality costs due to irregular operations. When any of these disruptions happen, operations personnel must make real-time decisions that return the aircraft and crew to their original schedule and deliver the passengers to their destinations as soon and as cost effectively as possible.

Aircraft recovery usually follows groundings and delays where the objective is to minimize the cost of reassigning aircraft and re-timing flights while taking into account available resources and other system constraints. Interested readers are referred to detailed reviews in Clarke and Smith [19], Rosenberger et al. [101], and Kohl et al. [102].

Crew recovery usually follows flight delay, flight cancellation, and aircraft rerouting decisions, but often flight dispatchers and crew schedulers coordinate to explore the best options. In crew recovery, the objective is to reassign crew to scheduled flights. It is desirable to return to the original crew schedule as soon as possible. Crew recovery is a hard problem because of complex crew work rules, complex crew cost computations, changing objectives, and having to explore many options. Often, airlines are forced to implement the first solution found due to time pressures.

Airline investment in technologies to expedite crew recovery and reduce crew costs has increased in recent years. Anbil et al. [103] develop a trip repair decision support system with a solution approach that involves iterating between column generation and optimization, based on a set partitioning problem formulation where the columns are trip proposals and the rows refer to both flights and non-reserve crews. The solution strategy is to determine whether the problem can be fixed by swapping flights among only the impacted crew, and bringing in reserves and other nondisrupted crews as needed. Heuristics are used to localize the search during column generation. Teodorović and Stojković [104] present a greedy heuristic based upon dynamic programming and employ a first-in-first-out approach for minimizing crew ground time. Wei et al. [105] and Song et al. [106] describe a heuristic-based framework and algorithms for crew management during irregular operations. Yu et al. [12] describe the CrewSolver system developed by CALEB Technologies for recovering crew schedules at Continental Airlines in real-time when disruptions occur. Abdelghany et al. [107] design a decision support tool that automates crew recovery for an airline with a hub-and-spoke network structure. Nissen and Haase [108] present a duty-period-based formulation for the airline crew rescheduling problem that is tailored to European carriers and labor rules. Schaefer et al. [109] develop approximate solution methods for crew scheduling under uncertainty. Once a crew schedule has been determined, Schaefer and Nemhauser [110] define a framework for perturbing the departure and arrival times that keeps the schedule legal without increasing the planned crew schedule cost. Yen and Birge [111] describe a stochastic integer programming model and devise a solution methodology for integrating disruptions in the evaluation of crew schedules.

In the past decade, there has been a trend toward integrated aircraft, crew, and passenger recovery from irregular operations. Clarke [100] provides a framework that simultaneously addresses flight delays and flight cancellations, and solves the mixed-integer linear programming aircraft recovery problem incorporating crew availability using an optimization-based solution procedure. Lettovský [112] and Lettovský et al. [113] discuss an integrated crew assignment, aircraft routing, and passenger flow mixed-integer linear programming problem, and solve it using Bender's decomposition. Stojković and Soumis [114,115] solve a nonlinear multicommodity flow model that considers both flight and crew schedule recovery using branch-and-price. Rosenberger et al. [101] develop a heuristic for rerouting aircraft and then revise the model to minimize crew and passenger disruptions. Kohl et al. [102] develop a disruption management system called Descartes (Decision Support for integrated Crew and AiRcrafT recovery), which is a joint effort between British Airways, Carmen Systems, and the Technical University of Denmark. Lan et al. [116] propose two new approaches for minimizing passenger disruptions and achieving more robust airline schedule plans by (a) intelligently routing aircraft to reduce the propagation of delays using a mixed-integer programming formulation with stochastically generated inputs and (b) minimizing the number of passenger misconnections by adjusting flight departure times within a small time window. Bratu and Barnhart [117] minimize both airline operating costs and estimated passenger delay and disruption costs to simultaneously determine aircraft, crew, and passenger recovery plans.

6.3 Revenue Management

The Airline Deregulation Act of 1978 is widely credited for having stimulated competition in the airline industry. For example, the year 1986, which was marked by economic growth and stable oil prices, saw the poor financial results of major airlines. This was just a symptom

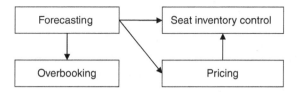

FIGURE 6.3   Interacting revenue management sub-problems.

of the increased competitive pressure faced by the carriers. In search of greater efficiencies, airlines began exploring ways to increase the average aircraft load and the revenue per seat. The discipline that has developed around this issue is usually referred to as "revenue management" (RM) although other terms are sometime used, such as "perishable asset revenue management" (PARM), "yield management," or "demand management." Given the complex and evolving nature of airline services, it does not consist of a single problem, with an unequivocal formulation, but rather of a collection of interacting sub-problems. The first such problem to be analyzed (before deregulation took place) was *overbooking*, which addresses the decision to sell more seats than currently available. The overbooking level depends on the likelihood that a customer will not show up at the gate, and on other factors, such as the lost revenue due to an unsold seat and the reimbursement to an eligible customer who is refused boarding. The former input to the overbooking problem is provided by forecasting methods applied to the reservation process and to the actual conversion rates of reservations in "go shows." The period prior to the 1978 Act saw another innovation: the promotional fare. "Early birds" would receive a discount over regular customers. From these early instances, *pricing* has evolved into a sophisticated discipline, with a deep impact in other industries as well. This simple yet effective innovation exploited the diversity of the customer population, namely the willingness of certain customers to trade off dollars for flexibility. However, the development of time-dependent fares posed a new challenge for airlines: *inventory seat control*, which consists of assigning inventory (number of seats) to a certain fare class. This is perhaps among the most important problems in the practice of revenue management. The earliest known implementation of inventory seat control is that of American Airlines in the first half of the 1980s. Since then, great progress has been made in improving the accuracy of inventory control systems. The relationship between the four sub-problems is shown in Figure 6.3.

The literature on revenue management is vast and growing rapidly. A complete bibliography is beyond the scope of this survey. Recent comprehensive treatments of revenue management are given by Phillips [118] and Van Ryzin and Talluri [119]. Yeoman and McMahon-Beattie [120] offer several real-world examples. The surveys of Weatherford and Bodily [121], McGill and Van Ryzin [122], and Barnhart et al. [9] contain comprehensive bibliographies. Several journals regularly publish RM articles, notably *Operations Research, Management Science, Interfaces*, the *European Journal of Operations Research*, and the recently founded *Journal of Pricing and Revenue Management*, which is entirely dedicated to this subject.

The rest of this section is organized as follows. First, the forecasting issue in revenue management is described. Then overbooking, inventory control, and pricing are reviewed. The section concludes with speculations on trends of the field.

6.3.1   Forecasting

Demand forecasting is a fundamental step in the RM exercise, as it influences all the others. Nevertheless, there are fewer published articles in this area than in any other area of RM. There are several possible reasons behind this relative scarcity. Actual booking data are not usually made available to academic researchers, and industry practitioners have no incentive in revealing internal methods or data. Another disincentive for researchers is that a majority of the articles published in flagship journals in operations research are of a theoretical, rather than empirical, nature. Finally, forecasting for RM purposes presents a formidable set of challenges. Next, the current approaches to this problem are reviewed, and the reader is pointed to the relevant literature.

Airlines sell a nonstorable product: a seat on an origin–destination itinerary fare. To maximize the financial metric of interest, it is vital for the airline not only to forecast the final demand of a given product, but also the process through which the demand accumulated over time. The former quantity is sometimes called the *historical booking forecast* and is a necessary input to overbooking decisions, while the latter, *the advanced booking forecast*, is used for pricing and inventory control, and aims at forecasting the number of bookings in a certain time interval prior to departure. Some of the relevant factors determining advanced booking behavior are:

1. Origin–destination pair
2. Class
3. Price
4. Time of the year
5. Time to departure
6. Presence of substitute products (e.g., a seat offered by the same company in a different class, or products offered by competing airlines).

Moreover, *group bookings* may be modeled explicitly, as their impact on inventory control policies is significant. Many methods have been proposed to explain observed reservation patterns; for a survey, see the book of Van Ryzin and Talluri [119] and the experimental comparison by Weatherford and Kimes [123]. Linear methods currently in place include autoregressive, moving-average models, Bayes and Hierarchical Bayes, and linear or multiplicative regressive models (sometime termed "pick-up" models). Model selection depends also on the chosen level of aggregation of the forecast. By choosing more aggregate models (e.g., aggregating similar classes) the modeler trades off greater bias for smaller variance; there is no "rule of thumb" or consensus on the ideal level of aggregation, although more sophisticated and application-specific models implemented in industrial applications usually provide very detailed forecasts.

It is important to remark that the data used in these statistical analyses are usually *censored*: the observed reservations are always bounded by the inventory made available on that day. Unconstraining demand is either based on mathematical methods (e.g., projection detruncation, or expectation maximization), or recorded reservation denials. For a comparison of popular unconstraining methods, see Weatherford and Pölt [124]. The availability of direct reservation channels (internet, airline-operated contact centers) has made reservation denials data both more abundant and reliable; demand correction based on denial rates is preferable, as it is usually more accurate and requires less computational effort. Figure 6.4 illustrates the relationships among the different components of the forecasting stage.

6.3.2 Overbooking

As mentioned before, research on overbooking predates deregulation. The original objective of overbooking was to maximize average load, subject to internally or externally

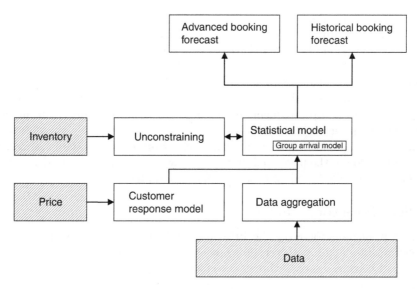

FIGURE 6.4 Forecasting components and relationships.

mandated maximum denial rates. More recently, airlines have been implementing revenue-maximizing overbooking policies. Overbooking can be roughly partitioned into three sub-problems: (1) identification of optimal policies (overbooking limits); (2) identification of "fair bumping" policies (i.e., to whom should one refuse boarding); and (3) estimation of no-show rates. Of the three problems, (1) is the better studied and understood. In its simplest, static, form (see, e.g., Beckmann [125]), it is akin to a newsvendor problem: the availability of a product in the face of underage and overage costs, and of uncertain demand, must be chosen. To better understand the similarity with this problem, consider a simple one-period model, in which random demand is observed in the first period, and departure occurs in the second period. The available capacity is denoted by C, and the overbooking limit must be set prior to the observation of demand to maximize expected revenue. Let p be the unit price per seat and a be the penalty incurred for bumping a passenger. If the number of reservations is r, a random number $Z(r)$, with cumulative distribution function $F_r(z)$, of passengers will show up. A popular behavioral model for "show" demand posits that customers show up with probability q, and that each individual behaves independently of each other; this "binomial" model is defined by:

$$F_r(x) = \binom{r}{x} q^x (1-q)^{r-x} \tag{6.17}$$

The problem faced by the airline is given by:

$$\max_r E(pZ(r) - a(Z(r) - C)^+) \tag{6.18}$$

It can be shown through simple algebra that the solution is given by:

$$1 - F_r(C) = \frac{p}{aq} \tag{6.19}$$

which closely resembles the solution of a newsvendor problem.

The static policy is robust and effective in practical applications; however, using dynamical models could yield revenue increases that, albeit small, can be of crucial importance to achieve profitability. Commonly used models are in discrete time, and assume Markov or independent and identically distributed (iid) demand, and binomial no-show rates, allowing a stochastic dynamic programming formulation. Rothstein [126] is an early contribution to this approach; Alstrup et al. [127] and Alstrup [128] present both theoretical and empirical results.

The problem of equitable bumping has been studied, among others, by Simon [129,130] and Vickrey [131]. Interestingly, the solution proposed in these early days—to conduct an ascending auction to bump passengers with the lowest disutility from boarding denial, has been implemented some three decades later.

Finally, the problem of estimating no-show rates from historical data has received relatively little attention in the published literature. Lawrence et al. [132] present a data-driven no-show forecasting rule based on customer/fare attributes.

6.3.3   Seat Inventory Control

One of the new entrants following the 1978 Act was PEOPLExpress. A nonunionized, barebone service airline carrier with very low fares, it grew phenomenally in the first half of the 1980s, and had become a serious threat to the incumbents. One of them, American Airlines, was competing directly with PEOPLExpress on its principal routes. Partly as a response to the threat posed by the new entrant, American Airlines introduced a seat inventory control system. Under this system, the availability of seats belonging to a certain class could be dynamically controlled. For example, a certain low-price class (e.g., non-refundable tickets) could be "closed" (i.e., made unavailable) at a certain date because the same seat had a high probability of being reserved under a higher-price class (e.g., refundable tickets). The first rule for inventory control dates back to the seminal work of Littlewood [133], which derives it for a two-class inventory. Some assumptions are necessary to obtain closed-form rules:

1. Over time classes are booked in a certain order
2. Low-valuation-before-high-valuation arrival pattern
3. Independent arrivals for different booking classes
4. Absence of no-shows
5. Single-leg reservations
6. No group reservations.

The formulation is reviewed here, as it is the basis for widely used revenue management heuristics in real-world applications. Let the classes be ranked from low valuation to high valuation $p_1 \leq p_2 \leq \cdots \leq p_N$, and denote their uncertain demands D_i, based on assumptions 1 and 2, where i also denotes the period during which the demand is observed. The decision of interest is the capacity x_i that should be allocated to class i. Let $V_i(y)$ denote the expected revenue from period i to period N when the available capacity at the beginning of that period is y. At stage $i-1$, the revenue is given by:

$$p_{i-1}x_{i-1} + V(C_{i-1} - x_{i-1}) \tag{6.20}$$

where x_i cannot exceed the available capacity in period $i-1$, C_{i-1}, and D_{i-1} the observed demand in that period. The Bellman equation is then

$$V_i(y) = E(\max\{p_{i-1}x_{i-1} + V_i(y - x_{i-1}) : x_{i-1} \in [0, \min\{D_{i-1}, x\}]\}), \quad 0 < i < N \tag{6.21}$$

$$V_N(y) = 0 \tag{6.22}$$

It can be shown that the optimal policy takes a simple form as shown in Van Ryzin and Talluri [119]: the optimal solution $\{x_i^*, i = 1, \ldots, N\}$ is such that $x_1^* \leq x_2^* \leq \cdots \leq x_N^*$; the value x_i^* does not depend on D_i, that is, the optimal inventory decision in a certain period can be taken *before* observing the demand in that period. According to the optimal policy, more and more capacity is reserved for classes with higher marginal value. This policy is sometimes termed *theft nesting*: at a given period, the set of higher-value classes is allocated at least a predetermined amount of inventory, but those classes can "steal" inventory from the lower-value class if seats are not sold by the end of the period; the opposite can never happen. An equivalent formulation of the problem uses as the relevant decision variables the booking limits for classes $i, \ldots, N$, defined as the sum of inventory allocations reserved to these classes. In the case of two classes (and two periods) only one decision is taken, and the solution to the recursive equation (6.1) takes the closed form for the booking limit for class 2: $C - F^{-1}(1 - p_1/p_2)$, where F is the cumulative distribution function of demand for class 2. This formula, also known as Littlewood's rule, has been used to formulate heuristics. Their popularity stems from their simplicity; from the fact that they predate the general formulation for the n-class problem; and finally because their sub-optimality, as observed from numerical experiments, is in the range 0–2%, which is considered low when compared to the vastly simplifying assumptions underlying the model. Two such heuristics are particularly popular in practice: expected marginal seat revenue version a (EMSR-a), and b (EMSR-b). Both originate from the work of Belobaba [134,135]. Both heuristics reduce the n-class problem to a 2-class problem. In the case of EMSR-a, pairwise comparisons of class i with higher-value classes are performed, and the booking limit for each such class in isolation is computed. The overall booking limit for classes $i + 1, \ldots, N$ is then set equal to the sum of the individual booking limits. By considering booking limits in isolation, EMSR-a ignores pooling effects among the demand of various classes, and as a result it is often more conservative than the optimal policy. EMSR-b alleviates this shortcoming by aggregating the demand of classes $i + 1, \ldots, N$ into an aggregated class, with demand equal to the sum of demands and marginal revenue equal to the weighted sum of marginal revenues: $\bar{p} = \sum_{k=i+1}^{N} p_k E[D_k] / \sum_{k=i+1}^{N} E[D_k]$. The booking limit in period i is then set by applying Littlewood's rule to class i and the aggregated class. EMSR-b is especially popular among practitioners, and has been extended in various directions. Notably, the case in which there is interaction between classes has been analyzed. In this setting, closing inventory of lower-value classes results in higher demand for higher-value classes (buy up); references on this subject are Belobaba and Weatherford [136] and Weatherford et al. [137].

Before concluding this section, we briefly describe the extension of single-leg capacity control mechanisms to a network setting. Indeed, the appropriate framework of analysis of RM is on an itinerary basis. However, the complexity of the problem increases significantly with size; moreover, network RM also poses organizational and data collection challenges, as responsibilities for revenue targets are assigned on a geographical basis, and data requirements for network RM are more demanding than for single-leg instances. Consequently, several exact formulations and heuristics are available. The heuristics rely on the concept of marginal network revenue value: when a seat is sold at a certain price, it generates revenue, but also results in lost revenues, as certain other itinerary fares will have one less seat available for sale. The network revenue value reflects this tradeoff, and can be effectively employed to make allocation decisions on a network. The advantage of such an approach lies in its robustness, and in the fact that it employs leg-based sub-problems, for which well-tested heuristics and tractable optimization formulations exist. As mentioned above, several network optimizations have been proposed as well. The topology of the network is captured through an incidence matrix, and the decision variables are the booking limits associated to an origin–destination class. The models differ in the way they capture uncertainty, in the intertemporal modeling of the decisions, and finally in the algorithmic approaches adopted to find a solution, exact or approximated. A comprehensive and current description of models is provided in Van Ryzin and Talluri [119]. There are no available data quantifying the financial benefits of network RM. Estimates rely on simulations, as those provided in the industry-standard platform PODS (Passenger Origin–Destination Simulator) and on the assessment of industry experts. The agreement [9] is that network RM can yield an improvement of 1% above and beyond single-leg RM.

6.3.4   Pricing

Most airlines condition passenger demand through seat inventory control. An alternative is to condition demand through pricing. The two practices are conceptually similar: if the only differentiating attribute among a set of classes is price, then closing lower-value classes is equivalent to posting a bid price for tickets equal to the cheapest class that is still open. Price controls, however, are more granular than inventory controls, as price points are not limited to the finite number of classes; moreover, dynamic price controls can effectively be used to limit the access to available seats. In its early days, commercial flight did not have the instruments to change prices dynamically, and to effectively advertise these changes. Moreover, customers were not used to observing changing prices for the same service; finally, a suitable framework of analysis was not available. In recent years, however, we are witnessing the evolution of revenue management systems in the direction of price-based controls. Low-cost carriers have grown in number and have thrived: Southwest, JetBlue in the U.S., EasyJet, RyanAir in Europe, TAM, and Go in South America. These carriers often offer a single product: an economy, nonrefundable ticket, and this product's availability is controlled through price. Despite the simplicity of the setting, dynamic pricing is a complex problem that has been attacked from many different angles, and is far from being solved. To obtain insightful results, researchers are forced to make simplifying assumptions that are either unrealistic or are difficult to test. Perhaps the single biggest challenge in dynamic pricing is understanding consumer behavior. Among the questions that have not received a definite answer are: how sensitive are the customers to a product's price changes, and to price changes of substitute products? Do customers take into account expected future prices when making their purchase decisions, and if they do, what is their learning process? How can we define the potential market for a certain product, and what is the impact of market size on optimal pricing? Some of these issues are studied in the marketing literature (for example, Nagle and Hogan [138]), but empirical validation in the airline industry is still missing. Perhaps the simplest model that captures the essential features of the airline industry is given by the following deterministic formulation for a single-product pricing problem with myopic customers:

$$\max \qquad \sum_{i=1}^{N} \pi_i(p_i, d_i(p_i)) \qquad\qquad (6.23)$$

$$\text{subject to:} \qquad \sum_{i=1}^{N} d_i(p_i) \leq C \qquad\qquad (6.24)$$

$$p_i \geq 0, \quad i = 1, \ldots, N \qquad\qquad (6.25)$$

The decision variables are the prices p_i of the product in period i, the market response is modeled by a deterministic demand function $d_i(\cdot)$, and profit is function of both price and demand. The above model assumes customers' myopic behavior, as demand in a certain period does not depend on the sequence of observed prices. Extensions of the model are possible in several directions. Gallego and Van Ryzin [139] considered a continuous-time, stochastic version of the pricing problem, and proved monotonicity and structural properties of the solution. The same authors analyze the multiperiod, multiproduct case (Gallego and Van Ryzin[140]). The case of non-myopic behavior has been analyzed by Stokey [141], Gale and Holmes [142], Gallego and Sahin [143], and Gallego et al. [144], who consider consumers with intertemporal valuations.

6.4 Aircraft Load Planning

The general problem of aircraft load planning can be described as transporting equipment (vehicles, palletized cargo, helicopters, ammunitions, etc.) and personnel from a set of departure points to a set of destinations with the objective of minimizing the number of loads used. The arrangement of the equipment and personnel affects the position of the center of gravity of the aircraft, which in turn has an impact on aircraft drag, which in turn has an impact on fuel consumption. The problem is further complicated by additional constraints such as hazardous material incompatibilities. The aircraft load planner must balance the load to ensure the safe and efficient use of the aircraft by complying with aircraft safety, weight and balance, and floor load restrictions for takeoff, in-flight, and landing. The speed and ease of on-loads and off-loads are of utmost importance too.

Aircraft load planning is both an important civilian and military application. The load planning process was done by manual procedures until 1977 when the first account of the use of computer-assisted approaches began within the U.S. Air Force, according to Martin-Vega [145], who provides a comprehensive review of work prior to 1985. Early semi-automated systems include CARLO (Cargo Loading system used by Scandinavian Airlines) by Larsen and Mikkelsen [146], AALPS (Automated Air Load Planning System by SRI International for the U.S. Army) as described in Frankel et al. [147,148] and Anderson and Ortiz [149], and DMES in Cochard and Yost [150]. Today AALPS is used at more than 300 sites throughout the continental United States Army community and by operations *Just Cause, Desert Shield, Desert Storm,* and *Restore Democracy.* Lockheed Corporation and the U.S. Air Force have utilized AALPS for designing cargo aircraft and configuring the C-17, respectively. Additional work in the area of aircraft load planning includes that of Eilon and Christofides [151], Hane et al. [152], Ng [153], Amiouny et al. [154], Heidelberg et al. [155], Thomas et al. [156], Gueret et al. [157], and Mongeau and Bes [158].

The aircraft loading problem belongs to the broad class of cutting and packing problems [159]. Even one-dimensional packing problems, such as the knapsack and bin-packing problems, are NP-hard. Typically 2-D bin-packing with the length and width of both the cargo and the aircraft's cargo hold are used. Height is not as significant because usually cargo is not stacked. Higher dimensional problems are much more combinatorial and cannot be solved exactly for large instances. A heuristic approach must be taken to achieve an acceptable result in a short amount of time, as discussed in Bischoff and Marriott [160] and Pisinger [161].

The challenge of balancing is generally focused on one dimension: the length of the plane, that is, positioning the center of gravity along the fore and aft axis of the aircraft. Balancing the center of gravity side-to-side is generally considered not significant.

Mongeau and Bes [158] take a mathematical programming approach to the aircraft loading problem for Airbus France, which can be solved optimally in a reasonable time.

To illustrate, the formulation is given as:

$$\max \quad M(x) \tag{6.26}$$

$$\text{subject to:} \quad CG(x) \le X_{\text{stab}} \tag{6.27}$$

$$M(x) \le M_{\text{max}} \tag{6.28}$$

$$\sum_{i=1}^{N_{\text{cont}}} \sum_{j \in H_k} M_i x_{ij} \le M_{\text{max}}^k, \quad k = 1, 2, \ldots, N_{\text{hold}} \tag{6.29}$$

$$\sum_{j=1}^{N_{\text{comp}}} x_{ij} \le 1, \quad\quad\quad \forall i = 1, 2, \ldots, N_{\text{cont}} \tag{6.30}$$

$$\sum_{j=1}^{N_{\text{comp}}} x_{ij} = 1, \quad\quad\quad \forall i \in I \tag{6.31}$$

$$x_{ik} = 1, \quad\quad\quad \text{for any given container } i \text{ required} \tag{6.32}$$
$$\text{to be in a specific compartment } k$$

$$X_{\text{target}} - \varepsilon \le CG(x) \le X_{\text{target}} + \varepsilon \tag{6.33}$$

The mathematical formulation requires the following notation with parameters:

$N_{\text{cont}} =$ number of containers on the ground,
$N_{\text{comp}} =$ number of compartments,
$N_{\text{hold}} =$ number of holds,
$M_i =$ mass of container i,
$CG(x) =$ center of gravity of the aircraft after loading,
$X_{\text{stab}} =$ maximal (longitudinal) position of the center of gravity of the aircraft after loading to satisfy stability requirements,
$M_{\text{max}} =$ maximal mass of freight that can be loaded,
$M_{\text{max}}^k =$ maximal mass of freight that can be loaded in hold k and $k = 1, 2, \ldots, N_{\text{hold}}$,
H_k where $H_k \subseteq \{1, 2, \ldots, N_{\text{comp}}\}$ are the compartments in hold k and $k = 1, 2, \ldots, N_{\text{hold}}$,
$I =$ a subset of the container list that is required to be loaded,
$X_{\text{target}} =$ ideal (longitudinal) position of the center of gravity of the aircraft after loading,
$\varepsilon =$ some positive allowable displacement of the center of gravity of the aircraft from its ideal position,
$x_{ij} \in \{0, 1\}$ where $x_{ij} = 1$, if container i is to be placed in container j, and 0 otherwise; and $i = 1, 2, \ldots, N_{\text{cont}}$; $j = 1, 2, \ldots, N_{\text{comp}}$.

The objective given in Equation 6.26 is to maximize the mass loaded $M(x)$ where

$$M(x) = \sum_{i=1}^{N_{\text{cont}}} \sum_{j=1}^{N_{\text{comp}}} M_i x_{ij} \tag{6.34}$$

Equation 6.27 ensures that the aircraft stability requirement is satisfied. Equations 6.28 and 6.29 limit the stress/mass capacity overall and for each hold, respectively. Equations 6.30 state that each container must be loaded at most once. Equations 6.31 enforce that a subset I of the container list is loaded. Equations 6.32 ensure, in the event that a specific container must be placed in a specific compartment, that the requirement is met. Equation 6.33 keeps the center of gravity within an allowable displacement from the ideal position. In the formulation, Mongeau and Bes [158] also include volume capacity constraints, which involve nonconvex, piecewise-linear functions that are transformed into simple linear constraints. These constraints are quite complex and vary according to both the aircraft and the container types. The interested reader may refer to the details in the article.

6.5 Future Research Directions and Conclusions

The application of Operations Research to airline planning processes has resulted in significant benefits to the industry through cost reductions, revenue improvements, operational efficiencies, and automated tools. The combination of advancements in computer hardware and software technologies with clever mathematical algorithms and heuristics has dramatically transformed the ability of operations researchers to solve large-scale, sophisticated airline optimization problems over the past 60 years. This trend will continue in the future; airline planning problems are rich with complexities that can lead to breakthroughs in methods and decision-support tools.

Many promising areas for future research exist in the airline industry. Some examples include:

1. Incorporating new forecasting techniques, overbooking strategies, and pricing strategies into revenue management decisions to address the increased market transparency of fares and fare classes on the Web.

2. Developing accurate consumer response models that take into account the multiple sales channels available to customers (e.g., travel agency, phone, online agencies, online airline Web sites), and their ability to compare fares.

3. Integrating schedule development to incorporate aircraft and crew planning decisions; for example, simultaneous consideration of aircraft routing, fleet assignment, and crew pairing.

4. Dynamically reassigning aircraft capacities to the flight network when improved passenger demand forecasts become available to maximize total revenue. This line of research of demand driven re-fleeting/demand driven dispatching has been recently explored in Sherali et al. [162] and Wang and Regan [163].

5. Exploiting the availability of real-time data to enable real-time decision-making; for example, in the area of recovery from irregular operations.

6. Developing real-time policies to assign locations and decide ordering quantities for spare parts. The problem formulation must take into account savings obtained from sharing the same warehouse for multiple service centers, and the service time requirements. Similar systems have already been implemented in military and industrial production environments.

7. Developing loyalty programs and customer contact programs that take into account the financial value associated to a customer, as well as his/her potential future evolution. This line of research, traditionally associated to the field of relationship marketing, has been recently explored in the context of airline campaign management [164].

8. Developing decision-support tools for fractional fleet services, which are growing in popularity, to address the stochastic nature of demand.

9. Developing new algorithms for aircraft fuel operational optimization and fuel hedging as fuel represents the second largest cost component of an airline's operations after labor [165].

References

1. The economic & social benefits of air transport, Air Transport Action Group brochure, 2004.
2. IATA Passenger and Freight Forecast 2005–2009, October 2005.
3. ICAO Annual Report of the Council, 2004.
4. Airports Council International (ACI) figure, 2005.
5. CANSO estimation, 2005.
6. Airbus: Global market forecast 2004–2023, http://www.airbus.com/en/myairbus/global_market_forcast.html.
7. Boeing: 2006 current market outlook, http://www.boeing.com/commercial/cmo.
8. Nemhauser, G. L., The age of optimization: solving large-scale real-world problems, *Operations Research*, 42, 5, 1994.
9. Barnhart, C., Belobaba, P., and Odoni, A. R., Applications of operations research in the air transport industry, *Transportation Science*, 37, 368, 2003.
10. Hicks, R. et al., Bombardier Flexjet significantly improves its fractional aircraft ownership operations, *Interfaces*, 35, 49, 2005.
11. Armacost, A. P. et al., UPS optimizes its air network, *Interfaces*, 34, 15, 2004.
12. Yu, G. et al., A new era for crew recovery at Continental Airlines, *Interfaces*, 33, 5, 2003.
13. Butchers, E. R. et al., Optimized crew scheduling at Air New Zealand, *Interfaces*, 31, 30, 2001.
14. Subramanian, R. et al., Coldstart: fleet assignment at Delta Air Lines, *Interfaces*, 24, 104, 1994.
15. Smith, B. C., Leimkuhler, J. F., and Darrow, R. M., Revenue management at American Airlines, *Interfaces*, 22, 8, 1992.
16. Etschmaier, M. and Mathaisel, D., Airline scheduling: an overview, *Transportation Science*, 19, 127, 1985.
17. Rushmeier, R. and Kontogiorgis, S. A., Advances in the optimization of airline fleet assignment, *Transportation Science*, 31, 159, 1997.
18. Barnhart, C. and Talluri, K., Airlines operations research, in C. ReVelle and A. E. McGarity (eds.), *Design and Operation of Civil and Environmental Engineering Systems*, John Wiley & Sons, Inc., New York, 435, 1997.
19. Clarke, M. and Smith, B., Impact of operations research on the evolution of the airline industry, *Journal of Aircraft*, 41, 62, 2004.
20. Klabjan, D., Large-scale models in the airline industry, in G. Desaulniers, J. Desrosiers, and M. M. Solomon (eds.), *Column Generation*, Springer, New York, 2005.
21. Barnhart, C. and Cohn, A., Airline schedule planning: accomplishments and opportunities, *Manufacturing & Service Operations Management*, 6, 3, 2004.
22. Rushmeier, R. A., Hoffman, K. L., and Padberg, M., Recent advances in exact optimization of airline scheduling problems, Technical Report, 1995.
23. Erdmann, A. et al., Modeling and solving an airline schedule generation problem, *Annals of Operations Research*, 107, 117, 2001.

24. Armacost, A., Barnhart, C., and Ware, K., Composite variable formulations for express shipment service network design, *Transportation Science*, 36, 1, 2002.

25. Armacost, A., Barnhart, C., Ware, K., and Wilson, A., UPS optimizes its air network, *Interfaces*, 34, 15, 2004.

26. Sherali, H. D., Bish, E. K., and Zhu, X., Airline fleet assignment concepts, models, and algorithms, *European Journal of Operational Research*, 172, 1, 2006.

27. Gopalan, R. and Talluri, K. T., Mathematical models in airline schedule planning: a survey, *Annals of Operations Research*, 76, 155, 1998.

28. Yu, G. and Yang, J., Optimization applications in the airline industry, in D. Z. Du and P. M. Pardalos (eds.), *Handbook of Combinatorial Optimization*, Kluwer Academic Publishers, Norwell, MA, 635, 1998.

29. Abara, J., Applying integer linear programming to the fleet assignment problem, *Interfaces*, 19, 20, 1989.

30. Daskin, M. S. and Panayotopoulos, N. D., A Lagrangian relaxation approach to assigning aircraft to routes in hub and spoke networks, *Transportation Science*, 23, 91, 1989.

31. Talluri, K. T., Swapping applications in a daily airline fleet assignment, *Transportation Science*, 30, 237, 1996.

32. Jarrah, A., Goodstein, J., and Narasimhan, R., An efficient airline re-fleeting model for the incremental modification of planned fleet assignments, *Transportation Science*, 34, 349, 2000.

33. Hane, C. A. et al., The fleet assignment problem: solving a large-scale integer program, *Mathematical Programming*, 70, 211, 1995.

34. Berge, M. A. and Hopperstad, C. A., Demand driven dispatch: a method for dynamic aircraft capacity assignment, models, and algorithms, *Operations Research*, 41, 153, 1993.

35. Desaulniers, G. et al., Daily aircraft routing and scheduling, *Management Science*, 43, 841, 1997.

36. Rexing, B. et al., Airline fleet assignment with time windows, *Transportation Science*, 34, 1, 2000.

37. Lohatepanont, M. and Barnhart, C., Airline scheduling planning: integrated models and algorithms for schedule design and fleet assignment, *Transportation Science*, 38, 19, 2004.

38. Yan, S. and Tseng, C. H., A passenger demand model for airline flight scheduling and fleet routing, *Computers and Operations Research*, 29, 1559, 2002.

39. Clarke, L. W. et al., Maintenance and crew considerations in fleet assignment, *Transportation Science*, 30, 249, 1996.

40. Barnhart, C. et al., Flight string models for aircraft fleeting and routing, *Transportation Science*, 32, 208, 1998.

41. Rosenberger, J. M., Johnson, E. L., and Nemhauser, G. L., A robust fleet-assignment model with hub isolation and short cycles, *Transportation Science*, 38, 357, 2004.

42. Belanger, N. et al., Weekly airline fleet assignment with homogeneity, *Transportation Research Part B—Methodological*, 40, 306, 2006.

43. Farkas, A., The influence of network effects and yield management on airline fleet assignment decisions, Ph.D. dissertation, MIT, Cambridge, MA, 1995.

44. Kniker, T. S., Itinerary-based airline fleet assignment, Ph.D. dissertation, Massachusetts Institute of Technology, Cambridge, MA, 1998.

45. Jacobs, T. L., Smith, B. C., and Johnson, E., O&D FAM: Incorporating passenger flows into the fleeting process, In: *Proceedings of the AGIFORS Symposium*, New Orleans, LA, 128, 1999.

46. Barnhart, C., Kniker, T. S., and Lohatepanont, M., Itinerary-based airline fleet assignment, *Transportation Science*, 36, 199, 2002.

47. Levin, A., Scheduling and fleet routing models for transportation system, *Transportation Science*, 5, 232, 1971.

48. Cordeau, J., Stojković, G., Soumis, F., and Desrosiers, J., Benders decomposition for simultaneous aircraft routing and crew scheduling, *Transportation Science*, 35, 375, 2001.

49. Feo, T. and Bard, J., Flight scheduling and maintenance base planning, *Management Science*, 35, 1415, 1989.

50. Clarke, L., Johnson, E., Nemhauser, G., and Zhu, Z., The aircraft rotation problem, in R.E. Burkard, T. Ibaraki, and M. Queyranne (eds.), *Annals of OR: Mathematics of Industrial Systems*, II, 33, 1997.

51. Gopalan, R. and Talluri, K., The aircraft maintenance routing problem, *Operations Research*, 46, 260, 1998.

52. Gabteni, S. and Gronkvist, M., A hybrid column generation and constraint programming optimizer for the tail assignment problem, *Computer Science*, 3990, 89, 2006.

53. Li, Y. H. and Wang, X. B., Integration of fleet assignment and aircraft routing, *Airports, Airspace, and Passenger Management Transportation Research Record*, 1915, 79, 2005.

54. Mercier, A., Cordeau, J. F., and Soumis, F., A computational study of Benders decomposition for the integrated aircraft routing and crew scheduling problem, *Computers & Operations Research*, 32, 1451, 2005.

55. Cohn, A. M. and Barnhart, C., Improving crew scheduling by incorporating key maintenance routing decisions, *Operations Research*, 51, 387, 2003.

56. Arabeyre, J. P. et al., The airline crew scheduling problem: a survey, *Transportation Science*, 3, 140, 1969.

57. Bornemann, D. R., The evolution of airline crew pairing optimization, *The 22nd AGIFORS Crew Management Study Group Proceeding*, Paris, France, 1982.

58. Barutt, J., and Hull, T., Airline crew scheduling: supercomputers and algorithms, *SIAM News*, 23, 20, 1990.

59. Desaulniers, G. et al., Crew scheduling in air transportation, in T. Cranic and G. Laporte (eds.), *Fleet Management and Logistics*, Kluwer Academic Publishers, Boston, MA, 1998.

60. Barnhart, C. et al., Airline crew scheduling, in Randolph W. Hall (ed.), *Handbook of Transportation Science*, 2nd ed., Kluwer Academic Publishers, Norwell, MA, 2003.

61. Gopalakrishnan, B. and Johnson, E. L., Airline crew scheduling: state-of-the-art, *Annals of Operations Research*, 140, 305, 2005.

62. Anbil, R., Forrest, J. J., and Pulleyblank, W. R., Column generation and the airline crew pairing problem, *Documenta Mathematica Journal der Deutschen Mathematiker, Extra Volume ICM*, 677, 1998.

63. Gershkoff, I., Optimizing flight crew schedules, *Interfaces*, 19, 29, 1989.

64. Rubin, J., A technique for the solution of massive set covering problems, with applications to airline crew scheduling, *Transportation Science*, 7, 34, 1973.

65. Anbil, R. et al., Recent advances in crew pairing optimization at American Airlines, *Interfaces*, 21, 62, 1991.

66. Anbil, R. et al., Crew-pairing optimization at American Airlines Decision Technologies, *Optimization in Industry*, in T. A. Ciriani and R. C. Leachman (eds.), John Wiley & Sons, 31, 1993.

67. Graves, G. W. et al., Flight crew scheduling, *Management Science*, 39, 736, 1993.

68. Hu, J. and Johnson, E., Computational results with a primal-dual subproblem simplex method, *Operations Research Letters*, 25, 149, 1999.

69. Minoux, M., Column generation techniques in combinatorial optimization: a new application to the crew pairing problems, *Proceedings of the XXIV AGIFORS Symposium*, France, 1984.

70. Baker, E. K., Bodin, L. D., and Fisher, M., The development of a heuristic set covering based system for air crew scheduling, *Transportation Policy Decision Making*, 3, 95, 1985.

71. Lavoie, S., Minoux, M., and Odier, E., A new approach of crew pairing problems by column generation and application to air transport, *European Journal of Operational Research*, 35, 45, 1988.

72. Barnhart, C. et al., A column generation technique for the long-haul crew assignment problem, in T. A. Ciriani and R. Leachman (eds.), *Optimization in Industry 2: Mathematical Programming and Modeling Techniques in Practice*, John Wiley & Sons, New York, 7, 1994.

73. Desaulniers, G. et al., Crew scheduling in air transportation, *Les Cahiers du GERAD, G97-26*, GERAD Inst. of Montreal, Quebec, Canada, 1997.

74. Stojković, M., Soumis, F., and Desrosiers, J., The operational airline crew scheduling problem, *Transportation Science*, 32, 232, 1998.

75. Galia, R. and Hjorring, C., Modeling of complex costs and rules in a crew pairing column generator, *Technical Report CRTR-0304*, Carmen Systems, 2003.

76. Barahona, F. and Anbil, R., The volume algorithm: producing primal solutions with a subgradient algorithm, *Mathematical Programming*, 87, 385, 2000.

77. Barahona, F. and Anbil, R., On some difficult linear programs coming from set partitioning, *Discrete Applied Mathematics*, 118, 3, 2002.

78. Hoffman, K. L. and Padberg, M., Solving airline crew scheduling problems by branch-and-cut, *Management Science*, 39, 657, 1993.

79. Vance, P. H., Barnhart, C., Johnson, E. L., and Nemhauser, G. L., Airline crew scheduling: a new formulation and decomposition algorithm, *Operations Research*, 45, 188, 1997.

80. Klabjan, D. et al., Solving large airline crew scheduling problems: random pairing generation and strong branching, *Computational Optimization and Applications*, 20, 73, 2001.

81. Makri, A. and Klabjan, D., Efficient column generation techniques for airline crew scheduling, *INFORMS Journal on Computing*, 16, 56, 2004.

82. Tran, V. H., Reinelt, G., and Bock, H. G., BoxStep methods for crew pairing problems, *Optimization and Engineering*, 7, 33, 2006.

83. Marchiori, E. and Steenbeck, A., An evolutionary algorithm for large scale set covering problems with application to airline crew scheduling, in *Real-World Applications of Evolutionary Computing*, LNCS1803, 367, 2000.

84. Ozdemir, H. T. and Mohan, C., Flight graph based genetic algorithm for crew scheduling in airlines, *Information Sciences*, 133, 165, 2001.

85. Kornilakis, H. and Stamatopoulos, P., Crew pairing optimization with genetic algorithms, *Methods and Applications of Artificial Intelligence, Lecture Notes in Artificial Intelligence*, 2308, 109, 2002.

86. Lagerholm, M., Peterson, C., and Söderberg, B., Airline crew scheduling using Potts mean field techniques, *European Journal of Operational Research*, 120, 81, 2000.

87. Kohl, N. and Karisch, S. E., Airline crew rostering: problem types, modeling, and optimization, *Annals of Operations Research*, 127, 223, 2004.

88. Cappanera, P. and Gallo, G., A multicommodity flow approach to the crew rostering problem, *Operations Research*, 52, 583, 2004.

89. Caprara, A. et al., Modeling and solving the crew rostering problem, *Operations Research*, 46, 820, 1998.

90. Christou, I. T. et al., A two-phase genetic algorithm for large-scale bidline-generation problems at Delta Air Lines, *Interfaces*, 29, 51, 1999.

91. Dawid, H., König, J., and Strauss, C., An enhanced rostering model for airline crews, *Computers and Operations Research*, 28, 671, 2001.

92. Day, P. R. and Ryan, D. M., Flight attendant rostering for short-haul airline operations, *Operations Research*, 45, 649, 1997.

93. Gamache, M. and Soumis, F., A method for optimally solving the rostering problem, in G. Yu (ed.), *Operations Research in the Airline Industry*, Kluwer Academic Publishers, Norwell, MA, 1998, 124.

94. Gamache, M. et al., The preferential bidding system at Air Canada, *Transportation Science*, 32, 246, 1998.

95. Gamache, M. et al., A column generation approach for large-scale aircrew rostering problems, *Operations Research*, 47, 247, 1999.

96. Shebalov, S. and Klabjan, D., Robust airline crew pairing: move-up crews, *Transportation Science*, 40, 300, 2006.

97. Sohoni, M. G., Johnson, E. L., and Bailey, T. G., Operational airline reserve crew planning, *Journal of Scheduling*, 9, 203, 2006.

98. Lučić, P. and Teodorvić, D., Simulated annealing for the multi-objective aircrew rostering problem, *Transportation Research Part A*, 33, 19, 1999.

99. Campbell, K., Durfee, R., and Hines, G., FedEx generates bid lines using simulated annealing, *Interfaces*, 27, 1, 1997.

100. Clarke, M., Development of heuristic procedures for flight rescheduling in the aftermath of irregular airline operations, Ph.D. dissertation, Massachusetts Institute of Technology, International Center for Air Transportation, Cambridge, MA, 1997.

101. Rosenberger, J. M., Johnson, E. L., and Nemhauser, G. L., Rerouting aircraft for airline recovery, *Transportation Science*, 37, 408, 2003.

102. Kohl, N. et al., Airline disruption management—perspectives, experiences, and outlook, forthcoming in *Journal of Air Transport Management*.

103. Anbil, R., Barahona, F., Ladanyi, L., Rushmeier, R., and Snowdon, J., Airline optimization, *OR/MS Today*, 26, 26, 1999.

104. Teodorvić, D. and Stojković, G., Model for operational daily airline scheduling, *Transportation Planning and Technology*, 14, 273, 1990.

105. Wei, G., Yu, G., and Song, M., Optimization model and algorithm for crew management during airline irregular operations, *Journal of Combinatorial Optimization*, 1, 305, 1997.

106. Song, M., Wei, G., and Yu, G., A decision support framework for crew management during airline irregular operations, in G. Yu (ed.), *Operations Research in the Airline Industry*, Kluwer Academic Publishers, Norwell, MA, 1998, 260.

107. Abdelghany, A., Ekollu, G., Narasimhan, R., and Abdelghany, K., A proactive crew recovery decision support tool for commercial airlines during irregular operations, *Annals of Operations Research*, 127, 309, 2004.

108. Nissen, R. and Haase, K., Duty-period-based network model of crew rescheduling in European airlines, *Journal of Scheduling*, 9, 255, 2006.

109. Schaefer, A. J. et al., Airline crew scheduling under uncertainty, *Transportation Science*, 39, 340, 2005.

110. Schaefer, A. J. and Nemhauser, G. L., Improving airline operational performance through schedule perturbation, *Annals of Operations Research*, 144, 3, 2006.

111. Yen, J. W. and Birge, J. R., A stochastic programming approach to the airline crew scheduling problem, *Transportation Science*, 40, 3, 2006.

112. Lettovský, L., Airline operations recovery: an optimization approach, Ph.D. thesis, Georgia Institute of Technology, 1997.

113. Lettovský, L., Johnson, E., and Nemhauser, G., Airline crew recovery, *Transportation Science*, 34, 337, 2000.

114. Stojković, M. and Soumis, F., An optimization model for the simultaneous operational flight and pilot scheduling problem, *Management Science*, 47, 1290, 2001.

115. Stojković, M. and Soumis, F., The operational flight and multi-crew scheduling problem, Technical Report G-2000-27, GERAD, 2003.

116. Lan, S., Clarke, J. P., and Barnhart, C., Planning for robust airline operations: optimizing aircraft routings and flight departure times to minimize passenger disruptions, *Transportation Science*, 40, 15, 2006.

117. Bratu, S. and Barnhart, C., Flight operations recovery: new approaches considering passenger recovery, *Journal of Scheduling*, 9, 279, 2006.

118. Phillips, R. L., *Pricing and Revenue Optimization*, Stanford University Press, Stanford, CA, 2005.

119. Van Ryzin, G. J., and Talluri, K. T., *The Theory and Practice of Revenue Management*, Kluwer Academic Publishers, Norwell, MA, 2004.

120. Yeoman, I. and McMahon-Beattie, U. (eds.), *Revenue Management and Pricing: Case Studies and Applications*, Int. Thomson Business Press, London, 2004.

121. Weatherford, L. R. and Bodily, S. E., A taxonomy and research overview of perishable-asset revenue management: yield management, overbooking, and pricing, *Operations Research*, 40, 831, 1992.

122. McGill, J. I. and Van Ryzin, G. J., Revenue management: research overview and prospects, *Transportation Science*, 33, 233, 1999.

123. Weatherford, L. R. and Kimes, S. E., A comparison of forecasting methods for hotel revenue management, *International Journal of Forecasting*, 19, 401, 2003.

124. Weatherford, L. R. and Pölt, S., Better unconstraining of airline demand data in revenue management systems for improved forecast accuracy and greater revenues, *Journal of Revenue Management and Pricing*, 1, 234, 2001.

125. Beckmann, J. M., Decision and team problems in airline reservations, *Econometrica*, 26, 134, 1958.

126. Rothstein, M., Stochastic models for airline booking policies, Ph.D. thesis, Graduate School of Engineering and Science, New York University, New York, 1968.

127. Alstrup, J., Boas, S., Madsen, O. B. G., Vidal, R., and Victor, V., Booking policy for flights with two types of passengers, *European Journal of Operational Research*, 27, 274, 1986.

128. Alstrup, J., Booking control increases profit at Scandinavian Airlines, *Interfaces*, 19, 10, 1989.

129. Simon, J., An almost practical solution to airline overbooking, *Journal of Transportation and Economic Policy*, 2, 201, 1968.

130. Simon, J., Airline overbooking: the state of the art—a reply, *Journal of Transportation and Economic Policy*, 6, 254, 1972.

131. Vickrey, W., Airline overbooking: some further solutions, *Journal of Transportation and Economic Policy*, 6, 257, 1972.

132. Lawrence, R. D., Hong, S. J., and Cherrier, J., Passenger-based predictive modeling of airline no-show rates, *ACM SIGKDD '03*, 2003.

133. Littlewood, K., Forecasting and control of passenger bookings, *AGIFORS Symposium Proceedings*, 1972, 95.

134. Belobaba, P., Air travel demand and airline seat inventory management, Ph.D. thesis, Flight Transportation Laboratory, Massachusetts Institute of Technology, Cambridge, MA, 1987.

135. Belobaba, P., Application of a probabilistic decision model to airline seat inventory control, *Operations Research*, 37, 183, 1989.

136. Belobaba, P., and Weatherford, L. R., Comparing decision rules that incorporate customer diversion in perishable asset revenue management situations, *Decision Sciences*, 27, 343, 1996.

137. Weatherford, L., Bodily, S. E., and Pfeifer, P. E., Modeling the customer arrival process and comparing decision rules in perishable asset revenue management situations, *Transportation Science*, 27, 239, 1993.

138. Nagle, T. T. and Hogan, J., *The Strategy and Tactics of Pricing: A Guide to Growing More Profitably* (4th ed.), Prentice Hall, Upper Saddle River, NJ, 2005.

139. Gallego, G. and Van Ryzin, G., Optimal dynamic pricing of inventories with stochastic demand over finite horizons, *Management Science*, 40, 999, 1994.

140. Gallego, G. and Van Ryzin, G., A multi-product dynamic pricing problem and its applications to network yield management, *Operations Research*, 45, 24, 1997.

141. Stokey, N., Intertemporal price discrimination, *Quarterly Journal of Economics*, 94, 355, 1979.

142. Gale, I. L. and Holmes, T. J., The efficiency of advance-purchase discounts in the presence of aggregate demand uncertainty, *International Journal of Industrial Organization*, Elsevier, 10, 413, 1992.

143. Gallego, G. and Sahin, O., Inter-temporal valuations, product design and revenue management, Working paper, Columbia University Department of Industrial Engineering and Operations Research, New York, 2006.

144. Gallego, G., Phillips, R., and Sahin, O., Strategic management of distressed inventory, Working paper, Columbia University Department of Industrial Engineering and Operations Research, New York, NY, 2004.

145. Martin-Vega, L., Aircraft load planning and the computer, *Computers & Industrial Engineering*, 9, 357, 1985.

146. Larsen, O. and Mikkelsen, G., An interactive system for the loading of cargo aircraft, *European Journal of Operational Research*, 4, 367, 1980.

147. Frankel, M. S. et al., Army data distributions system/packet radio testbed, Contract MDA903-80-C-0217, Quarterly Technical Reports 1–3, SRI International, Menlo Park, California, 1980.

148. Frankel, M. S. et al., Army data distributions system/packet radio testbed, Contract MDA903-80-C-0217, Final Technical Report, SRI International, Menlo Park, California, 1981.

149. Anderson, D. and Ortiz, C., AALPS: a knowledge-based system for aircraft loading, *IEEE Expert*, 1987, 71.

150. Cochard, D. D. and Yost, K. A., Improving utilization of air force cargo aircraft, *Interfaces*, 15, 53, 1985.

151. Eilon, S. and Christofides, N., The loading problem, *Management Science*, 17, 259, 1971.

152. Hane, C. et al., Plane loading algorithms, presented at the ORSA/TIMS Joint Meeting, Denver, 1988.

153. Ng, K. Y. K., A multicriteria optimization approach to aircraft loading, *Operations Research Technical Note*, 40, 1200, 1992.

154. Amiouny, S. V. et al., Balanced loading, *Operations Research*, 40, 238, 1992.

155. Heidelberg, K. R. et al., Automated air load planning, *Naval Research Logistics*, 45, 751, 1998.

156. Thomas, C. et al., Airbus packing at Federal Express, *Interfaces*, 28, 21, 1998.

157. Gueret, C. et al., Loading aircraft for military operations, *Journal of the Operational Research Society*, 54, 458, 2003.

158. Mongeau, M. and Bes, C., Optimization of aircraft container loading, *IEEE Transactions on Aerospace and Electronic Systems*, 39, 140, 2003.

159. Sweeney, P. E. and Paternoster, E. R., Cutting and packing problems: a categorized, application-oriented research bibliography, *Journal of the Operational Research Society*, 43, 691, 1992.

160. Bischoff, E. E. and Marriott, M. E., A comparative evaluation of heuristics for container loading, *European Journal of Operational Research*, 44, 267, 1990.
161. Pisinger, D., Heuristics for the container loading problem, *European Journal of Operational Research*, 141, 382, 2002.
162. Sherali, H. D., Bish, E. K., and Zhu, X. M., Polyhedral analysis and algorithms for a demand-driven refleeting model for aircraft assignment, *Transportation Science*, 39, 349, 2005.
163. Wang, X. B. and Regan, A., Dynamic yield management when aircraft assignments are subject to swap, *Transportation Research Part B—Methodological*, 40, 563, 2006.
164. Labbi, A., Tirenni, G., Berrospi, C., and Elisseeff A., Customer equity and lifetime management (CELM): Finnair case study, *Marketing Science*, submitted.
165. Subcommittee on Aviation Hearing on Commercial Jet Fuel Supply: impact and cost on the U.S. airline industry, http://www.house.gov/transportation/aviation/02-15-06/02-15-06memo.html, 2006.

7

Financial Engineering

Aliza R. Heching and
Alan J. King
IBM T. J. Watson Research Center

7.1   Introduction

Operations research provides a rich set of tools and techniques that are applied to financial decision making. The first topic that likely comes to mind for most readers is Markowitz's

Nobel Prize–winning treatment of the problem of portfolio diversification using quadratic programming techniques. This treatment, which first appeared in 1952, underlies almost all of the subsequent research into the pricing of risk in financial markets. Linear programming, of course, has been applied in many financial planning problems, from the management of working capital to formulating a bid for the underwriting of a bond issue. Less well known is the fundamental role that duality theory plays in the pricing of options and contingent claims, both in its discrete state and time formulation using linear programming and in its continuous time counterparts. This duality leads directly to the Monte Carlo simulation method for pricing and evaluating the risk of options portfolios for investment banks; this activity probably comprises the single greatest use of computing resources in any industry.

This chapter does not cover every possible topic in the applications of operations research (OR) to finance. We have chosen to highlight the main topics in investment theory and to give an elementary, mostly self-contained, exposition of each. A comprehensive perspective of the application of OR techniques to financial markets along with an excellent bibliography of the recent literature in this area can be found in the survey by Board et al. (2003). In this chapter, we chose not to cover the more traditional applications of OR to financial management for firms, such as the management of working capital, capital investment, taxation, and financial planning. For these, we direct the reader to consult Ashford et al. (1988). We also excluded financial forecasting models; the reader may refer to Campbell et al. (1997) and Mills (1999) for treatments of these topics. Finally, Board et al. (2003) provide a survey of the application of OR techniques for the allocation of investment budgets between a set of projects. Complete and up-to-date coverage of finance and financial engineering topics for readers in operations research and management science may be found in the handbooks of Jarrow et al. (1995) and the forthcoming volume of Birge and Linetsky (2007).

We begin this chapter by introducing some basic concepts in investment theory. In Section 7.2, we present the formulas for computing the return and variance of return on a portfolio. The formulas for a portfolio's mean and variance presume that these parameters are known for the individual assets in the portfolio. In Section 7.3, we discuss two methods for estimating these parameters when they are not known.

Section 7.4 explains how a portfolio's overall risk can be reduced by including a diverse set of assets in the portfolio. In Section 7.5, we introduce the risk–reward tradeoff efficient frontier and the Markowitz problem. Up to this point, we have assumed that the investor is able to specify a mathematical function describing his attitude toward risk. In Section 7.6, we consider utility theory that does not require an explicit specification of a risk function. Instead, utility theory assumes that investors specify a utility, or satisfaction, with any cash payout. The associated optimal portfolio selection problem will seek to maximize the investor's expected utility.

Section 7.7 discusses the Black–Litterman model for asset allocation. Black and Litterman use Bayesian updating to combine historical asset returns with individual investor views to determine a posterior distribution on asset returns that is used to make asset allocation decisions. Section 7.8 considers the challenges of risk management. We introduce the notion of coherent risk measures and conditional value-at-risk ($CVaR$), and show how a portfolio selection problem with a constraint on $CVaR$ can be formulated as a stochastic program.

In Sections 7.9 through 7.13, we turn to the problem of options valuation. Options valuation combines a mathematical model for the behavior of the underlying uncertain market factors with simulation or dynamic programming (or combinations thereof) to determine options prices. Section 7.14 considers the problem of asset-liability matching in a multi-period setting. The solution uses stochastic optimization based upon Monte Carlo simulation. Finally, in Section 7.15 we present some concluding remarks.

7.2 Return

Suppose that an investor invests in asset i at time 0 and sells the asset at time t. The rate of return (more simply referred to as the return) on asset i over time period t is given by:

$$r_i = \frac{\text{amount received at time } t - \text{amount invested in asset } i \text{ at time } 0}{\text{amount invested in asset } i \text{ at time } 0} \tag{7.1}$$

Now suppose that an investor invests in a portfolio of N assets. Let f_i denote the fraction of the portfolio that is comprised of asset i. Assuming that no short selling is allowed, $f_i \geq 0$. Clearly, $\sum_{i=1}^{N} f_i = 1$.

The portfolio return is given by the weighted sum of the returns on the individual assets in the portfolio:

$$r_p = \sum_{i=1}^{N} f_i r_i \tag{7.2}$$

We have described asset returns as if they are known with certainty. However, there is typically uncertainty surrounding the amount that will be received at the time that an asset is sold. We can use a probability distribution to describe this unknown rate of return. If return is normally distributed, then only two parameters—its expected return and its standard deviation (or variance)—are needed to describe this distribution. The expected return is the return around which the probability distribution is centered; it is the expected value of the probability distribution of possible returns. The standard deviation describes the dispersion of the distribution of possible returns.

7.2.1 Expected Portfolio Return

Suppose there are N assets with random returns $r_1, \ldots, r_N$. The corresponding expected returns are $E(r_1), \ldots, E(r_N)$. An investor wishes to create a portfolio of these N assets, by investing a fraction f_i of his wealth in asset i.

Using the properties of expectation, we may compute the expected portfolio return using Equation 7.2:

$$E(r_p) = \sum_{i=1}^{N} f_i E(r_i)$$

That is, an expected portfolio return is equal to the weighted sum of the expected returns of its individual asset components.

7.2.2 Portfolio Variance

The volatility of an asset's return can be measured by its variance. Variance is often adopted as a measure of an asset's risk. If σ_i^2 denotes the variance of asset i's return, then the variance of the portfolio's return is given by:

$$\sigma_p^2 = E\left[\left(\sum_{i=1}^{N} f_i r_i - \sum_{i=1}^{N} f_i \bar{r}_i\right)^2\right]$$

$$= E\left[\left(\sum_{i=1}^{N} f_i(r_i - \bar{r}_i)\right)\left(\sum_{j=1}^{N} f_j(r_j - \bar{r}_j)\right)\right]$$

$$= E\left[\sum_{i,j=1}^{N} f_i f_j(r_i - \bar{r}_i)(r_j - \bar{r}_j)\right] \qquad (7.3)$$

$$= \sum_{i,j=1}^{N} f_i f_j \sigma_{ij}$$

$$= \sum_{i=1}^{N} f_i^2 \sigma_i^2 + \sum_{i=1}^{N} \sum_{\substack{j=1 \\ j \neq i}}^{N} f_i f_j \sigma_{ij}$$

where $\bar{r}_i = E(r_i)$. Note that portfolio variance is a combination of the variance of the returns of each individual asset in the portfolio plus their covariance.

7.3   Estimating an Asset's Mean and Variance

Of course, asset i's rate of return and variance are not known and must be estimated. These values can be estimated based upon historical data using standard statistical methods. Alternatively, one can use a scenario-based approach. We describe the two methods below.

7.3.1   Estimating Statistics Using Historical Data

To estimate statistics using historical data, one must collect several periods of historical returns on the assets. The estimated average return on asset i is computed as the sample average of returns on asset i, $\overline{X}_i$, given by:

$$\overline{X}_i = \frac{\sum_t x_{it}}{T}$$

where x_{it} is the historical return on asset i in period t and there are T periods of historical data.

The variance of return on asset i is estimated by s_i^2, the historical sample variance of returns on investment i:

$$s_i^2 = \frac{\sum_t (x_{it} - \overline{X}_i)^2}{T - 1}$$

For example, Table 7.1 contains the monthly closing stock prices and monthly returns for Sun Microsystems and Continental Airlines for the months January 2004 through February 2006. The first column of this table indicates the month, the second and third columns contain the closing stock prices for SUN and Continental Airlines, respectively. The fourth and fifth columns contain the monthly stock returns for SUN ($X_{SUN,t}$) and Continental

TABLE 7.1 Monthly Closing Stock Prices and Returns for Sun Microsystems and Continental Airlines

Month	SUN Stock Price ($)	CAL Stock Price ($)	SUN Return (%)	CAL Return (%)
Jan-04	55.55	15.65	8.12	−5.04
Feb-04	61.66	14.97	11.00	−4.35
Mar-04	62.38	12.58	1.17	−15.97
Apr-04	63.00	10.66	0.99	−15.26
May-04	62.04	10.50	−1.52	−1.50
Jun-04	63.63	11.37	2.56	8.29
Jul-04	68.17	8.75	7.13	−23.04
Aug-04	61.60	9.65	−9.64	10.29
Sep-04	73.98	8.60	20.10	−10.88
Oct-04	74.76	9.24	1.05	7.44
Nov-04	82.45	11.16	10.29	20.78
Dec-04	81.71	13.75	−0.90	23.21
Jan-05	88.04	10.50	7.75	−23.64
Feb-05	98.80	11.14	12.22	6.10
Mar-05	105.00	12.00	6.28	7.72
Apr-05	98.74	12.00	−5.96	0.00
May-05	103.00	13.70	4.31	14.17
Jun-05	114.14	13.32	10.82	−2.77
Jul-05	126.33	15.81	10.68	18.69
Aug-05	75.35	13.28	−40.35	−16.00
Sep-05	78.20	9.70	3.78	−26.96
Oct-05	73.76	13.20	−5.68	36.08
Nov-05	78.10	15.59	5.88	18.11
Dec-05	79.34	21.27	1.59	36.43
Jan-06	95.20	20.50	19.99	−3.62
Feb-06	74.35	23.79	−21.90	16.05

TABLE 7.2 Expected Historical Monthly Returns, Variances, and Standard Deviations of Returns

	Expected Monthly Return	Variance of Return	Standard Deviation
SUN	2.30%	1.54	12.40%
Continental Airlines	2.86%	3.04	17.43%

Airlines $(X_{CAL,t})$, respectively; these columns were populated using Equation 7.1. In this example, $T = 26$.

Table 7.2 shows the mean and standard deviation of returns for these two stocks, based upon the 26 months of historical data. The average monthly return for SUN, $\overline{X}_{SUN}$, and the average monthly return for Continental, $\overline{X}_{CAL}$, is computed as the arithmetic average of the monthly returns in the fourth and fifth columns, respectively. An estimate of the variance of monthly return on SUN's (Continental's) stock is computed as the variance of the returns in the fourth (fifth) column of Table 7.1. If variance is used as a measure of risk, then Continental is a riskier investment as it has a higher volatility (its variance is higher).

7.3.2 The Scenario Approach to Estimating Statistics

Sometimes, historical market conditions are not considered a good predictor of future market conditions. In this case, historical data may not be a good source for estimating expected returns or risk. When historical estimates are determined to be poor predictors of the future, one can consider a scenario approach.

The scenario approach proceeds as follows:

Define a set of S future economic scenarios and assign likelihood $p(s)$ that scenario s will occur. $\sum_{s \in S} p(s) = 1$, since in the future the economy must be in exactly one of these

TABLE 7.3   Definition of Future Possible Scenarios for States of the Economy

Scenario (s)	Likelihood $(p(s))$	Return $(r_i(s))$	$p(s)^* r_i(s)$	$p(s)^* (r_i(s) - r_i)^2$
Weak economy	0.30	-15.00%	-4.50%	1.03
Stable economy	0.45	9.00%	4.05%	0.13
Strong economy	0.25	16.00%	4.00%	0.39
			3.55%	1.55

economic conditions. Next, define each asset's behavior (its return) under each of the defined economic scenarios. Asset i's expected return is computed as:

$$r_i = \sum_s p(s)r_i(s) \tag{7.4}$$

where $r_i(s)$ is asset i's return under scenario s.

Similarly, we compute the variance of return on asset i as:

$$v_i = \sum_s p(s)(r_i(s) - r_i)^2 \tag{7.5}$$

For example, suppose we use the scenario approach to predict expected monthly return on SUN stock. We have determined that the economy may be in one of three states: weak, stable, or strong, with a likelihood of 0.3, 0.45, and 0.25, respectively. Table 7.3 indicates the forecasted monthly stock returns under each of these future economic conditions.

The first and second columns in Table 7.3 indicate the economic scenario and likelihood that the scenario will occur, respectively. The third column contains the expected return under each of the defined future scenarios. The fourth and fifth columns contain intermediate computations needed to calculate the expected return and standard deviation of returns on SUN stock, based upon Equations 7.3 and 7.4. Using these equations we find that the expected monthly return on SUN stock is 3.55%, variance of monthly return is 1.55, with corresponding standard deviation of 12.46%.

Comparing the estimates of mean and standard deviation of SUN's monthly return using the historical data approach versus the scenario-based approach we find that while the estimates of volatility of return are close in value, the estimates of monthly return differ significantly (2.3% versus 3.55%). In Section 7.7, we discuss the negative impact that can result from portfolio allocation based upon incorrect parameter estimation. Thus, care must be taken to determine the correct method and assumptions when estimating these values.

7.4   Diversification

We now explore how a portfolio's risk, as measured by the variance of the portfolio return, can be reduced when stocks are added to the portfolio. This phenomenon, whereby a portfolio's risk is reduced when assets are added to the portfolio, is known as diversification.

Portfolio return is the weighted average of the returns of the assets in the portfolio, weighted by their appearance in the portfolio. However, portfolio variance (as derived in Equation 7.3) is given by:

$$\sum_{i=1}^{N} f_i^2 \sigma_i^2 + \sum_{i=1}^{N} \sum_{\substack{j=1 \\ j \neq i}}^{N} f_i f_j \sigma_{ij}$$

Namely, portfolio variance is comprised of two components. One component is the variances of the individual assets in the portfolio and the other component is the covariance between the returns on the different assets in the portfolio. Covariance of return between asset i and j is the expected value of the product of the deviations of each of the assets from their respective means. If the two assets deviate from their respective means in identical fashions (i.e., both are above their means or below their means at the same time) then the covariance is highly positive, and its contribution to portfolio variance is highly positive. If the return on one asset deviates below its mean at the time that the return on the other asset deviates above its mean, then the covariance is highly negative, which reduces the overall portfolio variance.

Let us explore the impact of the covariance term on overall portfolio variance. If all of the assets are independent, then the covariance terms equal zero and only the variance of the individual assets in the portfolio contribute to overall portfolio variance. In this case, the variance formula is:

$$\sum_{i=1}^{N} f_i^2 \sigma_i^2$$

If the investor invests an equal amount in each of the N independent assets, then the portfolio variance is:

$$\sum_{i=1}^{N} \left(\frac{1}{N}\right)^2 \sigma_i^2 = \frac{1}{N}\sum_{i=1}^{N}\frac{1}{N}\sigma_i^2 = \frac{1}{N}\overline{\sigma}_i^2$$

where $\overline{\sigma}_i^2$ is the average variance of the assets in the portfolio. As N gets larger, the portfolio variance goes to zero. Thus, for a portfolio comprised only of independent assets, when the number of assets in the portfolio is large enough the variance of the portfolio return is zero.

Now consider N assets that are not independent. Without loss of generality, assume the assets appear in the portfolio with equal weight. Then, variance of portfolio return is:

$$\frac{1}{N}\overline{\sigma}_i^2 + \sum_{i=1}^{N}\sum_{\substack{j=1\\j\neq i}}^{N} \left(\frac{1}{N}\right)^2 \sigma_{ij}$$

$$= \frac{1}{N}\overline{\sigma}_i^2 + \frac{N-1}{N}\sum_{i=1}^{N}\sum_{\substack{j=1\\j\neq i}}^{N} \frac{\sigma_{ij}}{N(N-1)} \tag{7.6}$$

$$= \frac{1}{N}\overline{\sigma}_i^2 + \frac{N-1}{N}\overline{\sigma}_{ij}$$

The final equality in Equation 7.6 reveals that as the number of assets in the portfolio increases, the contribution of the first component (variance) becomes negligible—it is diversified away—and the second term (covariance) approaches average covariance. Thus, while increasing the number of assets in a portfolio will diversify away the individual risk of the assets, the risk attributed to the covariance terms cannot be diversified away.

As a numerical example, consider the SUN and Continental Airlines monthly stock returns from Section 7.3.1. Average monthly return on Sun Microsystems (Continental Airlines) stock from January 2004 through February 2006 was 2.3% (2.86%). A portfolio comprised of equal investments in Sun Microsystems and Continental Airlines yields an average monthly return of 2.58%. The variance of the monthly returns on Sun Microsystems stock over the 26 months considered is 1.54. The variance of the monthly returns on Continental Airlines stock over that same time period is 3.04. However, for the portfolio consisting of equal investments in Sun Microsystems and Continental Airlines, the variance of portfolio returns is 1.09. This represents a significant reduction in risk from the risk associated with either of the stocks alone. The explanation lies in the value of the covariance between the monthly returns on these two stocks; the value of the covariance is -0.12. Because the returns on the stocks have a negative covariance, diversification reduces the portfolio risk.

7.5   Efficient Frontier

The discussion in Section 7.4 illustrates the potential benefits of combining assets in a portfolio. For a *risk averse* investor, diversification provides the opportunity to reduce portfolio risk while maintaining a minimum level of return. An investor can consider different combinations of assets, each of which has an associated risk and return.

This naturally leads one to question whether there is an *optimal* way to combine assets. We address this question within the context of assuming that: (i) for a fixed return investors prefer the lowest possible risk, that is, investors are risk averse, (ii) for a given level of risk, investors always prefer the highest possible return, (this property is referred to as nonsatiation), and (iii) the first two moments of the distribution of an asset's return are sufficient to describe the asset's character; there is no need to gather information about higher moments such as skew.

Given these assumptions, the risk-return trade-off of portfolios of assets can be graphically displayed by constructing a plot with risk (as measured by standard deviation) on the horizontal axis and return on the vertical axis. We can plot every possible portfolio on this risk–return space. The set of all portfolios plotted form a *feasible region*.

By the risk averse assumption, for a given level of return investors prefer portfolios that lie as far to the left as possible, as these have the lowest risk. Similarly, by the nonsatiation property for a given level of risk investors prefer portfolios that lie higher on the graph as these yield a greater return. The upper left perimeter of the feasible region is called the *efficient frontier*. It represents the least risk combination for a given level of return. The efficient frontier is concave.

Figure 7.1 shows a mean-standard deviation plot for 10,000 random portfolios created from the 30 stocks in the Dow from 1986 through 1991. It represents the feasible region of portfolios. Figure 7.2 is the efficient frontier for the stocks from the Dow Industrials from 1985 through 1991. (Both figures were taken from the NEOS Server for Optimization website.)

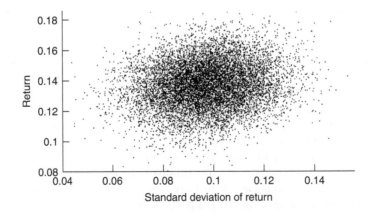

FIGURE 7.1   Return vs. standard deviation for 10,000 random portfolios from the Dow Industrials.

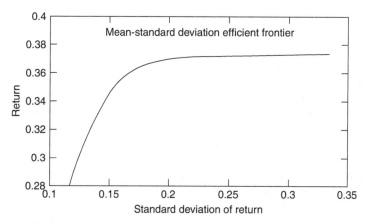

FIGURE 7.2 Efficient frontier for Dow Industrials from 1985–1991.

7.5.1 Risk-Reward Portfolio Optimization

Markowitz (1952) developed a single period portfolio choice problem where the objective is to minimize portfolio risk (variance) for a specified return on the portfolio. For this model, it is assumed that all relevant information required by investors to make portfolio decisions is captured in the mean, standard deviation, and correlation of assets. This method for portfolio selection is often referred to as mean-variance optimization as it trades off an investor's desire for higher mean return against an aversion to greater risk as measured by portfolio variance.

The Markowitz model for portfolio optimization is given by:

$$\text{Minimize} \quad \sum_{i,j} f_i f_j \sigma_{ij}$$

$$\text{Subject to:} \quad \sum_{i=1}^{N} f_i r_i = \underline{R} \tag{7.7}$$

$$\sum_{i=1}^{N} f_i = 1$$

The objective is to minimize variance subject to two constraints: (i) portfolio return must equal the targeted return $\underline{R}$ and (ii) total allocation must equal 1. Negative values for f_i correspond to short selling.

One method that can be employed to solve this constrained optimization problem is to form an auxiliary function L called the Lagrangian by (i) rearranging each constraint so that the right hand side equals to zero and (ii) introducing a single Lagrange multiplier for each constraint in the problem as follows:

$$L = \sum_{i,j} f_i f_j \sigma_{ij} - \lambda \left(\sum_{i=1}^{N} f_i r_i - \underline{R} \right) - \mu \left(\sum_{i=1}^{N} f_i - 1 \right) \tag{7.8}$$

We now treat the Lagrangian Equation 7.8 as an unconstrained minimization problem. A necessary condition for a point to be optimal is that the partial derivative of the Lagrangian with respect to each of the variables must equal zero. Thus, we take the partial derivative of L with respect to each of the i asset weights f_i, μ, and λ and set each partial derivative equal to zero. (Notice that the partial derivative of L with respect to λ yields the portfolio return constraint, and the partial derivative of L with respect to μ yields the constraint on the asset weights.) The result is a system of $i + 2$ constraints. We use this set of constraints to solve for the $i + 2$ unknowns: f_i, μ, and λ.

The Lagrangian method can often only be successfully implemented for small problems. For larger problems with many variables it may be virtually impossible to solve the set of equations for the $i + 2$ unknowns. Instead, one can find the solution to Problem 7.7 using methods developed for optimizing quadratic programs. A *quadratic program* is a mathematical optimization model where the objective function is quadratic and all constraints are linear equalities or inequalities. The constraints of the optimization problem define the *feasible region* within which the optimal solution must lie. Quadratic programs are, in general, difficult to solve. Quadratic programming solution methods work in two phases. In the first phase a feasible solution is found. In the second phase, the method searches along the edges and surfaces of the feasible region for another solution that improves upon the current feasible solution. Unless the objective function is convex, the method will often identify a local optimal solution.

The solution to the Markowitz problem yields a point that lies on the efficient frontier. By varying the value of $\underline{R}$ in Problem 7.7, one can map out the entire efficient frontier.

7.6   Utility Analysis

The solution to the Markowitz problem provides one means for making investment decisions in the mean-variance space. It requires the investor to define a measure of risk and a measure of value and then utilizes an explicitly defined trade-off between these two measures to determine the investor's preferences. Utility theory provides an alternative way to establish an investor's preferences without explicitly defining risk functions.

Utility describes an investor's attitude toward risk by translating the investor's satisfaction associated with different cash payouts into a utility value. The application of utility to uncertain financial situations was first introduced by von Neumann and Morgenstern (1944). Utility functions can be used to explain how investors make choices between different portfolios.

Utility theory is often introduced by way of the concept of *certainty equivalent*. Certainty equivalent is the amount of wealth that is equally preferred to an uncertain alternative. Risk averse investors will prefer a lower certain cash payout to a higher risky cash payout. That is, their certainty equivalent is lower than the expected value of uncertain alternatives. Risk seeking individuals have a certainty equivalent that is higher than the expected value of the uncertain alternatives. A utility function assigns different weights to different outcomes according to the risk profile of the individual investor. The shape of a utility function is defined by the risk profile of the investor.

Each individual investor may have a different utility function, as each investor may have a different attitude toward risk. However, all utility functions U satisfy the following:

- *Nonsatiation*: Utility functions must be consistent with more being preferred to less. If x and y are two cash payouts and $x > y$, then $U(x) > U(y)$. This property is equivalent to stating that the first derivative of the utility function, with respect to cash payout, is positive.

- *Risk preference*: Economic theory presumes that an investor will seek to maximize the utility of his investment. However, all investors will not make identical investment decisions because they will not all share the same attitudes toward risk. Investors can be classified into three classes, according to their willingness to accept risk: risk averse, risk neutral, or risk taking. Risk-averse investors invest in investments where the utility of the return exceeds the risk-free rate. If no such investment exists, the investor invests in risk-free investments. The utility function of a risk-averse investor is increasing and concave in the cash payout $(U''(x) < 0)$; the value assigned to each additional dollar received decreases due to the risk-averse nature of the individual. Risk-neutral investors ignore risk when making investment decisions. They seek investments with a maximum return, irrespective of the risk involved. The utility function of a risk-neutral individual is linear increasing in the cash payout; the same value is assigned to each additional dollar. Risk-taking investors are more likely to invest in investments with a higher risk involved. The utility function of a risk-taking individual is convex $(U''(x) > 0)$.

- *Changes in wealth*: Utility functions also define how an investor's investment decisions will be affected by changes in his wealth. Specifically, if an investor has a larger amount of capital will this change his willingness to accept risk? The Arrow–Pratt absolute risk aversion coefficient, given by

$$A(x) = -\frac{U''(x)}{U'(x)}$$

is a measure of an investor's absolute risk aversion. $A'(x)$ measures changes in an investor's absolute risk aversion as a function of changes in his wealth. If $A'(x) > 0$ $(A'(x) < 0)$, the investor has increasing (decreasing) absolute risk aversion and he will hold fewer (more) dollars in risky assets as his wealth increases. If $A'(x) = 0$, the investor has constant absolute risk aversion and he will hold the same dollar amount of risky assets as his wealth increases.

A utility function only *ranks* alternatives according to risk preferences; its numerical value has no real meaning. Thus, utility functions are unique up to a positive affine transformation. More specifically, if $U(x)$ is a utility function then $V(x) = a + bU(x)$ $(b > 0)$ will yield the same rankings (and hence the same investment decisions) as $U(x)$.

7.6.1 Utility Functions

Although each investor can define his own utility function, there are a number of predefined utility functions that are commonly used in the finance and economics literature. In this section, we describe the exponential utility function. We also reconcile between mean-variance optimization and utility optimization.

The exponential utility function is often adopted for financial analysis. The exponential utility function is defined as:

$$U(x) = 1 - e^{-x/R}, \quad \text{for all } x \tag{7.9}$$

where x is the wealth and $R > 0$ represents the investor's risk tolerance. Greater values of R mean that the investor is less risk averse. (In fact, as $R \to \infty$ the investor becomes risk neutral.) $U'(x) = (1/R)e^{-x/R}$ and $U''(x) = -(1/R)^2 e^{-x/R}$. The second derivative of Equation 7.9 is strictly negative so the exponential utility function is concave; the exponential utility function describes the behavior of a risk-averse investor. The absolute risk aversion coefficient is $(1/R)$, which is constant with wealth; the investor invests constant dollar amounts in risky investments as his wealth increases.

We now demonstrate the relationship between utility and mean-variance for concave (i.e., risk-averse) investors, following King (1993). Suppose $U(X)$ is a strictly concave utility function. The Taylor expansion of $U(X)$ is an approximation of the function at a particular point, say M, using its derivatives. The second-order Taylor expansion of $U(X)$ about point M is given by:

$$U(X) = U(M) + U'(M)(X - M) + \frac{1}{2}U''(M)(X - M)^2 \qquad (7.10)$$

Now suppose that the point M is equal to expected wealth, that is, $E(X) = M$. Then the expected value of the second-order Taylor expansion expression (Equation 7.10) is equal to

$$E[U(X)] = U(M) + \frac{1}{2}U''(M)E(X - M)^2$$

$$= U(M) + \frac{1}{2}U''(M)\sigma^2$$

(The middle term drops out as $E(X) = M$.) The second derivative is negative for a strictly concave utility function. This implies that maximizing the second-order Taylor expansion is equivalent to

$$\min \quad E(X - M)^2 = \sigma^2$$

for all X with $E(X) = M$, which is a mean-variance problem with a given mean return M.

It follows that mean-variance is a second-order approximation to utility maximization. Of course, due to the two-sided nature of the variance, eventually this approximation will become negatively-sloped—and hence not really valid as a utility function—as X increases in value. The range over which the approximation is valid can be pretty wide. The upper bound of the range is the point where $U'(X) + U''(X)(X - M) = 0$; which is the maximum point of the approximating quadratic. For the logarithmic utility $U(X) = \log(X)$, the upper bound of the range where the mean-variance approximation remains quadratic is $X = 2M$. In other words, the mean-variance approximation for the logarithm is valid for a range that includes twice the mean value of the return!

7.6.2   Utility in Practice

In practice, utility is not often used as an objective criterion for investment decision making because utility curves are difficult to estimate. However, Holmer (1998) reports that Fannie Mae uses expected utility to optimize its portfolio of assets and liabilities. Fannie Mae faces a somewhat unique set of risks due to the specific nature of its business. Fannie Mae buys mortgages on the secondary market, pools them, and then sells them on the open market to investors as mortgage backed securities. Fannie Mae faces many risks such as: prepayment risk, risks due to potential gaps between interest due and interest owed, and long-term asset and liability risks due to interest rate movements. Utility maximization allows Fannie Mae to explicitly consider degrees of risk aversion against expected return to determine its risk-adjusted optimal investment portfolio.

7.7 Black–Litterman Asset Allocation Model

Markowitz mean-variance portfolio optimization requires mean and covariance as input and outputs optimal portfolio weights. The method has been criticized because:

  i. The optimal portfolio weights are highly dependent upon the input values. However, it is difficult to accurately estimate these input values. Chopra and Ziemba (1993), Kallberg and Ziemba (1981, 1984), and Michaud (1989) use simulation to demonstrate the significant cash-equivalent losses due to incorrect estimates of the mean. Bengtsson (2004) showed that incorrect estimates of variance and covariance also have a significant negative impact on cash returns.

 ii. Markowitz mean-variance optimization requires the investor to specify the universe of return values. It is unreasonable to expect an investor to know the universe of returns. On the other hand, mean-variance optimization is sensitive to the input values so incorrect estimation can significantly skew the results.

iii. Black and Litterman (1992) and He and Litterman (1999) have studied the optimal Markowitz model portfolio weights assuming different methods for estimating the assets' means and found that the resulting portfolio weights were unnatural. Unconstrained mean-variance optimization typically yields an optimal portfolio that takes many large long and short positions. Constrained mean-variance optimization often results in an extreme portfolio that is highly concentrated in a small number of assets. Neither of these portfolio profiles is typically considered acceptable to investors.

 iv. Due to the intricate interaction between mean and variance, the optimal weights determined by Markowitz's mean-variance estimation are often non-intuitive. A small change in an estimated mean of a single asset can drastically change the weights of many assets in the optimal portfolio.

Black and Litterman observed the potential benefit of using mathematical optimization for portfolio decision making, yet understood an investment manager's hesitations in implementing Markowitz's mean-variance optimization model. Black and Litterman (1992) developed a Bayesian method for combining individual investor subjective views on asset performance with market equilibrium returns to create a mixed estimate of expected returns. The Bayes approach works by combining a prior belief with additional information to create an updated "posterior" distribution of expected asset returns. In the Black–Litterman framework the equilibrium returns are the prior and investor subjective views are the additional information. Together, these form a posterior distribution on expected asset returns. These expected returns can then be used to make asset allocation decisions. If the investor has no subjective views on asset performance, then the optimal allocation decision is determined solely according to the market equilibrium returns. Only if the investor expresses opinions on specific assets will the weights for those assets shift away from the market equilibrium weights in the direction of the investor's beliefs.

The Black–Litterman model is based on taking a market equilibrium perspective on asset returns. Asset "prior" returns are derived from the market capitalization weights of the optimal holdings of a mean-variance investor, given historical variance. Then, if the investor has specific views on the performance of any assets, the model combines the equilibrium returns with these views, taking into consideration the level of confidence that the investor associates with each of the views. The model then yields a set of updated expected asset returns as well as updated optimal portfolio weights, updated according to the views expressed by the investor.

The key inputs to the Black–Litterman model are market equilibrium returns and the investor views. We now consider these inputs in more detail.

7.7.1 Market Equilibrium Returns

Black and Litterman use the market equilibrium expected returns, or capital asset pricing model (CAPM) returns, as a neutral starting point in their model. (See, e.g., Sharpe, 1964.) The basic assumptions are that (i) security markets are frictionless, (ii) investors have full information relevant to security prices, and (iii) all investors process the information as if they were mean-variance investors. The starting point for the development of the CAPM is to form the efficient frontier for the market portfolios and to draw the capital market line (CML). The CML begins at the risk-free rate on the vertical axis (which has risk 0) and is exactly tangent to the efficient frontier. The point where the CML touches the efficient frontier is the pair (σ_m, r_m), which is defined to be the "market" standard deviation and "market" expected return. By changing the relative proportions of riskless asset and market portfolio, an investor can obtain any combination of risk and return that lies on the CML. Because the CML is tangent to the efficient frontier at the market point, there are no other combinations of risky and riskless assets that can provide better expected returns for a given level of risk. Now, consider a particular investor portfolio i with expected return $E(r_i)$ and standard deviation σ_i. For an investor to choose to hold this portfolio it must have returns comparable to the returns that lie on the CML. Thus, the following must hold:

$$E(r_i) = r_f + \frac{\sigma_i}{\sigma_m}(r_m - r_f)$$

where r_f is the risk-free rate. This equation is called the CAPM.

The interpretation of the CAPM is that investors' portfolios have an expected return that includes a reward for taking on risk. This reward, by the CAPM hypothesis, must be equal to the return that would be obtained from holding a portfolio on the CML that has an equivalent risk. Any remaining risk in the portfolio can be diversified away (by, for example, holding the market portfolio) so the investor does not gain any reward for the non-systematic, or diversifiable, risk of the portfolio. The CAPM argument applied to individual securities implies that the holders of individual securities will be compensated only for that part of the risk that is correlated with the market, or the so-called systematic risk. For an individual security j the CAPM relationship is

$$E(r_j) = r_f + \beta_j(r_m - r_f)$$

where $\beta_j = (\sigma_j/\sigma_m)\rho_{mj}$ and ρ_{mj} is the correlation between asset j returns and the market returns.

The Black–Litterman approach uses the CAPM in reverse. It assumes that in equilibrium the market portfolio is held by mean-variance investors and it uses optimization to back out the expected returns that such investors would require given their observed holdings of risky assets. Let N denote the number of assets and let the excess equilibrium market returns (above the risk-free rate) be defined by:

$$\prod = \lambda \sum w \tag{7.11}$$

where
$\prod = N \times 1$ vector of implied excess returns
$\sum = N \times N$ covariance matrix of returns

$w = N \times 1$ vector of market capitalization weights of the assets

$\lambda =$ risk aversion coefficient that characterizes the expected risk-reward tradeoff.

λ is the price of risk as it measures how risk and reward can be traded off when making portfolio choices. It measures the rate at which an investor will forego expected return for less variance. λ is calculated as $\lambda = (r_m - r_f)/\sigma_m^2$, where σ_m^2 is the variance of the market return. The elements of the covariance matrix are computed using historical correlations and standard deviations. Market capitalization weights are determined by measuring the dollar value of the global holdings of all equity investors in the large public stock exchanges. The capitalization weight of a single equity name is the dollar-weighted market-closing value of its equity share times the outstanding shares issued. Later we will show that Equation 7.11 is used to determine the optimal portfolio weights in the Black–Litterman model.

7.7.2 Investor Views

The second key input to the Black–Litterman model is individual investor views.

Assume that an investor has K views, denoted by a $K \times 1$ vector Q. Uncertainty regarding these views is denoted by an error term ε, where ε is normally distributed with mean zero and $K \times K$ covariance matrix Ω. Thus, a view has the form:

$$Q + \varepsilon = \begin{bmatrix} Q_1 \\ \vdots \\ Q_K \end{bmatrix} + \begin{bmatrix} \varepsilon_1 \\ \vdots \\ \varepsilon_K \end{bmatrix}$$

$\varepsilon = 0$ means that the investor is 100% confident about his view; in the more likely case that the investor is uncertain about his view, ε takes on some positive or negative value.

ω denotes the variance of each error term. We assume that the error terms are independent of each other. (This assumption can be relaxed.) Thus, the covariance matrix Ω is a diagonal matrix where the elements on the diagonal are ω, the variances of each error term. A higher variance indicates greater investor uncertainty with the associated view. The error terms ε do not enter directly into the Black–Litterman formula; only their variances enter via the covariance matrix Ω.

$$\Omega = \begin{bmatrix} \omega_1 & 0 & \cdots & & 0 \\ 0 & \ddots & & & \vdots \\ \vdots & & & & 0 \\ 0 & \cdots & & 0 & \omega_K \end{bmatrix}$$

The Black–Litterman model allows investors to express views such as:

View 1: Asset A will have an excess return of 5.5% with 40% confidence.

View 2: Asset B will outperform asset C by 3% with 15% confidence.

View 3: Asset D will outperform assets E and F by 1% with 20% confidence.

The first view is called an *absolute* view, while the second and third views are called *relative* views. Notice that the investor assigns a level of confidence to each view.

Each view can be seen as a portfolio of long and short positions. If the view is an absolute view then the portfolio position will be long. If the view is a relative view, then the portfolio will take a long position in the asset that is expected to "overperform" and a short position in the asset that is expected to "underperform." In general, the impact on the optimal portfolio weights is determined by comparing the equilibrium difference in the performance of these assets to the performance expressed by the investor view. If the performance expressed in the view is better than the equilibrium performance, the model will tilt the portfolio toward the outperforming asset. More specifically, consider View 2, which states that asset B will outperform asset C by 3%. If the equilibrium returns indicate that asset B will outperform asset C by more than 3% then the view represents a weakening view in the performance of asset B and the model will tilt the portfolio away from asset B.

One of the most challenging questions in applying the Black–Litterman model is how to populate the covariance matrix Ω and how to translate the user specified expressions of confidence into uncertainty in the views. We will discuss this further below.

7.7.3   An Example of an Investor View

We now illustrate how one builds the inputs for the Black–Litterman model, given the three views expressed. Suppose that there are $N = 7$ assets: Assets A $-$ G. The Q matrix is given by:

$$Q = \begin{bmatrix} 5.5 \\ 3 \\ 1 \end{bmatrix}$$

Note that the investor only has views on six of the seven assets. We use the matrix P to match the views to the individual assets. Each view results in a $1 \times N$ vector so that P is a $K \times N$ matrix. In our case, where there are seven assets and three views, P is a 3×7 matrix. Each column corresponds to one of the assets; column 1 corresponds to Asset A, column 2 corresponds to Asset B, and so on. In the case of absolute views, the sum of the elements in the row equals 1. In our case, View 1 yields the vector:

$$P_1 = \begin{bmatrix} 1 & 0 & 0 & 0 & 0 & 0 & 0 \end{bmatrix}$$

In the case of relative views, the sum of the elements equals zero. Elements corresponding to relatively outperforming assets have positive values; elements corresponding to relatively underperforming assets take negative values. We determine the values of the individual elements by dividing 1 by the number of outperforming and underperforming assets, respectively. For View 2, we have one outperforming asset and one underperforming asset. Thus, View 2 yields the vector:

$$P_2 = \begin{bmatrix} 0 & 1 & -1 & 0 & 0 & 0 & 0 \end{bmatrix}$$

View 3 has one outperforming asset (Asset D) and two relatively underperforming assets (Assets E and F). Thus, Asset D is assigned a value of $+1$ and Assets E and F are assigned values of -0.5 each. View 3 yields the vector:

$$P_3 = \begin{bmatrix} 0 & 0 & 0 & 1 & -0.5 & -0.5 & 0 \end{bmatrix}$$

Matrix P is given by:

$$\begin{bmatrix} 1 & 0 & 0 & 0 & 0 & 0 & 0 \\ 0 & 1 & -1 & 0 & 0 & 0 & 0 \\ 0 & 0 & 0 & 1 & -0.5 & -0.5 & 0 \end{bmatrix}$$

The variance of the kth view portfolio can be calculated according to the formula $p_k \sum p'_k$, where p_k is the kth row of the P matrix and $\sum$ is the covariance matrix of the excess equilibrium market returns. (Recall, these form the neutral starting point of the Black–Litterman model.) The variance of each view portfolio is an important source of information regarding the confidence that should be placed in the investor's view k.

7.7.4 Combining Equilibrium Returns with Investor Views

We now state the Black–Litterman equation for combining equilibrium returns with investor views to determine a vector of expected asset returns that will be used to determine optimal portfolio weights. The vector of combined asset returns is given by:

$$E[R] = \left[\left(\tau \sum \right)^{-1} + P'\Omega^{-1}P \right]^{-1} \left[\left(\tau \sum \right)^{-1} \prod + P'\Omega^{-1}Q \right] \tag{7.12}$$

where:

$E[R] = N \times 1$ vector of combined returns

$\quad \tau =$ scalar, indicating uncertainty of the CAPM prior

$\sum = N \times N$ covariance matrix of equilibrium excess returns

$\quad \mathrm{P} = K \times N$ matrix of investor views

$\quad \Omega = K \times K$ diagonal covariance matrix of view error terms (uncorrelated view uncertainty)

$\quad \Pi = N \times 1$ vector of equilibrium excess returns

$\quad Q = K \times 1$ vector of investor views.

Examining this formula, we have yet to describe how the value of τ should be set and how the matrix Ω should be populated. Recall that if an investor has no views, the Black–Litterman model suggests that the investor does not deviate from the market equilibrium portfolio. Only weights on assets for which the investor has views should change from their market equilibrium weights. The amount of change depends upon τ, the investor's confidence in the CAPM prior, and ω, the uncertainty in the views expressed.

The literature does not have a single view regarding how the value of τ should be set. Black and Litterman (1992) suggest a value close to zero. He and Litterman (1999) set τ equal to 0.025 and populate the covariance matrix Ω so that

$$\Omega = \begin{bmatrix} \left(p_1 \sum p_1' \right)\tau & 0 & \cdots & & 0 \\ 0 & & \ddots & & \vdots \\ \vdots & & & & \\ & & & & 0 \\ 0 & & \cdots & 0 & \left(p_K \sum p_K' \right)\tau \end{bmatrix}$$

We note that the implied assumption is that the variance of the view portfolio is the information that determines an investor's confidence in his view. There may be other information that contributes to the level of an investor's confidence but it is not accounted for in this method for populating Ω.

Formula 7.12 uses Bayes approach to yield posterior estimates of asset returns that reflect a combination of the market equilibrium returns and the investor views. These updated returns are now used to compute updated optimal portfolio weights.

In the case that the investor is unconstrained, we use Formula 7.11. Using Formula 7.11 w^*, the optimal portfolio weights, are given by:

$$w^* = \left(\lambda \sum \right)^{-1} \mu \tag{7.13}$$

where μ is the vector of combined returns given by Equation 7.12. Equation 7.13 is the solution to the unconstrained maximization problem $\max_w w'\mu - \lambda w' \sum w/2$.

In the presence of constraints (e.g., risk, short selling, etc.) Black and Litterman suggest that the vector of combined returns be input into a mean-variance portfolio optimization problem.

We note two additional comments on the updated weights w^*:

i. Not all view portfolios necessarily have equal impact on the optimal portfolio weights derived using the Black–Litterman model. A view with a higher level of uncertainty is given less weight. Similarly, a view portfolio that has a covariance with the market equilibrium portfolio is given less weight. This is because such a view represents less new information and hence should have a smaller impact in moving the optimal portfolio weights away from the market equilibrium weights. Finally, following the same reasoning, a view portfolio that has a covariance with another view portfolio has less weight.

ii. A single view causes all returns to change, because all returns are linked via the covariance matrix $\sum$. However, only the weights for assets for which views were expressed change from their original market capitalization weights. Thus, the Black–Litterman model yields a portfolio that is intuitively understandable to the investor. The optimal portfolio represents a combination of the market portfolio and a weighted sum of the view portfolios expressed by the investor.

7.7.5 Application of the Black–Litterman Model

The Black–Litterman model was developed at Goldman Sachs in the early 1990s and is used by the Quantitative Strategies group at Goldman Sachs Asset Management. This group develops quantitative models to manage portfolios. The group creates views and then uses the Black–Litterman approach to transform these views into expected asset returns. These expected returns are used to make optimal asset allocation decisions for all of the different portfolios managed by the group. Different objectives or requirements (such as liquidity requirements, risk aversion, etc.) are incorporated via constraints on the portfolio. The Black–Litterman model has gained widespread use in other financial institutions.

7.8 Risk Management

Risk exists when more than one outcome is possible from the investment. Sources of risk may include business risk, market risk, liquidity risk, and the like. Variance or standard deviation of return is often used as a measure of the risk associated with an asset's return. If variance is small, there is little chance that the asset return will differ from what is expected; if variance is large then the asset returns will be highly variable.

Financial institutions manage their risk on a regular basis both to meet regulatory requirements as well as for internal performance measurement purposes. However, while variance is a traditional measure of risk in economics and finance, in practice it is typically not the risk measure of choice. Variance assumes symmetric deviations above and below expected return. In practice, one does not observe deviations below expected return as often as deviations above expected return due to positions in options and options-like instruments in portfolios. Moreover, variance assigns equal penalty to deviations above and below the mean return. However, investors typically are not averse to receiving higher than anticipated returns. Investors are more interested in shortfall risk measures. These are risk measures that measure either the distance of return below a target or measure the likelihood that return will fall below a threshold.

One measure of shortfall risk is downside risk. Downside risk measures the expected amount by which the return falls short of a target. Specifically, if z is the realized return and X is the target then downside risk is given by $E[\max(X - z, 0)]$. Semivariance is another measure of shortfall risk. Semivariance measures the variance of the returns that fall below a target value. Semivariance is given by $E[\max(X - z, 0)^2]$.

7.8.1 Value at Risk

Value at risk (*VaR*) is a risk measure that is used for regulatory reporting. Rather than measuring risk as deviation from a target return, *VaR* is a quantile of the loss distribution of a portfolio. Let $L(f, \tilde{r})$ be the random loss on a portfolio with allocation vector f and random return vector $\tilde{r}$. Let F be its distribution function so that $F(f, u) = \Pr\{L(f, \tilde{r}) \le u\}$. *VaR* is the α-quantile of the loss distribution and is defined by:

$$VaR_\alpha(f) = \min\{u : F(f, u) \ge \alpha\} \tag{7.14}$$

Thus, *VaR* is the smallest amount u such that with probability α the loss will not exceed u.

The first step in computing *VaR* is to determine the underlying market factors that contribute to the risk (uncertainty) in the portfolio. The next step is to simulate these sources of uncertainty and the resulting portfolio loss. Monte Carlo simulation is used largely because many of the portfolios under management by the more sophisticated banks include a preponderance of instruments that have optional features. As we shall see in Section 7.10, the price changes of these instruments can best be approximated by simulation. *VaR* can then be calculated by determining the distribution of the portfolio losses. The key question is what assumptions to make about the distributions of these uncertain market factors. Similar to the two methods that we discuss for estimating asset returns and variances, one can use historical data or a scenario approach to build the distributions.

Using historical data, one assumes that past market behavior is a good indicator of future market behavior. Take T periods of historical data. For each period, simulate the change in the portfolio value using the actual historical data. Use these T data points of portfolio profit/loss to compute the loss distribution and hence *VaR*. The benefit of this method is that there is no need for artificial assumptions about the distribution of the uncertainty of the underlying factors that impact the value of the portfolio. On the other hand, this method assumes that future behavior will be identical to historical behavior.

An alternative approach is to specify a probability distribution for each of the sources of market uncertainty and to then randomly generate events from those distributions. The events translate into behavior of the uncertain factors, which result in a change in the portfolio value. One would then simulate the portfolio profit/loss assuming that these randomly generated events occur and construct the loss distribution. This distribution is used to compute *VaR*.

7.8.2 Coherent Risk Measures

VaR is a popular risk measure. However, *VaR* does not satisfy one of the basic requirements of a good risk measure: *VaR* is not subadditive for all distributions (i.e., it is not always the case that $VaR(A + B) < VaR(A) + VaR(B)$), a property one would hope to hold true if risk is reduced by adding assets to a portfolio. This means that the *VaR* of a diversified portfolio may exceed the sum of the *VaR* of its component assets.

Artzner, Delbaen, Eber, and Heath (1999) specify a set of axioms satisfied by all *coherent* risk measures. These are:

- Subadditivity: $\rho(A + B) \leq \rho(A) + \rho(B)$
- Positive homogeneity: $\rho(\lambda A) = \lambda \rho(A)$ for $\lambda \geq 0$
- Translation invariance: $\rho(A + c) = \rho(A) - c$ for all c
- Monotonicity: $A \leq B$ then $\rho(B) \leq \rho(A)$

Subadditivity implies that the risk of a combined position of assets does not exceed the combined risk of the individual assets. This allows for risk reduction via diversification, as we discuss in Section 7.4.

Conditional value-at-risk (*CVaR*), also known as expected tail loss, is a coherent risk measure. *CVaR* measures the expected losses conditioned on the fact that the losses exceed *VaR*. Following the definition of *VaR* in Equation 7.14, if F is continuous then *CVaR* is defined as:

$$CVaR(f, \alpha) = E\{L(f, \tilde{r}) | L(f, \tilde{r}) \geq VaR(f, \alpha)\} \qquad (7.15)$$

An investment strategy that minimizes *CVaR* will minimize *VaR* as well.

An investor wishing to maximize portfolio return subject to a constraint on maximum *CVaR* would solve the following mathematical program:

$$\max \quad E\left(\sum_{i=1}^{N} f_i r_i\right)$$

$$\text{subject to:} \quad CVaR_\alpha(f_1, \ldots, f_N) \leq C \qquad (7.16)$$

$$\sum_{i=1}^{N} f_i = 1$$

$$0 \leq f_i \leq 1$$

where f_i is the fraction of wealth allocated to asset i, r_i is the return on asset i, and C is the maximum acceptable risk. Formulation 7.16 is a nonlinear formulation due to the constraint on *CVaR*, and is a hard problem to solve.

Suppose, instead of assuming that the loss distribution F is continuous, we discretize the asset returns by considering a finite set of $s = 1, \ldots, S$ scenarios of the portfolio performance. Let $p(s)$ denote the likelihood that scenario s occurs; $0 \leq p(s) \leq 1$; $\sum_{s \in S} p(s) = 1$. $\varphi(s)$ denotes the vector of asset returns under scenario s and $\rho(f, \varphi(s))$ denotes the portfolio return under scenario s assuming asset allocation vector f. Then, using this discrete set of asset returns, the expected portfolio return is given by

$$\sum_{s \in S} p(s)\rho(f, \varphi(s)) \qquad (7.17)$$

The investor will wish to maximize Equation 7.17. Definition 7.15 applies to continuous loss distributions. Rockafellar and Uryasev (2002) have proposed an alternative definition of *CVaR* for any general loss distribution, where $u = VaR_{\alpha(f)}$:

$$CVaR(f, \alpha) = \frac{1}{1 - \alpha}\left(\sum_{\{s \in S | \rho(f, \varphi(s)) \leq u\}} p(s) - \alpha\right)u + \frac{1}{1 - \alpha}\sum_{\{s \in S | \rho(f, \varphi(s)) > u\}} p(s)\rho(f, \varphi(s))$$

$$(7.18)$$

Using Definition 7.18, we can solve Problem 7.16 using a stochastic programming approach. First, we define an auxiliary variable $z(s)$ for each scenario s, which denotes the shortfall in portfolio return from the target u:

$$z(s) = \max(0, u - \rho(f, \varphi(s))) \qquad (7.19)$$

Following Rockafellar and Uryasev (2002), *CVaR* can be expressed using the shortfall variables (Equation 7.19) as:

$$CVaR(f, \alpha) = \min \left[u - \frac{1}{1 - \alpha} \sum_{s=1}^{S} p(s) z(s) \right]$$

The linear program is given by:

$$
\begin{aligned}
\max \quad & \sum_{s \in S} p(s) \rho(f, \varphi(s)) \\
\text{subject to:} \quad & \sum_{i=1}^{N} f_i = 1 \\
& 0 \leq f_i \leq 1 \\
& u - \frac{1}{1 - \alpha} \sum_{s=1}^{S} p(s) z(s) \geq C \\
& z(s) \geq u - \rho(f, \varphi(s)) \quad s = 1, \ldots, S \\
& z(s) \geq 0 \qquad\qquad\quad\; s = 1, \ldots, S
\end{aligned}
\qquad (7.20)
$$

where the last two inequalities follow from the definition of the shortfall variables (Equation 7.19) and the maximization is taken over the variables (f, u, z).

Formulation 7.20 is a linear stochastic programming formulation of the *CVaR* problem. To solve this problem requires an estimate of the distribution of the asset returns, which will be used to build the scenarios. If historical data are used to develop the scenarios then it is recommended that as time passes and more information is available, the investor reoptimize (Equation 7.20) using these additional scenarios. We direct the reader to Rockafellar and Uryasev (2000, 2002) for additional information on this subject.

7.8.3   Risk Management in Practice

VaR and *CVaR* are popular risk measures. *VaR* is used in regulatory reporting and to determine the minimum capital requirements to hedge against market, operational, and credit risk. Financial institutions may be required to report portfolio risks such as the 30-day 95% *VaR* or the 5% quantile of 30-day portfolio returns and to hold reserve accounts in proportion to these calculated amounts. *VaR* is used in these contexts for historical reasons. But even though as we saw above it is not a coherent risk measure, there is possibly some justification in continuing to use it. *VaR* is a frequency measure, so regulators can easily track whether the bank is reporting *VaR* accurately; *CVaR* is an integral over the tail probabilities and is likely not as easy for regulators to track.

In addition to meeting regulatory requirements, financial institutions may use *VaR* or *CVaR* to measure the performance of its business units that control financial portfolios by comparing the profit generated by the portfolio actions versus the risk of the portfolio itself.

For purposes of generating risk management statistics, banks will simulate from distributions that reflect views of the market and the economy. Banks will also incorporate some probability of extreme events such as the wild swings in correlations and liquidity that occur in market crashes.

7.9   Options

In this section, we discuss options. An option is a derivative security, which means that its value is derived from the value of an underlying variable. The underlying variable may or may not be a traded security. Stocks or bonds are examples of traded securities; interest rates or the weather conditions are examples of variables upon which the value of an option may be contingent but that are not traded securities. Derivative securities are sometimes referred to as contingent claims, as the derivative represents a claim whose payoff is contingent on the history of the underlying security.

The two least complex types of option contracts for individual stocks are calls and puts. Call options give the holder the right to buy a specified number of shares (typically, 100 shares) of the stock at the specified price (known as the exercise or strike price) by the expiration date (known as the exercise date or maturity). A put option gives the holder the right to sell a specified number of shares of the stock at the strike price by maturity. American options allow the holder to exercise the option at any time until maturity; European options can only be exercised at maturity. The holder of an option contract may choose whether or not he wishes to exercise his option contract. However, if the holder chooses to exercise, the seller is obligated to deliver (for call options) or purchase (for put options) the underlying securities.

When two investors enter into an options contract, the buyer pays the seller the option price and takes a *long* position; the seller takes a *short* position. The buyer has large potential upside from the option, but his downside loss is limited by the price that he paid for the option. The seller's profit or loss is the reverse of that of the buyer's. The seller receives cash upfront (the price of the option) but has a potential future liability in the case that the buyer exercises the option.

7.9.1   Call Option Payoffs

We first consider European options. We will define European option payoffs at their expiration date.

Let C represent the option cost, S_T denote the stock price at maturity, and K denote the strike price. An investor will only exercise the option if the stock price exceeds the strike price, that is, $S_T > K$. If $S_T > K$, the investor will exercise his option to buy the stock at price K and gain $(S_T - K)$ for each share of stock that he purchases. If $S_T < K$, then the investor will not exercise the option to purchase the stock at price K. Thus, the option payoff for each share of stock is

$$\max(S_T - K, 0) \tag{7.21}$$

The payoff for the option writer (who has a short position in a call option) is the opposite of the payoff for the long position. If, at expiration, the stock price is below the strike price the holder (owner) will not exercise the option. However, if the stock price is above the strike price the owner will exercise his option. The writer must sell the stock to the owner at the strike price. For each share of stock that he sells, the writer must purchase the stock in the open market at a price per share of S_T and then sell it to the owner at a price of K. Thus, the writer loses $(S_T - K)$ for each share that he is obligated to sell to the owner.

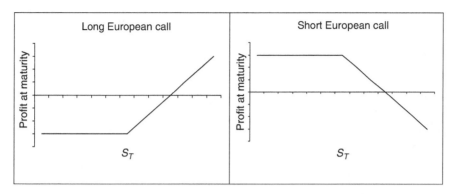

FIGURE 7.3   Profit from European call options.

The writer thus has an unlimited potential loss depending upon the final stock price of the underlying asset; the writer's payoff is

$$-\max(S_T - K, 0) = \min(K - S_T, 0)$$

Thus, an investor with a short position in a European call option has potentially unlimited loss depending upon the final stock price of the underlying stock. This risk must be compensated by the price of the option C.

The graph on the left in Figure 7.3 shows the profit at maturity for the owner of a call option. The point of inflection occurs when the ending stock price equals the strike price. The negative payoff is the price the investor paid for the option. As mentioned, the option holder's downside loss is limited by the price paid for the option. The payoff for the call option writer is shown in the graph on the right in Figure 7.3 and is the reverse of the payoff to the call option buyer.

When the price of the underlying stock is above the strike price, we say that the option is "in the money." If the stock price is below the strike price we say that the option is "out of the money." If the stock price is exactly equal to the strike price we say that the call option is "at the money."

American options can be exercised at any time prior to maturity. The decision of whether to exercise hinges upon a comparison of the value of exercising immediately (the *intrinsic value* of the option) against the expected future value of the option if the investor continues to hold it. We will discuss this in further detail in Section 7.12, where we discuss pricing American options.

7.9.2   Put Option Payoffs

An investor with a long position in a put option profits if the price of the underlying stock drops below the option's strike price. Similar to the definitions for a call option, we say that a put option is "in the money" if the stock price at maturity is lower than the strike price. The put option is "out of the money" if the stock price exceeds the strike price. The option is "at the money" if the strike price equals the stock price at maturity.

Let P denote the cost of the put option, K is its strike price, and S_T the stock price at expiration. The holder of a put option will only sell if the stock price at expiration is lower than the strike price. In this case, the owner can sell the shares to the writer at the strike price and will gain $(K - S_T)$ per share. Thus, payoff on a put option is $\max(K - S_T, 0)$; we note that positive payoff on a put is limited at K. If the stock price is higher than the strike price at maturity then the holder will not exercise the put option as he can sell his shares on the open market at a higher price. In this case, the option will expire worthless.

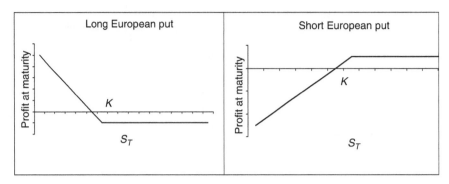

FIGURE 7.4 Profit from European put options.

A put writer has opposite payoffs. If the stock price exceeds the strike price, the put holder will never exercise his option; however, if the stock price declines, the writer will lose money. The price P paid by the owner must compensate the writer for this risk.

Figure 7.4 shows the option payoff for a put holder and writer.

7.10   Valuing Options

The exact value of a stock option is easy to define at maturity. Valuing options prior to expiration is more difficult and depends upon the distribution of the underlying stock price, among other factors. Black and Scholes (1973) derived a differential equation that can be used to price options on non-dividend paying stocks. We discuss the Black–Scholes formula in Section 7.10.1. However, in general, exact formulas are not available for valuing options. In most cases, we rely on numerical methods and approximations for options valuation. In Sections 7.10.3 and 7.12, we discuss two numerical methods for pricing options: Monte Carlo simulation and dynamic programming. Dynamic programming is useful for pricing American options, where the holder has the ability to exercise prior to the expiration date. Monte Carlo simulation is useful for pricing a European option, where the option holder cannot exercise the option prior to the maturity date. In the following sections, we describe Monte Carlo simulation and dynamic programming and show how they can be used to price options. We first will lay out some background and basic assumptions that are required.

7.10.1   Risk Neutral Valuation in Efficient and Complete Markets

We base our discussion of options pricing on the assumption that markets are efficient and complete. Efficient markets are arbitrage-free. Arbitrage provides an investor with a riskless investment opportunity with unlimited return, without having to put up any money. We assume that if any such opportunities exist there would be infinite demand for such assets. This would immediately raise the price of the investments and the arbitrage opportunity would disappear.

Black and Scholes derived a differential equation that describes the value of a trader's portfolio who has a short position in the option and who is trading in the underlying asset and a cash-like instrument. Efficiency of markets is one of the key assumptions required in their derivation. In addition, they assume that instantaneous and continuous trading opportunities exist, no dividends, transaction costs, or taxes are paid, and that short selling is permitted. Finally, they assume that the price of the underlying stock follows a specific stochastic process called Brownian motion. (See Section 7.10.2 for discussion of Brownian motion.) In this Black–Scholes framework it turns out that there is a trading strategy (called "delta-hedging") that makes the portfolio's return completely riskless. In an efficient market, a riskless portfolio will return the risk-free rate. This arbitrage reasoning combined with the delta-hedging strategy leads to a partial differential equation that resembles the heat equation of classical physics. Its solution provides the option's value at any point in time. In the case of European-style options (those that have a fixed exercise date) the solution can be achieved in closed form—this is the famous Black–Scholes formula. The Black–Scholes formulas for the values of a European call C or put P are:

$$C = S\Phi(d_1) - Ke^{-rT}\Phi(d_2)$$

$$P = Ke^{-rT}\Phi(-d_2) - S\Phi(-d_1)$$

where:

$$d_1 = [\log(S/K) + T(r + \sigma^2/2)]/\sigma\sqrt{T}$$

$$d_2 = d_1 - \sigma\sqrt{T}$$

Here, r is the risk-free rate (the rate of return of a riskless security such as a U.S. Treasury security over time T), log denotes the natural logarithm, and $\Phi()$ is the cumulative distribution function for the standard normal distribution $N(0, 1)$.

Exotic option contracts, especially those with exercise rules that give the owner the discretion of when to exercise, or options with underlying assets that are more complicated than equity stocks with no dividends, or options that depend on multiple assets, are very difficult to solve using the Black–Scholes partial differential equation.

Harrison and Pliska (1981) developed a more general perspective on options pricing that leads to a useful approach for these more general categories of options. The existence of the riskless trading strategy in the Black–Scholes framework is mathematically equivalent to the existence of a dual object called a "risk-neutral measure," and the options price is the integral of the option payouts with this risk-neutral measure. When the risk-neutral measure is unique, the market is called "complete." This assumption means that there is a single risk-neutral measure that can be used to price all the options.

This perspective leads to the following methodology for options pricing. Observe the prices of the traded options. Usually these are of a fairly simple type (European or American calls and puts), for which closed-form expressions like the Black–Scholes formula can be used. Then invert the formula to find the parameters of the risk-neutral distribution. This distribution can then be used to simulate values of any option—under the assumption that the market is efficient (arbitrage-free) and complete.

7.10.2   Brownian Motion

A key component of valuing stock options is a model of the price process of the underlying stock. In this section, we describe the Brownian motion model for the stock prices.

The efficient market hypothesis, which states that stock prices reflect all history and that any new information is immediately reflected in the stock prices, ensures that stock prices follow a Markov process so the next stock price depends only upon the current stock price and does not depend upon the historical stock price process. A Markov process is a stochastic process with the property that only the current value of the random variable is relevant for the purposes of determining the next value of the variable. A Wiener process is a type of Markov process. A Wiener process $Z(t)$ has normal and independent increments with variance proportional to the square root of time, that is, $Z(t) - Z(s)$ has a normal distribution with mean zero and variance $\sqrt{t - s}$. It turns out that $Z(t), t > 0$ will be a continuous function of t. If Δt represents an increment in time and ΔZ represents the change in Z over that increment in time then the relationship between ΔZ and Δt can be expressed by:

$$\Delta Z(t) = \varepsilon \sqrt{\Delta t} \tag{7.22}$$

where ε is drawn from a standard normal distribution. A Wiener process is the limit of the above stochastic process as the time increments get infinitesimally small, that is, as $\Delta t \to 0$. Equation 7.22 is expressed as

$$\mathrm{d}Z(t) = \varepsilon \sqrt{\mathrm{d}t} \tag{7.23}$$

If $x(t)$ is a random variable and Z is a Wiener process, then a generalized Wiener process is defined as

$$\mathrm{d}x(t) = a\,\mathrm{d}t + b\,\mathrm{d}Z$$

where a and b are constants. An Ito process is a further generalization of a generalized Wiener process. In an Ito process, a and b are not constants rather, they can be functions of x and t. An Ito process is defined as

$$\mathrm{d}x(t) = a(x,t)\,\mathrm{d}t + b(x,t)\,\mathrm{d}Z$$

Investors are typically interested in the rate of return on a stock, rather than the absolute change in stock price. Let S be the stock price and consider the change in stock price dS over a small period of time dt. The rate of return on a stock, dS/S, is often modeled as being comprised of a deterministic and stochastic component. The deterministic component, μdt, represents the contribution of the average growth rate of the stock. The stochastic component captures random changes in stock price due to unanticipated news. This component is often modeled by σdZ, where Z is a Brownian motion. Combining the deterministic growth rate (also known as drift) with the stochastic contribution to rate of change in stock price yields the equation:

$$\frac{\mathrm{d}S}{S} = \mu\,\mathrm{d}t + \sigma\,\mathrm{d}Z \tag{7.24}$$

an Ito process. μ and σ can be estimated using the methods described in Section 7.3.

The risk-neutral measure Q of Harrison and Pliska as applied to the Black–Scholes framework is induced by the risk-neutral process X that satisfies the modified Brownian motion

$$\frac{\mathrm{d}X}{X} = (r - \sigma^2/2)\,\mathrm{d}t + \sigma\,\mathrm{d}Z \tag{7.25}$$

It is important to note that this process is not the same as the original process followed by the stock—the drift has been adjusted. This adjustment is required to generate the probability measure that makes the delta-hedging portfolio process into a martingale. According to the theory discussed in Section 7.10.1, we price options in the Black–Scholes framework by integrating their cash flows under the risk-neutral measure generated by Equation 7.25. In the following section, we discuss how efficient markets and risk neutral valuation are used to compute an option's value using Monte Carlo simulation.

7.10.3   Simulating Risk-Neutral Paths for Options Pricing

In this section, we discuss how simulation can be used to price options on a stock by simulating the stock price under the risk-neutral measure over T periods, each of length $\Delta t = 1/52$. At each time interval, we simulate the current stock price and then step the process forward so that there are a total of T steps in the simulation. To simulate the path of the stock price over the T week period, we consider the discrete time version of Equation 7.25: $\Delta S/S = (r - \sigma^2/2)\Delta t + \sigma dZ = (r - \sigma^2/2)\Delta t + \sigma \varepsilon \sqrt{\Delta t}$. As ε is distributed like a standard normal random variable, $\Delta S/S \sim N((r - \sigma^2/2)\Delta t, \sigma\sqrt{\Delta t})$.

Each week, use the following steps to determine the simulated stock price:

   i. Set $i = 0$.

   ii. Generate a random value v_1 from a standard normal distribution. (Standard spreadsheet tools include this capability.)

   iii. Convert v_1 to a sample v_2 from a $N((r - \sigma^2/2)\Delta t, \; \sigma\sqrt{\Delta t})$ by setting $v_2 = (r - \sigma^2/2)\Delta t + \sigma\sqrt{\Delta t}v_1$.

   iv. Set $\Delta S = v_2 S$. ΔS represents the incremental change in stock price from the prior period to the current period.

   v. $S' = S + \Delta S$, where S' is the simulated updated value of the stock price after one period.

   vi. Set $S = S'$, $i = i + 1$.

   vii. If $i = T$ then stop. S is the simulated stock price at the end of 6 months. If $i < T$, return to step (i).

Note that randomness only enters in step (ii) when generating a random value v_1. All other steps are mere transformations or calculations and are deterministic. The payoff of a call option at expiration is given by Equation 7.21. Further, in the absence of arbitrage opportunities (i.e., assuming efficient markets) and by applying the theory of risk neutral valuation we know that the value of the option is equal to its expected payoff discounted by the risk-free rate. Using these facts, we apply the Monte Carlo simulation method to price the option. The overall methodology is as follows:

   i. Define the number of time periods until maturity, T.

   ii. Use Monte Carlo simulation to simulate a path of length T describing the evolution of the underlying stock price, as described above. Denote the final stock price at the end of this simulation by S^F.

   iii. Determine the option payoff according to Equation 7.21, assuming S^F, the final period T stock price determined in step (ii).

   iv. Discount the option payoff from step (iii) assuming the risk-free rate. The resulting value is the current value of the option.

   v. Repeat steps (ii)–(iv) until the confidence bound on the estimated value of the option is within an acceptable range.

7.10.4   A Numerical Example

A stock has expected annual return of $\mu = 15\%$ per year and standard deviation of $\sigma = 20\%$ per year. The current stock price is $S = \$42$. An investor wishes to determine the value of a European call option with a strike price of \$40 that matures in 6 months. The risk-free rate is 8%.

We will use Monte Carlo simulation to simulate the path followed by the stock price and hence the stock price at expiration which determines the option payoff. We consider weekly time intervals, that is, $\Delta t = 1/52$. Thus $T = 24$ assuming, for the sake of simplicity, that there are 4 weeks in each month.

To simulate the path of the stock price over the 24-week period, we follow the algorithm described in Section 7.10.3. We first compute the risk-neutral drift $(r - \sigma^2/2)\Delta t$, which with these parameter settings works out to be 0.0012. The random quantity ε is distributed like a standard normal random variable, so $\Delta S/S \sim N(.0012, .0277)$.

The starting price of the stock is \$42. Each week, use the following steps to determine the simulated updated stock price:

    i. Generate a random value v_1 from a standard normal distribution.

    ii. Convert v_1 to a sample v_2 from a $N(.0012, .0277)$ by setting $v_2 = .0012 + .0277v_1$.

    iii. Set $\Delta S = v_2 S$.

    iv. Set $S = S + \Delta S$

Steps (i)–(iv) yield the simulated updated stock price after 1 week. Repeat this process $T = 24$ times to determine S^F, the stock price at the end of 6 months. Then, the option payoff equals $P = \max(S^F - 40, 0)$. P is the option payoff based upon a single simulation of the ending stock price after 6 months, that is, based upon a single simulation run. Perform many simulation runs and after each run compute the arithmetic average and confidence bounds of the simulated values of P. When enough simulation runs have been performed so that the confidence bounds are acceptable, the value of the option can be computed based upon the expected value of P: $V = e^{-.08(.5)}E(P)$.

7.11   Dynamic Programming

Dynamic programming is a formal method for performing optimization over time. The algorithm involves breaking a problem into a number of subproblems, solving the smaller subproblems, and then using those solutions to help solve the larger problem. Similar to stochastic programming with recourse, dynamic programming involves sequential decision making where decisions are made, information is revealed, and then new decisions are made. More formally, the dynamic programming approach solves a problem in *stages*. Each stage is comprised of a number of possible *states*. The optimal solution is given in the form of a policy that defines the optimal action for each stage. Taking action causes the system to transition from one stage to a new state in the next stage.

There are two types of dynamic programming settings: deterministic and stochastic. In a deterministic setting, there is no system uncertainty. Given the current state and the action taken, the future state is known with certainty. In a stochastic setting, taking action will select the probability distribution for the next state. For the remainder we restrict our discussion to a stochastic dynamic programming setting, as finance problems are generally not deterministic. If the current state is the value of a portfolio, and possible actions are allocations to different assets, the value of the portfolio in the next stage is not known with certainty (assuming that some of the assets under consideration contain risk).

A dynamic program typically has the following features:

    i. The problem is divided into $t = 1, \ldots, T$ stages. x_t denotes the state at the beginning of stage t and $a_t(x_t)$ denotes the action taken during stage t given state x_t.

Taking action transitions the system to a new state in the next stage so that $x_{t+1} = f(x_t, a_t(x_t), \varepsilon_t)$, where ε_t is a random noise term. The initial state x_0 is known.

ii. The cost (or profit) function in period t is given by $g_t(x_t, a_t(x_t), \varepsilon_t)$. This cost function is additive in the sense that the total cost (or profit) over the entire T stages is given by:

$$g_T(x_T, a_T(x_T), \varepsilon_T) + \sum_{t=1}^{T-1} g_t(x_t, a_t(x_t), \varepsilon_t) \tag{7.26}$$

The objective is to optimize the expected value of Equation 7.26.

iii. Given the current state, the optimal solution for the remaining states is independent of any previous decisions or states. The optimal solution can be found by backward recursion. Namely, the optimal solution is found for the period T subproblem, then for the periods $T-1$ and T subproblem, and so on. The final period T subproblem must be solvable.

The features of dynamic programming that define the options pricing problem differ somewhat from the features described here. In Section 7.12.1 we note these differences.

7.12 Pricing American Options Using Dynamic Programming

Monte Carlo simulation is a powerful tool for options pricing. It performs well even in the presence of a large number of underlying stochastic factors. However, at each step simulation progresses forward in time. On the other hand, options that allow for early exercise must be evaluated backward in time where in each period the option holder must compare the intrinsic value of the option against the expected future value of the option.

Dealing with early exercise requires one to go backward in time, as at each decision point the option holder must compare the value of exercising immediately against the value of holding the option. The value of holding the option is simply the price of the option at that point.

In this section we will show how one can use dynamic programming to price an American option. The method involves two steps:

i. Build a T stage tree of possible states. The states correspond to points visited by the underlying stock price process.

ii. Use dynamic programming and backward recursion to determine the current value of the option.

7.12.1 Building the T Stage Tree

Cox et al. (1979) derived an exact options pricing formula under a discrete time setting. Following their analysis, we model stock price as following a multiplicative binomial distribution: if the stock price at the beginning of stage t is S then at the beginning of stage $t+1$

the stock price will be either uS or dS with probability q and $(1-q)$, respectively. Each stage has length Δt. We will build a "tree" of possible stock prices in any stage. When there is only one stage remaining, the tree looks like:

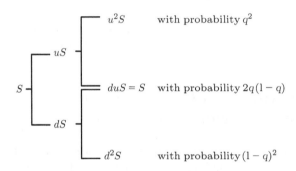

Suppose there are two stages. In each stage, the stock price will move "up" by a factor of u with probability q and "down" by a factor of d with probability $(1-q)$. In this case, there are three possible final stock prices and the tree is given by:

The tree continues to grow according to this method. In general, at stage t there are $t+1$ possible stock prices (states) that will appear on the tree. These are given by:

$$u^j d^{t-j}S, \quad \text{for } j = 0,\ldots,t$$

The values of u, d, and q are determined based upon the assumptions of efficient markets, risk neutral valuation, and the fact the variance of the change in stock price is given by $\sigma^2 \Delta t$ (from Section 7.10.3). These values are:

$$u = e^{\sigma\sqrt{\Delta t}}$$

$$d = e^{-\sigma\sqrt{\Delta t}}$$

$$q = \frac{a-d}{u-d}$$

where $a = e^{r\Delta t}$.

7.12.2 Pricing the Option Using the Binomial Tree

We now use backward enumeration through the binomial tree to determine the current stage 0 value of the option. We will illustrate the concept using the trees developed in

Section 7.12.1. Let K denote the strike price. With one period remaining, the binomial tree had the form:

$$S \left[\begin{array}{l} uS \quad \text{with probability } q \\ \\ \\ dS \quad \text{with probability } 1 - q \end{array} \right.$$

The corresponding values of the call at the terminal points of the tree are $C_u = \max(0, uS - K)$ with probability q and $C_d = \max(0, dS - K)$ with probability $(1 - q)$. The current value of the call is given by the present value (using the risk-free rate) of $qC_u + (1 - q)C_d$.

When there is more than one period remaining, each node in the tree must be evaluated by comparing the intrinsic value of the option against its continuation value. The intrinsic value is the value of the option if it is exercised immediately; this value is determined by comparing the current stock price to the option strike price. The continuation value is the discounted value of the expected cash payout of the option under the risk neutral measure, assuming that the optimal exercise policy is followed in the future. Thus, the decision is given by:

$$g_t = \max\left\{\max(0, x_t - K), E[e^{-r\Delta t} g_{t+1}(x_{t+1})|x_t]\right\} \tag{7.27}$$

The expectation is taken over the risk neutral probability measure. x_t, the current state, is the current stock price. Notice that the action taken in the current state (whether to exercise) does not affect the future stock price. Further, this value function is not additive. However, its recursive nature makes a dynamic programming solution method useful.

7.12.3 A Numerical Example

We illustrate the approach using the identical setting as that used to illustrate the Monte Carlo simulation approach to options pricing. However, here we will assume that we are pricing an American option. The investor wishes to determine the value of an American call option with a strike price of $40 that matures in 1 month. (We consider only 1 month to limit the size of the tree that we build.) The current stock price is $S = \$42$. The stock return has a standard deviation of $\sigma = 20\%$ per year. The risk-free rate is 8%.

We first build the binomial tree and then use the tree to determine the current value of the option. We consider weekly time intervals, that is, $\Delta t = 1/52$. Thus $T = 4$ assuming, for the sake of simplicity, that there are 4 weeks in each month.

$$u = e^{\sigma\sqrt{\Delta t}} = 1.028$$

$$d = e^{-\sigma\sqrt{\Delta t}} = 0.973$$

$$q = \frac{e^{r\Delta t} - d}{u - d} = 0.493$$

The tree of stock movements over the 4-week period looks like:

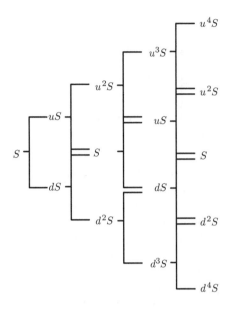

We will evaluate each node in the tree by backward evaluation starting at the fourth time period and moving backward in time. For each node we will use the Equation 7.27 to compare the intrinsic value of the option against its continuation value to determine the value of that node. The binomial tree for the value of the American call option is:

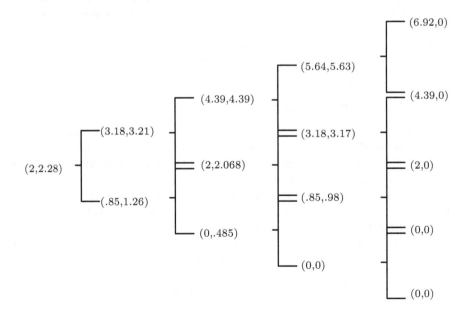

Every node in the tree contains two numbers in parenthesis. The first number is the intrinsic value of the option. The second number is the discounted expected continuation value, assuming that optimal action is followed in future time periods. The option value at time zero (current time) is 2.28. Note that although the option is currently in the money, it is not optimal to exercise even though it is an American option and early exercise is allowed. By using the binomial tree to evaluate the option we find that the expected continuation value of the option is higher than its current exercise value.

7.13 Comparison of Monte Carlo Simulation and Dynamic Programming

Dynamic programming is a powerful tool that can be used for pricing options with early exercise features. However, dynamic programming suffers from the so-called curse of dimensionality. As the number of underlying variables increases, the time required to solve the problem grows significantly. This reduces the practical use of dynamic programming as a solution methodology. The performance of Monte Carlo simulation is better in the sense that its convergence is independent of the state dimension. On the other hand, as we have discussed, simulation has traditionally been viewed as inappropriate for pricing options with early exercise decisions as these require estimates of future values of the option and simulation only moves forward in time. Recent research has focused on combining simulation and dynamic programming approaches to pricing American options to gain the benefits of both techniques. See, for example, Broadie and Glasserman (1997).

7.14 Multi-Period Asset Liability Management

The management of liability portfolios of relatively long-term products, like pensions, variable annuities, and some insurance products requires a perspective that goes beyond a single investment period. The portfolio optimization models of Sections 7.5 through 7.7 are short-term models. Simply rolling these models over into the next time horizon can lead to suboptimal results. First, the models may make an excessive number of transactions. Transactions are not free, and realistic portfolio management models must take trading costs into consideration. Second, the models depend only on second moments. Large market moves, such as during a market crash, are not part of the model assumptions. Finally, policies and regulations place conditions on the composition of portfolios.

Academic and finance industry researchers have, over the past few decades, been exploring the viability of using multi-period balance sheet modeling to address the issues of long-term asset liability management. A multi-period balance sheet model treats the assets and liabilities as generating highly aggregated cash flows over multiple time periods. The aggregations are across asset classes, so that individual securities in an asset class, say stocks, are accumulated into a single asset class, say S&P 500. Other asset classes are corporate bonds of various maturity classes, and so forth. The asset class cash flows are aggregated over time periods, typically 3 or 6 months, so that cash flows occurring within a time interval, say, $(t-1, t]$, are treated as if they all occur at the end-point t. The model treats the aggregate positions in each asset category as variables in the model. There is a single decision to be made for each asset class at the beginning of each time period, which is the change in the holdings of each asset class. The asset holdings and liabilities generate cash flows, which then flow into account balances. These account balances are assigned targets, and the objective function records the deviation from the targets. The objective of the model is to maximize the sum of the expected net profits and the expected penalties for missing the targets over the time horizon of the model.

A simplified application of such a model to a property-casualty insurance problem is as follows. Let A_t denote a vector of asset class values at time t and i_t denote their cash flows (e.g., interest payment, dividends, etc.) at time t. Let x_t denote the portfolio of holdings in the asset classes. Cash flows are generated by the holdings and by asset sales:

$$C_t = A_t \Delta x_t + i_t x_{t-1}$$

where $\Delta x_t := x_t - x_{t-1}$. The cash flows are subject to market and economic variability over the time horizons of interest, say $t = 1, \ldots, T$.

Liability flows from the property-casualty portfolio are modeled by aggregating and then forecasting the aggregated losses minus premium income. Loss events are classified by frequency of occurrence and intensity given loss. These can be simulated over the time horizon T using actuarial models. The evolution of the liability portfolio composition (say, by new sales, lapses of coverage, and so forth) can also be modeled. The key correlation to capture in the liability model is the relationship between the liability flows and the state of the economy. For example, high employment is indicative of strong economic activity, which can lead to increases in the number and size of the policies in the insurance portfolio; high inflation will lead to higher loss payouts given losses; and so forth.

Various performance, accounting, tax, and regulatory measurements are computed from the net cash flows. For example, one measurement could be shareholder's equity at the time horizon S_T. Another could be annual net income N_t. Yet another could be premium-surplus ratio P_t—a quantity used in the property-casualty industry as a proxy for the rating of the insurance company.

In these aggregated models, we try to model the change in performance measurements as linear computations from the cash and liability flows and the previous period quantities. For example, net income is computed as cash flow minus operating expenses. If $O_t \Delta x_t$ is a proxy for the contribution of portfolio management to expenses, for example, the cost of trading, then net income can be modeled by the following equation

$$N_t = C_t - L_t - O_t \Delta x_t$$

Shareholder's equity can change due to a number of influences; here we just capture the change due to the addition of net income:

$$S_t = S_{t-1} + N_t$$

Finally, premium-surplus ratio can be approximated by fixing premium income to a level L and assuming (this is a major simplification!) that the surplus is equivalent to shareholders equity:

$$P_t = \frac{L}{S_t}$$

A typical objective for a multi-period balance sheet model/an asset-liability matching problem is to create performance targets for each quantity and to penalize the shortfall. Suppose that the targets are $\overline{N}_t$ for annual net income, $\overline{S}_T$ for shareholder's equity, and $\overline{P}_t$ for premium-surplus ratio. Then the model to be optimized is:

$$\text{Maximize} \quad E\{S_T - \lambda \sum_t [\overline{N}_t - N_t]^+ - \mu \sum [L - S_t \overline{P}_t]^+\}$$

$$\text{Subject to} \quad N_t = A_t \Delta x_t + i_i x_{t-1} - L_t - O_t \Delta x_t \qquad (7.28)$$

$$S_t = N_t + S_{t-1}$$

where the parameters λ and μ are used to balance the various contributions in the objective function, the premium-surplus ratio target relationship has been multiplied through by the denominator to make the objective linear in the decision variables, and the objective is integrated over the probability space represented by the discrete scenarios.

The objective function in formulation 7.28 can be viewed as a variation of the Markowitz style, where we are modeling "expected return" through the shareholder's equity at the end of the horizon, and "risks" through the shortfall penalties relative to the targets for net income and premium-surplus ratio.

7.14.1 Scenarios

In multi-period asset liability management the probability distribution is modeled by discrete scenarios. These scenarios indicate the values, or states, taken by the random quantities at each period in time. The scenarios can branch so that conditional distributions given a future state can be modeled. The resulting structure is called a "scenario tree." Typically there is no recombining of states in a scenario tree, so the size of the tree grows exponentially in the number of time periods. For example, in the property-casualty model, the scenarios are the values and cash flows of the assets and the cash flows of the liabilities. The values of these quantities at each time point t and scenario s is represented by the triple (A_t^s, i_t^s, L_t^s). The pair (s, t) is sometimes called a "node" of the scenario tree. The scenario tree may branch at this node, in which case the conditional distribution for the triple $(A_{t+1}, i_{t+1}, L_{t+1})$ given node (s, t) is the values of the triples on the nodes that branch from this node.

It is important to model the correlation between the asset values and returns and the liability cash flows in these conditional distributions. Without the correlations, the model will not be able to find positions in the asset classes that hedge the variability of the liabilities. In property-casualty insurance, for example, it is common to correlate the returns of the S&P 500 and bonds with inflation and economic activity. These correlations can be obtained from historical scenarios, and conditioned on views as discussed in Section 7.7.

The scenario modeling framework allows users to explicitly model the probability and intensity of extreme market movements and events from the liability side. One can also incorporate "market crash" scenarios in which the historical correlations are changed for some length of time that reflects unusual market or economic circumstances—such as a stock market crash or a recession. Finally, in these models it is usual to incorporate the loss event scenarios explicitly rather than follow standard financial valuation methodology, which would tend to analyze the expected value of loss distributions conditional on financial return variability. Such methodology would ignore the year-to-year impact of loss distribution variability on net income and shareholder's equity. However, from the asset liability management (ALM) perspective, the variability of the liability cash flows is very important for understanding the impact of the hedging program on the viability of the firm.

7.14.2 Multi-Period Stochastic Programming

The technology employed in solving an asset liability management problem such as this is multi-period stochastic linear programming. For a recent survey of stochastic programming, see Shapiro and Ruszczynski (2003).

The computational intensity for these models increases exponentially in the number of time periods, so the models must be highly aggregated and strategic in their recommendations. Nevertheless, the models do perform reasonably well in practice, usually generating 300 basis points of excess return over the myopic formulations based on the repetitive application of one period formulations, primarily through controlling transaction costs and because the solution can be made more robust by explicitly modeling market crash scenarios. A recent collection of this activity is in the volume edited by Ziemba and Mulvey (1998). See also Ziemba's monograph (Ziemba, 2003) for an excellent survey of issues in asset liability management.

7.15   Conclusions

In this chapter, we saw the profound influence of applications of Operations Research to the area of finance and financial engineering. Portfolio optimization by investors, Monte Carlo simulation for risk management, options pricing, and asset liability management, are all techniques that originated in OR and found deep application in finance. Even the foundations of options pricing are based on deep applications of duality theory. As the name financial engineering suggests, there is a growing part of the body of financial practice that is regarded as a subdiscipline of engineering—in which techniques of applied mathematics and OR are applied to the understanding of the behavior and the management of the financial portfolios underpinning critical parts of our economy: in capital formation, economic growth, insurance, and economic-environmental management.

References

1. Artzner, P., F. Delbaen, J.-M. Eber, D. Heath, 1999, Coherent measures of risk, *Mathematical Finance*, 203–228.
2. Ashford, R.W., R.H. Berry, and R.G. Dyson, 1988, "Operational Research and Financial Management," *European Journal of Operational Research*, 36(2), 143–152.
3. Bengtsson, C., 2004, "The Impact of Estimation Error on Portfolio Selection for Investors with Constant Relative Risk Aversion." Working paper, Department of Economics, Lund University.
4. Bertsekas, D.P. and J.N. Tsitsiklis, 1996, *Neuro-Dynamic Programming*, Athena Scientific, Belmont, MA, 1996.
5. Bevan, A. and K. Winkelmann, 1998, "Using the Black Litterman Global Asset Allocation Model: Three Years of Practical Experience," Fixed Income Research, Goldman Sachs & Company, December.
6. Birge, J. and V. Linetsky, "Financial Engineering," in: *Handbook in Operations Research and Management Science*, Elsevier, Amsterdam, In press.
7. Black, F. and R. Litterman, 1992, "Global Portfolio Optimization," *Financial Analysts Journal*, September/October, 28–43.
8. Black, F. and M. Scholes, 1973, "The Price of Options and Corporate Liabilities," *Journal of Political Economy*, 81, 637–654.
9. Board, J., C. Sutcliffe, and W.T. Ziemba, 2003, "Applying Operations Research Techniques to Financial Markets," *Interfaces*, 33, 12–24.
10. Broadie, M. and P. Glasserman, 1997, "Pricing American-Style Securities Using Simulation," *Journal of Economic Dynamics and Control*, 21, 1323–1352.
11. Campbell, J.Y., A.W. Lo, and A.C. MacKinlay, 1997, *The Econometrics of Financial Markets*, Princeton University Press, Princeton, NJ.
12. Chopra, V.K. and W.T. Ziemba, 1993, "The Effect of Errors in Mean, Variance, and Covariance Estimates on Optimal Portfolio Choice," *Journal of Portfolio Management*, 19(2), 6–11.
13. Cox, J.C. and S.A. Ross, 1976, "The Valuation of Options for Alternative Stochastic Processes," *Journal of Financial Economics*, 3, 145–166.

14. Cox, J.C., S.A. Ross, and M. Rubinstein, (1979), "Option Pricing: A Simplified Approach," *Journal of Financial Economics*, 7, 229–263.

15. Fama, E.F., 1965, "The Behavior of Stock Prices," *Journal of Business*, 38, 34–105.

16. Fusai, G. and A. Meucci, 2003, "Assessing Views," *Risk*, March 2003, s18–s21.

17. Glasserman, P., 2004, *Monte Carlo Methods in Financial Engineering*, Springer-Verlag, New York.

18. Harrison, J.M. and S.R. Pliska, 1981, "Martingales and Stochastic Integrals in the Theory of Continuous Trading," *Stochastic Processes and their Applications*, 11, 215–260.

19. He, G. and R. Litterman, 1999, "The Intuition Behind Black-Litterman Model Portfolios," *Investment Management Research*, Goldman Sachs & Company, December.

20. Holmer, M.R., 1998, "Integrated Asset-Liability Management: An Implementation Case Study," in *Worldwide Asset and Liability Modeling*, W.T. Ziemba and J.R. Mulvey (eds.), Cambridge University Press.

21. Jarrow, R.V. Maksimovic and W. Ziemba (eds.), 1995, "Finance," In: *Handbooks in Operations Research and Management Science*, vol. 9, North-Holland, Amsterdam.

22. Kallberg, J.G. and W.T. Ziemba, 1981, "Remarks on Optimal Portfolio Selection," In: *Methods of Operations Research*, G. Bamberg, O. Optin, Oelgeschlager, Gunn and Hain (eds.), Cambridge, MA, pp. 507–520.

23. Kallberg, J.G. and W.T. Ziemba, 1984, "Mis-Specifications in Portfolio Selection Problems," In: *Risk and Capital*, G. Bamberg and K. Spremann (eds.), Springer-Verlag, New York, pp. 74–87.

24. King, A.J. 1993, "Asymmetric Risk Measures and Tracking Models for Portfolio Optimization Under Uncertainty, *Annals of Operations Research*, 45, 165–177.

25. Michaud, R.O., 1989, "The Markowitz Optimization Enigma: is 'Optimized' Optimal?" *Financial Analysts Journal*, 45(1), 31–42.

26. Mills, T.C., 1999, *The Econometric Modelling of Financial Time Series*, 2nd ed., Cambridge University Press, Cambridge, UK.

27. Pedersen, C.S. and S.E. Satchell, 1998, "An Extended Family of Financial Risk Measures," *Geneva Papers on Risk and Insurance Theory*, 23, 89–117.

28. Rockafellar, R.T. and S. Uryasev, 2000, "Optimization of Conditional Value-at-Risk," *The Journal of Risk*, 2(3), 21–41.

29. Rockafellar, R.T. and S. Uryasev, 2002, "Conditional Value-at-Risk for General Distributions," *Journal of Banking and Finance*, 26(7), 1443–1471.

30. Samuelson, P.A., 1965, "Rational Theory of Warrant Pricing," *Industrial Management Review*, 6, 13–31

31. Shapiro, A. and A. Ruszczynski, 2003, "Stochastic Programming," In: *Handbooks in Operations Research and Management Science*, vol. 10, Elsevier.

32. Sharpe, W.F., 1964, "Capital Asset Prices: A Theory of Market Equilibrium," *Journal of Finance*, September, 425–442.

33. Uryasev, S. 2001, "Conditional Value-at-Risk: Optimization Algorithms and Applications," *Financial Engineering News*, 14, 1–5.

34. von Neumann, J. and O. Morgenstern, 1944, *Theory of Games and Economic Behaviour*, Princeton University Press, Princeton, NJ.

35. Ziemba, W.T. 2003, *The Stochastic Programming Approach to Asset, Liability, and Wealth Management*, AIMR Research Foundation Publications, No. 3: 1–192.

36. Ziemba, W.T. and J.R. Mulvey (eds.), 1998, *Worldwide Asset and Liability Modeling*, Cambridge University Press, Cambridge, UK.

8

Supply Chain Management

Donald P. Warsing
North Carolina State University

8.1 Introduction

Perhaps the best way to set the tone for this chapter is to consider a hypothetical encounter that an industrial engineer or operations manager is likely to experience. It's not quite the "guy walks into a bar" setup, but it's close. Consider the following scenario: A software salesperson walks into the office of an industrial engineer and says, "I have some software to sell you that will optimize your supply chain." Unfortunately, the punch line—which I'll leave to the reader's imagination—is not as funny as many of the "guy walks into a bar" jokes you're likely to hear. In fact, there are some decidedly not-funny examples of what could happen when "optimizing" the supply chain is pursued via software implementations without careful consideration of the structure and data requirements of the software system. Hershey Foods' enterprise resource planning (ERP) system implementation (Stedman, 1999) and Nike's advanced planning system (APS) implementation (Koch, 2004) are frequently cited examples of what could go wrong. Indeed, without some careful thought and the appropriate perspective, the industrial engineer (IE) in our hypothetical scenario might as well be asked to invest in a bridge in Brooklyn, to paraphrase another old joke.

Encounters similar to the one described above surely occur in reality countless times every day as of this writing in the early twenty-first century, and in this author's opinion it happens all too often because those magical words "optimize your supply chain" seem to be used frivolously and in a very hollow sense. The question is, what could it possibly mean to optimize the supply chain? Immediately, an astute IE or operations manager must ask himself or herself a whole host of questions, such as: What part of my supply chain? All of it? What does "all of it" mean? Where does my supply chain begin and where does it end? What is the objective function being optimized?...minimize cost?...which costs?...maximize profit?...whose profit? What are the constraints? Where will the input data behind the formulation come from? Is it even conceivable to formulate the problem? Provided I could formulate the problem, could I solve it?...to optimality?

Now, in fairness, this chapter is not technically about supply chain *optimization*, but supply chain *management*. Still, we should continue our line of critical questioning. What does it mean to "manage the supply chain"? Taken at face value, it's an *enormous* task. In the author's view, the term "manage the supply chain," like "optimize your supply chain," is thrown around much too loosely. A better approach, and the one taken in this chapter, is to consider the various aspects of the supply chain that must be managed. Our challenge at hand, therefore, is to figure out how to identify and use the appropriate tools and concepts to lay a good foundation for the hard work that comprises "managing the supply chain."

In the first edition of their widely-cited text book on supply chain management (SCM), Chopra and Meindl (2001) define a *supply chain* as "all stages involved, directly or indirectly, in fulfilling a customer request" (p. 3). (In the more recent edition of the text, Chopra and Meindl 2004, the authors substitute the word "parties" for stages. Since this is inconsistent with the definition of SCM that is to follow, I have chosen the earlier definition.) This definition is certainly concise, but perhaps a bit too concise. A more comprehensive definition of the term "supply chain" is given by Bozarth and Handfield (2006), as follows: "A network of manufacturers and service providers that work together to convert and move goods from the raw materials stage through to the end user" (p. 4). In a more recent edition of their text, Chopra and Meindl (2004) go on to define *supply chain management* as "the management of flows between and among supply chain stages to maximize total supply chain profitability" (p. 6). The nice aspects of this definition are that it is, once again, concise and that it clearly emphasizes managing the *flows* among the stages in the chain. The problematic aspect of the definition, for this writer, is that its objective, while worthy and theoretically the right one (maximizing the net difference between the revenue generated by the chain and the total costs of running the chain), would clearly be difficult to measure in reality. Would all parties in the supply chain willfully share their chain-specific costs with all others in the chain? This brings us back to the criticisms lodged at the outset of this chapter related to "optimizing the supply chain"—a noble goal, but perhaps not the most practical one.

The author is familiar with a third-party logistics (3PL) company that derives a portion of its revenue from actually carrying out the goals of optimizing—or at least *improving*—its clients' supply chains. Figure 8.1 is taken from an overview presentation by this 3PL regarding its services and represents a cyclic perspective on what it means to manage the supply chain—through design and redesign, implementation, measurement, and design (re-) assessment. Figure 8.2 provides an idea of the kinds of tools this 3PL uses in carrying out its SCM design, evaluation, and redesign activities. What is striking about this company's approach to doing this is that it involves neither a single software application nor a simple series of software applets integrated into a larger "master program," but a whole host of discrete tools applied to various portions of the overarching problem of "managing the supply chain." The software tools listed are a mix of off-the-shelf software and proprietary code, customized in an attempt to solve more integrated problems. The result is an approach that views the overall challenge of managing the supply chain as one of understanding how the various problems that comprise "supply chain management" interact with one another, even as they are solved discretely and independently, many times for the sheer purpose of tractability.

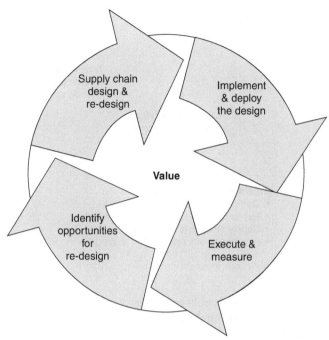

FIGURE 8.1   Supply chain design cycle. Presentation by unnamed 3PL company.

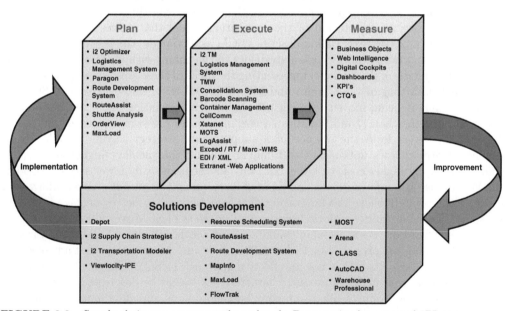

FIGURE 8.2   Supply chain management tasks and tools. Presentation by unnamed 3PL company.

Coming back to definitions of SCM, the definition from Bozarth and Handfield (2006) offers an objective different from Chopra and Meindl's profit-maximizing one—more customer focused, but perhaps no less problematic in practice. Their definition is as follows: "The *active* management of supply chain activities and relationships in order to maximize

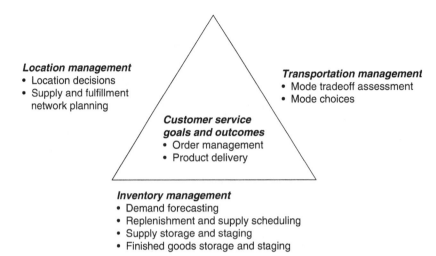

FIGURE 8.3   Drivers of supply chain management strategy. (Adapted from Ballou, 2004.)

customer value and achieve a sustainable competitive advantage" (p. 8). The idea of a sustainable competitive advantage through the management of SCM-related activities is also emphasized by Ballou (2004), who does a commendable job basing the whole of his textbook on logistics and SCM on the various management activities that must be understood and integrated to derive positive customer outcomes and thereby achieve a sustainable advantage. This will be the approach taken in the remainder of this chapter.

Figure 8.3 shows an adaptation of the conceptual model that serves as the structural basis for Ballou's (2004) text. Ballou describes SCM as a triangle, the legs of which are the inventory, transportation, and location strategies of the chain, with customer service goals located in the middle of the triangle, representing the focus of these strategies. This approach is consistent with that of the Chopra and Meindl (2004) text, whose *drivers* of SCM are inventory (*what* is being moved through the chain), transportation (*how* the inventory is moved through the chain), and facilities (*where* inventories are transformed in the chain). These SCM drivers form the essence of a supply chain strategy, laying out the extent to which a firm chooses approaches to fulfilling demand that achieve higher service, but almost necessarily higher cost, or lower cost, but almost necessarily lower service. (Chopra and Meindl also list *information* as an SCM driver, but in the author's view, information is more appropriately viewed as an *enabler* that may be used to better manage inventory, transportation, and facilities. More discussion on *enablers* of SCM will follow.)

The idea of SCM drivers is useful in thinking about another theme in SCM emphasized by many authors and first proposed by Fisher (1997) in an important article that advanced the notion that "one size fits all" is not an effective approach to SCM. Fisher (1997) cogently lays out a matrix that matches *product* characteristics—what he describes as a dichotomy between *innovative* products like technology-based products and *functional* products like toothpaste or other staple goods—and *supply chain* characteristics—another dichotomy between *efficient* (cost-focused) supply chains and *responsive* (customer service-focused) supply chains. Chopra and Meindl (2004) take this conceptual model a step further, first by pointing out that Fisher's product characteristics and supply chain strategies are really continuous spectrums, and then by superimposing the Fisher model, as it were, on a frontier that represents the natural tradeoff between responsiveness and efficiency. Clearly, it stands to reason that a firm, or a supply chain, *cannot* maximize cost efficiency and customer responsiveness simultaneously; some aspects of each of these objectives necessarily

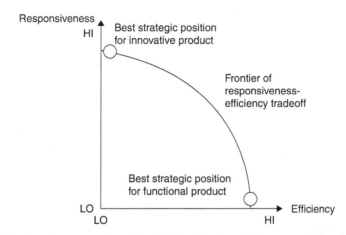

FIGURE 8.4 Overlay of conceptual models of Fisher (1997) and Chopra and Meindl (2004).

work at cross purposes. Chopra and Meindl's frontier and Fisher's product dichotomy are presented in Figure 8.4. The value of this perspective is that it clearly identifies a *market-driven* basis for strategic choices regarding the three SCM drivers: Should our inventory management decisions be focused more on efficiency—minimizing inventory levels—or on responsiveness—maximizing product availability? Should our transportation choices be focused more on efficiency—minimizing transportation costs, perhaps through more extensive economies of scale—or on responsiveness—minimizing delivery lead times and maximizing reliability? Should our facilities (network design) decisions be focused more on efficiency—minimizing the number of locations and maximizing their size and scale—or on responsiveness—seeking high levels of customer service by choosing many, focused locations closer to customers?

Clearly, the legs of the Ballou (2004) triangle frame the SCM strategy for any firm. What is it, however, that allows a firm to implement that strategy and execute those driver-related decisions, effectively? Marien (2000) presents the results of a survey of approximately 200 past attendees of executive education seminars, asking these managers first to provide a rank-ordering of four *enablers* of effective SCM, and then to rank the various attributes that support each enabler. The enablers in Marien's survey were identified through an extensive search of the academic and trade literature on SCM, and the four resulting enablers were (1) technology, (2) strategic alliances, (3) organizational infrastructure, and (4) human resources management. Marien's motivation for the survey was similar to the criticism laid out at the outset of this chapter—the contrast between what he was hearing from consultants and software salespeople, that effective SCM somehow flowed naturally from information technology implementations, and from practicing managers, who, at best, were skeptical of such technology-focused approaches and, at worst, still bore the scars of failed, overzealous IT implementations. Indeed, the result of Marien's (2000) survey indicate that managers, circa 1998, viewed organizational infrastructure as, far and away, the most important enabler of effective SCM, with the others—technology, alliances, and human resources—essentially in a three-way tie for second place. The most important attributes of organizational infrastructure, according to Marien's survey results, were a coherent business strategy, formal process flows, commitment to cross-functional process management, and effective process metrics.

The process management theme in Marien's (2000) results is striking. It is also consistent with what has arguably become the most widely-accepted structural view of supply chain management among practitioners, the supply chain operations reference model, or

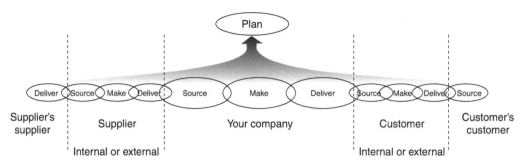

FIGURE 8.5   The Supply Chain Operations Reference (SCOR®) model. (From Supply-Chain Council, 2006.)

SCOR* (Supply-Chain Council, 2006). Figure 8.5 presents the structural framework for the SCOR model, based on the primary business processes that comprise SCM according to the model—plan, source, make, deliver, and return. Underlying these primary processes is an extensive checklist, as it were, of the sub-processes and tasks that, according to the model, should be in place to enable effective SCM. Bozarth and Warsing (2006) offer an academic perspective that attempts to tie the underlying detail of the SCOR model back to company performance and offers a research agenda for understanding this link; in fact, this theme has already begun to emerge in the literature (Lockamy and McCormack, 2004).

   An important theme that must be addressed in this chapter, however—and one that is clear from Figure 8.5—is that supply chain management is encapsulated in managing *internal* business processes and their interfaces with the firms *immediately* upstream and downstream from your company. While the common lingo claiming to manage the chain from the "supplier's suppliers to the customer's customers" is catchy, it is not clear how this is to be done. What is more important, from an external perspective, is managing the most important *dyadic* relationships in the chain, the links immediately upstream and downstream. These dyads form the business relationships that a company can directly influence, and *direct* influence is what really matters in dealing with a system as complex and dynamic as a supply chain. Vollmann et al. (2005, chapter 3) offer an excellent discussion of this key idea, and one that, hopefully, will help bring practitioners of SCM down from the clouds and back to the hard work of making their links in the chain more effective.

   In the remainder of this chapter, therefore, I discuss each of the drivers of SCM separately—managing inventories, managing transportation, and managing locations (i.e., the network design problem)—identifying important enablers along the way. Then, I present the important issues in managing dyads in the chain. Finally, a discussion follows to tie these themes together, and the chapter ends with a few concluding remarks.

8.2   Managing Inventories in the Supply Chain

Inventory management can be concisely captured in the following policy, of sorts: "Before you run out, get more."† Though concise, the statement captures the essence of the basic problem of inventory management, and immediately begs for an answer to two

---

   *SCOR® is a registered trademark of the Supply-Chain Council in the United States and Canada.

   †I owe this concise statement, with only slight alterations applied, to my good friend and colleague Douglas J. Thomas.

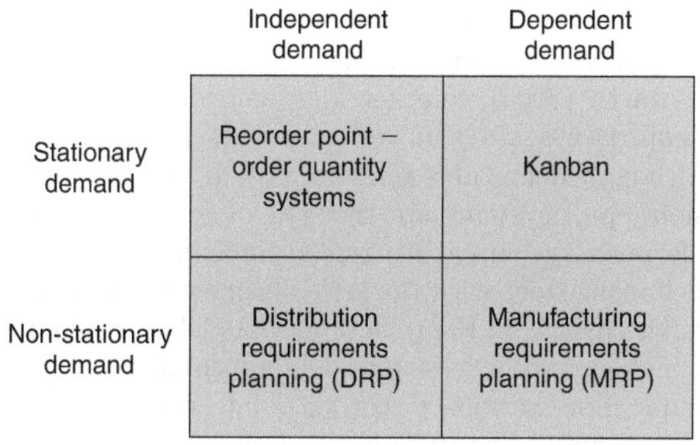

	Independent demand	Dependent demand
Stationary demand	Reorder point – order quantity systems	Kanban
Non-stationary demand	Distribution requirements planning (DRP)	Manufacturing requirements planning (MRP)

FIGURE 8.6   Framework for choice of inventory management tools. (Based on Bozarth, 2005.)

questions—"How long before you run out (i.e., *when*)?" and "*How much* more do you get?" Thus, managing inventory boils down to answering those two questions, "when?" and "how much?"

Before I dive into tools and techniques for managing inventories—and consistent with the theme I laid out in the Introduction—let us first step back and make sure that we are clear about the characteristics of the items to be managed. These characteristics, it stands to reason, should have a significant impact on the kinds of tools that are appropriate to the management task. Again, my overarching theme is to beware of (buying or) applying a single, elaborate, sophisticated tool to every problem. Sometimes you really might be better off with the proverbial "hammer" to solve the simple, "nail-like" aspects of your problem. Sometimes, I will admit, an advanced power-tool might be warranted.

Bozarth (2005) offers an effective framework for placing the inventory management problem in the context of two important aspects of demand for the products being managed. This framework has been reproduced in Figure 8.6. The first of the two context-shaping aspects of demand is whether demand is driven externally by forces outside the firm's direct control or whether demand is driven by internal management decisions. The former situation describes *independent demand* items, those items whose demand is driven by customer tastes, preferences, and purchasing patterns; typically, these are finished goods. Inventory management in these cases is influenced greatly by the extent to which the firm can effectively describe the random variations in these customer purchasing patterns. The latter situation describes *dependent demand* items, those items whose demand is driven by the demand of other items; a good example of such items would be component parts, whose demand is driven by the production schedules for the finished goods of which they are components. Note that the production schedule for finished items is solely under the control of the manufacturing firm; its management chooses the production plan in response to projected consumer demand, and therefore, inventory management for dependent demand items is largely an issue of ensuring that component parts are available in sufficient quantities to execute the production plan.

The second important aspect of demand that helps determine which inventory management tools to apply is the stability of demand over time. Though Bozarth (2005) describes the dichotomy as "stable" versus "variable," I have chosen to characterize demand as *stationary* or *non-stationary*—the former being the case in which the expected value of demand remains constant over time, or at least over the planning horizon, and the latter being the case in which it does not. In either case, stationary or non-stationary, demand could be uncertain or known with certainty. Obviously, then, the simplest case would be one in which demand is stationary and known with certainty, and the most challenging would be one in which demand is non-stationary and uncertain.

As indicated by Bozarth (2005), another idea embedded in his framework is that the stationary demand approaches are *pull systems*, whereas the non-stationary approaches are *push systems*. Although definitions tend to vary (see Hopp and Spearman, 2004, for an excellent discussion), pull systems are those that generate orders only in response to actual demand, while push systems drive replenishments from the schedule of projected future demands, which, consistent with Bozarth's framework, will vary over time. Finally, I should point out that the matrix in Figure 8.6 presents what would appear to be *reasonable* approaches to managing demand in these four demand contexts, but not necessarily the *only* approaches. For example, non-stationary reorder point-order quantity systems are certainly possible, but probably not applied very often in practice due to the significant computational analysis that would be required. The effort required to manage the system must clearly be considered in deciding which system to apply. While the "optimal" solution might indeed be a non-stationary reorder point-order quantity system, the data required to systematically update the policy parameters of such a system and the analytical skill required to carry out—and *interpret*—the updates may not be warranted in all cases.

As this chapter deals with managing the supply chain, I will focus only on managing those items that move between firms in the chain—that is, the independent demand items. Management of dependent demand items is an issue of production planning, more of a "microfocus" than is warranted in this chapter. An excellent discussion of production planning problems and comparisons between MRP and Kanban systems can be found in Vollmann et al. (2005). Thus, I will spend the next few sections describing the important issues in managing independent demand inventories in the supply chain. First, however, I will present an inventory management model that does not appear, *per se*, in the framework of Figure 8.6. This model is presented, however, to give the reader an appreciation for the important tradeoffs in *any* inventory management system—stationary or non-stationary, independent or dependent demand.

8.2.1 Newsvendor and Base-Stock Models

An important starting point with all inventory management decision making is an understanding of the nature of demand. At one extreme, demand for an item can be a "spike," a single point-in-time, as it were, at which there is demand for the item in question, after which the item is never again demanded by the marketplace. This is the situation of the classic "newsvendor problem." Clearly, demand for today's newspaper is concentrated solely in the current day; there will be, effectively, no demand for today's newspaper tomorrow (it's "yesterday's news," after all) or at any point in future. Thus, if I run a newsstand, I will stock papers only to satisfy this spike in demand. If I don't buy enough papers, I will run out and lose sales—and potentially also lose the "goodwill" of those customers who wanted to buy, but could not. If I buy too many papers, then I will have to return them or sell them to a recycler at the end of the day, in either case at pennies on the dollar, if I'm lucky. The other extreme for demand is for it to occur at a stationary rate per unit time infinitely into the future. That situation will be presented in the section that follows.

I present the newsvendor model in this chapter for the purpose of helping the reader understand the basic tension in inventory decision making—namely, balancing the cost of having too much inventory versus the cost of having too little.* It does not take too much

---

*For an excellent treatment of the newsvendor problem, the reader is referred to chapter 9 in Cachon and Terwiesch (2006).

imagination to see that this too much–too little knife edge can be affected significantly by transportation and network design decisions. The important issue before us at this point, however, is to understand how to evaluate the tradeoff. Note also that in the newsvendor case, one only needs to answer the question of "how much" as the answer to the "when" question, in the case of a demand spike, is simply "sufficiently in advance of the first sale of the selling season." The newsvendor model, however, is also the basis for a base-stock inventory policy, which applies to the case of recurring demand with no cost to reorder, in which case, the answer to "when" is "at every review of the inventory level."

To illustrate the costs and tradeoffs in the newsvendor model, let us imagine a fictional retailer—we'll call them "The Unlimited"—who foresees a market for hot-pink, faux-leather biker jackets in the Northeast region for the winter 2007–08 season. Let's also assume that The Unlimited's sales of similar products in Northeastern stores in recent winter seasons— for example, flaming-red, fake-fur, full-length coats in 2005–06 and baby-blue, faux-suede bomber jackets in 2004–05—were centered around 10,000 units. The Unlimited's marketing team, however, plans a stronger ad campaign this season and estimates that there is a 90% chance that demand will be at least 14,000 units. The Unlimited also faces fierce competition in the Northeast from (also fictional) catalog retailer Cliff's Edge, who has recently launched an aggressive direct-mail and internet ad campaign. Thus, there is concern that sales might come in three to four thousand units below historical averages for similar products. Jackets must be ordered from an Asian apparel producer in March for delivery in October; only one order can be placed and it cannot be altered once it is placed. The landed cost of each jacket is $100 (the cost to purchase the item, delivered). They sell for $200 each. In March 2008, any remaining jackets will be deeply discounted for sale at $25 each.

Thus, this case gives us all the basics we will need to determine a good order quantity: a distribution of demand (or at least enough information to infer a distribution), unit sales price, unit cost, and end-of-season unit salvage value. In general, if an item sells for p, can be purchased for c, and has end-of-season salvage value of s, then the cost of missing a sale is $p - c$, the profit margin that we would have made on that sale, and the cost of having a unit in excess at the end of the selling season is $c - s$, the cost we paid to procure the item less its salvage value. If demand is a random variable D with probability distribution function (pdf) f, then the optimal order quantity maximizes expected profit, given by

$$\pi(Q) = \int\limits_{-\infty}^{Q} [(p - c)D - (c - s)(Q - D)]f(D)\mathrm{d}D + \int\limits_{Q}^{\infty} Q(p - c)f(D)\mathrm{d}D \qquad (8.1)$$

(i.e., if the order quantity exceeds demand, we get the unit profit margin on the D units we sell, but we suffer a loss of $c - s$ on the $Q - D$ units in excess of demand; however, if demand exceeds the order quantity, then we simply get the unit profit margin on the Q units we are able to sell).

One approach to finding the best order quantity would be to minimize Equation 8.1 directly. Another approach, and one that is better in drawing out the intuition of the model, is to find the best order quantity through a marginal analysis. Assume that we have ordered Q units. What are the benefits and the risks of ordering another unit versus the risks and benefits of not ordering another unit? If we order another unit, the probability that we will not sell it is $\Pr\{D \leq Q\} = F(Q)$, where F is obviously the cdf of D. The cost in this case is $c - s$, meaning that the expected cost of ordering another unit beyond Q is $(c - s)F(Q)$. If we do not order another unit, the probability that we could have sold it is $\Pr\{D > Q\} = 1 - F(Q)$ (provided that D is a continuous random variable). The cost in this case is $p - c$, the opportunity cost of the profit margin foregone by our under-ordering, meaning that the expected cost of not ordering another unit beyond Q is $(p - c)[1 - F(Q)]$.

Another way of expressing the costs is to consider $p - c$ to be the *underage cost*, or $c_u = p - c$, and to consider $c - s$ the *overage cost*, or $c_o = c - s$. A marginal analysis says that we should, starting from $Q = 1$, continue ordering more units until that action is no longer profitable. This occurs where the net difference between the expected cost of ordering another unit—the cost of overstocking—and the expected cost of not ordering another unit—the cost of understocking—is zero, or where $c_o F(Q) = c_u[1 - F(Q)]$. Solving this equation, we obtain the optimal order quantity,

$$Q^* = F^{-1} \left(\frac{c_u}{c_u + c_o} \right) \tag{8.2}$$

where the ratio $c_u/(c_u + c_o)$ is called the *critical ratio*, which we denote by CR, and which specifies the optimal order quantity at a *critical fractile* of the demand distribution.

Returning to our fictional retailer example, we have $c_u = \$100$ and $c_o = \$75$; therefore, $CR = 0.571$. The issue now is to translate this critical ratio into a critical fractile of the demand distribution. From the information given above for "The Unlimited," let's assume that the upside and downside information regarding demand implies that the probabilities are well-centered around the mean of 10,000. Moreover, let's assume that demand follows a normal distribution. Thus, we can use a normal look-up table (perhaps the one embedded in the popular Excel spreadsheet software) and use the 90th-percentile estimate from Marketing to compute the standard deviation of our demand distribution. Specifically, $z_{0.90} = 1.2816 = (14,000 - 10,000)/\sigma$, meaning that $\sigma = 3,121$. Thus, another table look-up (or Excel computation) gives us $Q^* = 10,558$, the value that accumulates a total probability of 0.571 under the normal distribution pdf with mean 10,000 and standard deviation 3,121.

As it turns out, the optimal order quantity in this example is not far from the expected value of demand. This is perhaps to be expected as the cost of overage is only 25% less than the cost of underage. One might reasonably wonder what kind of service outcomes this order quantity would produce. Let us denote α as the *in-stock probability*, or the probability of not stocking out of items in the selling season, and β as the *fill rate*, or the percentage of overall demand that we are able to fulfill. For the newsvendor model, let us further denote $\alpha(Q)$ as the in-stock probability given that Q units are ordered, and let $\beta(Q)$ denote the fill rate if Q units are ordered. It should be immediately obvious that $\alpha(Q^*) = CR$. Computing the fill rate is a little more involved. As fill rate is given by (Total demand $-$ Unfulfilled demand) $\div$ (Total demand), it follows that

$$\beta(Q) = 1 - \frac{S(Q)}{\mu} \tag{8.3}$$

where $S(Q)$ is the expected number of units short (i.e., the expected unfulfilled demand) and μ is the expected demand. Computing $S(Q)$ requires a *loss function*, which also is typically available in look-up table form for the standard normal distribution.* Thus, for normally distributed demand

$$\beta(Q) = 1 - \frac{\sigma L(z)}{\mu} \tag{8.4}$$

where $L(z)$ is the standard normal loss function[†] and $z = (Q - \mu)/\sigma$ is the standardized value of Q. Returning once again to our example, we can compute $\beta(Q^*) = 0.9016$, meaning

---

*Cachon and Terwiesch (2006) also provide loss function look-up tables for Erlang and Poisson distributions.

[†]That is, $L(z) = \int_z^\infty (x - z)\phi(x)\mathrm{d}x$, where $x \sim N(0, 1)$ and ϕ is the standard normal pdf. This can also be computed directly as $L(z) = \phi(z) - z[1 - \Phi(z)]$, where Φ is the standard normal cdf.

TABLE 8.1 Comparative Outcomes of Newsvendor Example for Increasing $g = k(p-c)$

k	$\alpha(Q^*)$	$\beta(Q^*)$	Q^*	Q^*/μ	z^*	Expected Overage
0	0.571	0.901	10,558	1.06	0.179	1543.9
0.2	0.615	0.916	10,913	1.09	0.293	1754.5
0.4	0.651	0.927	11,211	1.12	0.388	1943.2
0.6	0.681	0.935	11,468	1.15	0.470	2114.3
0.8	0.706	0.942	11,691	1.17	0.542	2269.0
1	0.727	0.948	11,884	1.19	0.604	2407.3
2	0.800	0.965	12,627	1.26	0.842	2975.4
4	0.870	0.980	13,515	1.35	1.126	3718.3
6	0.903	0.986	14,054	1.41	1.299	4196.4
8	0.923	0.989	14,449	1.44	1.426	4557.2
10	0.936	0.991	14,750	1.48	1.522	4837.0

Note: Q^* is rounded to the nearest integer, based on a critical ratio rounded to three decimal places.

that we would expect to fulfill about 90.2% of the demand for biker jackets next winter season if we order 10,558 of them.

A reasonable question to ask at this point is whether those service levels, a 57% chance of not stocking out and a 90% fill rate are "good" service levels. One answer to that question is that they are neither "good" nor "bad"; they are *optimal* for the overage and underage costs given. Another way of looking at this is to say that, if the decision maker believes that those service levels are too low, then the cost of underage must be understated, for if this cost were larger, then optimal order quantity would increase and the in-stock probability and fill rate would increase commensurately. Therefore, newsvendor models like our example commonly employ another cost factor, what we will call g, the cost of losing customer *goodwill* due to stocking out. This cost inflates the underage cost, making it $c_u = p - c + g$. The problem with goodwill cost, however, is that it is not a cost that one could pull from the accounting books of any company. Although one could clearly argue that it is "real"—customers do indeed care, in many instances, about the inconvenience of not finding items they came to purchase—it would indeed be hard to evaluate for the "average" customer. The way around this is to allow the service outcome to serve as a proxy for the actual goodwill cost. The decision maker probably has a service target in mind, one that represents the in-stock probability or fill rate that the firm believes is a good representation of their commitment to their customers, but does not "break the bank" in expected overstocking costs. Again, back to our example, Table 8.1 shows the comparative values of Q, α, and β for goodwill costs $g = k(p - c)$, with k ranging from 0 to 10. Thus, from Table 8.1, as g ranges from 0 to relatively large values, service—measured both by α and β—clearly improves, as does the expected unsold inventory.

For comparison, Table 8.1 displays a ratio of the optimal order quantity to the expected demand, Q^*/μ, and also shows the standardized value of the optimal order quantity, $z^* = (Q^* - \mu)/\sigma$. These values are, of course, specific to our example and would obviously change as the various cost factors and demand distribution parameters change. It may be more instructive to consider a "parameter-free" view of the service level, which is possible if we focus on the in-stock probability. Figure 8.7 shows a graph similar to Figure 12.2 from Chopra and Meindl (2004), which plots $\alpha(Q^*) = CR$, the optimal in-stock probability, versus the ratio of overage and underage costs, c_o/c_u.* This graph shows clearly how service levels increase as c_u increases relative to c_o, in theory increasing to 100% in-stock probability at $c_0 = 0$ (obviously an impossible level for a demand distribution that is unbounded in the positive direction). Moreover, for high costs of overstocking, the optimal in-stock probability could be well below 50%.

---

*Note that $CR = c_u/(c_u + c_o)$ can also be expressed as $CR = 1/(1 + c_o/c_u)$.

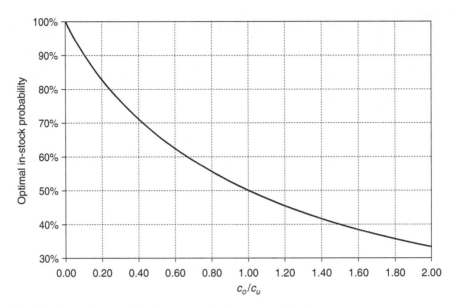

FIGURE 8.7   Optimal service level versus ratio of overage and underage costs.

These ideas lead to a natural perspective on *safety stock*, which is the amount of inventory held in excess of expected demand to protect against stockouts. In the case of the newsvendor model, where demand is a spike, the safety stock is simply given by $Q - \mu$. Note also that at $Q = \mu$, the safety stock is zero and the expected in-stock probability is 50%.

We also pointed out above that the newsvendor model is the basis for the simplest inventory system to manage items with repeating demands, namely a base-stock policy. A base-stock policy is appropriate when the cost of placing orders is zero (or negligible). In a *periodic review* setting, one in which the inventory level is reviewed on a regular cycle, say every T days, a base-stock policy is one in which, at each review, an order is placed for $B - x$ units, where B is the base-stock level and x is the *inventory position* (inventory on-hand plus outstanding orders). In a *continuous review* setting, where the inventory level is reviewed continuously (e.g., via a computerized inventory records system), a base-stock policy is sometimes called a "sell one-buy one" system as an order will be placed every time an item is sold (i.e., as soon as the inventory position dips below B). One can see how a newsvendor model serves as the basis for setting a base-stock level. If one can describe the distribution of demand over the replenishment lead time, with expected value $E[D_{DLT}]$ and variance $\text{var}[D_{DLT}]$, then the base-stock level could be set by choosing a fractile of this distribution, in essence, by setting the desired in-stock probability. For normally distributed demand over the lead time, with mean μ_{DLT} and standard deviation σ_{DLT}, this gives $B = \mu_{DLT} + z_\alpha \sigma_{DLT}$, where z_α is the fractile of the standard normal distribution that accumulates α probability under the pdf. Moreover, if demand is stationary, this base stock level is as well.

A Supply-Chain Wide Base-Stock Model

In a supply-chain context, one that considers managing base-stock levels across multiple sites in the chain, Graves and Willems (2003) present a relatively simple-to-compute approximation for the lead times that tend to result from the congestion effects in the network caused by demand uncertainty and occasional stockouts at various points in the network.

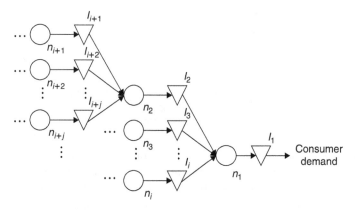

FIGURE 8.8   A general production network.

We will focus on what is called the *stochastic service formulation*, in which the *actual* time to replenish supply at any node in the network is stochastic, even if *nominal* replenishment times are deterministic. (Graves and Willems cite Lee and Billington, 1993, as the motivation for this approach). This results from each node occasionally stocking out, causing congestion in the system as other orders build up while supply is unavailable at some point in the network. The other alternative is a *guaranteed-service formulation*, in which each node holds sufficient inventory to guarantee meeting its service time commitment; this, however, requires an assumption of bounded demand. Graves and Willems trace this model back to the work of Kimball, from 1955, but published in Kimball (1988). For the stochastic service formulation, the congestion-based adjustment to lead time employed by Graves and Willems comes from Ettl et al. (2000), and leads to a fairly straightforward non-linear optimization problem to minimize the cost of holding inventories throughout the network, subject to an in-stock probability constraint at the network node farthest downstream.

In this case, the supply chain network is defined as a graph $G = (N, E)$, comprised of a set of nodes $N = \{1, \ldots, n\}$ and edges (arcs) $E = \{(i, j) : i, j \in N, i \neq j\}$. Figure 8.8 shows a pure production network, in which all nodes have exactly one successor, but could have multiple predecessors. (This is done for simplicity of exposition. The optimization method applies to more general arborescent networks as well. Graves and Willems give examples of networks for battery packaging and distribution and for bulldozer assembly.) Node 1 (n_1) is the node farthest downstream (i.e., the node that fulfills demand for the finished item), and downstream inventory I_i is controlled by node n_i. The *fulfillment time* of any node j in this network, which we denote by L_j, is the nominal amount of time required for node j to fulfill an order from a successor node. We denote the *replenishment time* of node j by τ_j. Thus, as indicated by Graves and Willems (following the model of Ettl et al., 2000), the worst-case replenishment time of node j is

$$\tau_j^{\max} = L_j + \max_{i:(i,j)\in E}\{\tau_i\} \tag{8.5}$$

but the expected replenishment time can be expressed as

$$E[\tau_j] = L_j + \sum_{i:(i,j)\in E} \pi_{ij} L_i \tag{8.6}$$

where π_{ij} is the probability that supply node i causes a stockout at j. The values of π_{ij} can be estimated using equation 3.4 in Graves and Willems (2003), derived from Ettl et al. (2000), which is

$$\pi_{ij} = \frac{1 - \Phi(z_i)}{\Phi(z_i)} \left[1 + \sum_{h:(h,j)\in E} \frac{1 - \Phi(z_h)}{\Phi(z_h)} \right]^{-1} \tag{8.7}$$

where z_i is the safety stock factor at node i, as in our discussion above.

Using the equations above—and assuming that demand over the replenishment lead time at each node j is normally distributed with mean $\mu_j E[\tau_j]$ and standard deviation $\sigma_j \sqrt{E[\tau_j]}$—leads to a relatively straightforward non-linear optimization problem to minimize the total investment in safety stock across the network. Specifically, that optimization problem is to

$$\min C = \sum_{j=1}^{N} h_j \sigma_j \sqrt{E[\tau_j]} \left(z_j + \int_{z_j}^{\infty} (x - z_j)\phi(x)\mathrm{d}x \right) \tag{8.8}$$

subject to $z_1 = \Phi^{-1}(\alpha)$, where α is the target service level at the final inventory location, Φ is the standard normal cdf, ϕ is the standard normal pdf, and h_i is the annual inventory holding cost at node i. The safety stock factors z_i $(i = 2, \ldots, n)$ are the decision variables and are also used in computing π_{ij} in Equation 8.6 via Equation 8.7. Feigin (1999) provides a similar approach to setting base-stock levels in a network, but one that uses fill rate targets—based on a fill rate approximation given by Ettl et al. (1997), an earlier version of Ettl et al. (2000)—as constraints in a non-linear optimization.

Base-Stock Model with Supply Uncertainty

Based on the work of Bryksina (2005), Warsing, Helmer, and Blackhurst (2006) extend the model of Graves and Willems (2003) to study the impact of disruptions or uncertainty in supply on stocking levels and service outcomes in a supply chain. Warsing et al. consider two possible fulfillment times, a normal fulfillment time L_j, as in Equation 8.6, and worst possible fulfillment time K_j, the latter of which accounts for supply uncertainty, or disruptions in supply. Using p_j to denote the probability that node j is disrupted in any given period, Warsing et al. extend the estimate of Equation 8.6 further by using $(1 - p_j)L_j + p_j K_j$ in place of L_j. They then compare the performance of systems using the original Graves and Willems base-stock computations, base-stock levels adjusted for supply uncertainty, and a simulation-based adjustment to safety stock levels that they develop.

Table 8.2 shows the base case parameters from a small, three-node example from Warsing et al. (2006), and Table 8.3 shows the results, comparing the Graves and Willems (2003) solution (GW), the modification of this solution that accounts for supply uncertainty (Mod-GW), and the algorithmic adjustment to the safety stock levels developed by Warsing et al. (WHB). An interesting insight that results from this work is the moderating effect of safety stock placement on the probability of disruptions. Specifically, as the ratio of downstream inventory holding costs become less expensive relative to upstream holding costs, the system places more safety stock downstream, which buffers the shock that results from a 10-fold increase in the likelihood of supply disruptions at an upstream node in the network.

TABLE 8.2 Base Case Parameters for Example from Warsing et al. (2006)

Node	L_i	K_i	p_i	h_i	μ_i	σ_i
1	1	2	0.01	10	200	30
2	3	10	0.01	1	200	30
3	3	10	0.01	1	200	30

TABLE 8.3 Results from Warsing et al. (2006) with Target In-Stock Probability of 95%

Scenario	Model	Base Stock Levels			In-Stock Probability		Fill Rate	
		Node 1	Node 2	Node 3	Mean (%)	Std Dev (%)	Mean (%)	Std Dev (%)
Base case	GW	296	694	694	91.5	2.1	93.1	2.0
	Mod-GW	299	709	709	92.9	1.8	94.3	1.8
	WHB	299	1233	709	94.5	2.2	95.8	2.0
$h_3 = 5$	GW	549	655	613	94.9	2.8	95.7	2.6
	Mod-GW	556	670	628	95.4	1.9	96.3	1.7
	WHB	556	670	628	95.4	1.9	96.3	1.7
$p_2 = 0.10$	GW	296	694	694	71.1	4.9	75.4	4.3
	Mod-GW	301	847	709	77.5	5.4	81.3	4.8
	WHB	301	1488	800	91.6	2.8	93.8	2.3
$h_3 = 5$	GW	549	655	613	80.2	3.2	83.6	2.7
$p_2 = 0.10$	Mod-GW	554	810	629	84.3	4.4	87.3	3.9
	WHB	554	1296	629	94.4	2.4	95.9	1.8

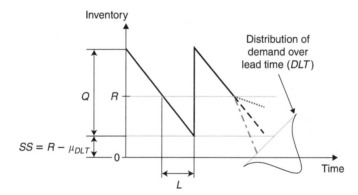

FIGURE 8.9 Continuous-review, perpetual-demand inventory model.

8.2.2 Reorder Point—Order Quantity Models

As discussed above, when there is no cost to place a replenishment order, a base-stock model that reorders at every periodic review—or whenever inventory is depleted by one or more units in a continuous review system—is appropriate. With costly ordering, however, this is likely not to be the cost-minimizing policy. Therefore, in this case, the question of *when* to get more is now relevant in addition to the question of *how much* to get.

In addition to costly ordering, let us now consider a *perpetual demand* setting, one in which demand is assumed to continue infinitely far into the future, and clearly one that is the antithesis of the spike demand that characterized the newsvendor setting. The easiest case to analyze in this setting would be a continuous-review inventory model. A pictorial view of a continuous-review, perpetual-demand model with stationary demand is shown in Figure 8.9. In this setting, an order of size Q is placed whenever the inventory level reaches, or dips below, reorder point R. The picture also provides a visual representation of safety stock for this model. From the visual we see that safety stock SS is not only the amount of inventory held in excess of demand—in this case demand over the replenishment lead time—to protect against stockouts, but it can also be seen to be the amount of inventory that is expected to be on hand when a replenishment order arrives. As indicated by the three dashed lines in the second cycle of the figure, if demand over the replenishment lead time is uncertain, the demand rate could track to expectations over the replenishment lead time (middle line), or it could experience a slow down (upper line) or an increase (lower line), in the latter case leaving the system at risk of a stockout. Setting the reorder point sufficiently high is the means of preventing such stockouts, and therefore, the reorder point, as indicated by the figure is given by

$$R = \mu_{DLT} + SS \qquad (8.9)$$

The problem in this case, then, is to find good values of two decision variables, the reorder point R and the order quantity Q, or the "when" and "how much" decisions we discussed above. If D now denotes the expected value of annual demand, A denotes the fixed cost to place a replenishment order, and h denotes the annual cost of holding a unit in inventory, then the total annual costs of ordering and holding in this model are given by

$$TAC(R,Q) = \frac{AD}{Q} + h\left(\frac{Q}{2} + R - \mu_{DLT}\right) \qquad (8.10)$$

since D/Q obviously gives the number of orders placed annually, $Q/2$ gives the cycle stock (or the average inventory—over and above any safety stock—given a stationary demand rate over time), and $R - \mu_{DLT}$ gives the safety stock.* Some time-honored texts—for example, Hax and Candea (1984) and Silver et al. (1998)—include explicit stockout costs in the annual cost equation. Our approach, however, will be to consider a service constraint in finding the (Q, R) solution that minimizes Equation 8.10.

Thus, an important issue in our formulation will be how we specify the service constraint on our optimization problem. The most general case is the one in which *both* demand per unit time period (typically days or weeks) *and* replenishment lead time (correspondingly expressed in days or weeks) are random variables. Let us, therefore, assume that the lead time L follows a normal distribution with mean μ_L and standard deviation σ_L. Further, let us assume that the distribution of the demand rate per unit time d also is normal, and let μ_d and σ_d denote its mean and standard deviation, respectively.[†] The normal-distribution assumption allows a well-defined convolution of demand over the lead time, which is therefore also normally distributed, with mean $\mu_{DLT} = \mu_d \mu_L$ and standard deviation[‡] $\sigma_{DLT} = \sqrt{\mu_L \sigma_d^2 + \mu_d^2 \sigma_L^2}$. Moreover, given a value of the reorder point R, the expected number of units short in an order cycle can be expressed as $S(R) = \sigma_{DLT} L(z)$, where $L(z)$, as above, is the standard normal loss function evaluated at z, which in this case is given by $z = (R - \mu_{DLT})/\sigma_{DLT}$. If unmet demand in a replenishment cycle is fully backlogged, it should be apparent that the expected demand met in a replenishment cycle is Q, meaning

---

*Note that this model implies that expected inventory is $Q/2 + SS$. Technically, however, we should be careful about how demand is handled in the event of a stockout. If demand is backlogged, then expected inventory at any point in time is actually $Q/2 + SS + E[BO]$, where $E[BO]$ is the expected number of backorders. Silver et al. (1998, p. 258), however, indicate that since $E[BO]$ is typically assumed to be small relative to inventory, it is reasonable to use $Q/2 + SS$ for expected inventory. If demand is lost in the event of a stockout, then expected inventory is exactly $Q/2 + SS$.

†As indicated by Silver et al. (1998), the assumption of the normal distribution for lead time demand is common for a number of reasons, particularly that it is "convenient from an analytic standpoint" and "the impact of using other distributions is usually quite small" (p. 272).

‡The general form of this equation, from the variance of a random-sized sum of random variables, can be found in example 4 in Appendix E of Kulkarni (1995, pp. 577–578).

that the fill rate in this continuous-review, perpetual demand model is given by

$$\beta(R,Q) = 1 - \frac{S(R)}{Q} = 1 - \frac{\sigma_{DLT}L(z)}{Q} \text{*} \tag{8.11}$$

Note, therefore, that the problem to minimize Equation 8.10 subject to a fill rate constraint based on Equation 8.11 is non-linear in both the objective function and the constraint. The objective (Equation 8.10), however, is convex in both Q and R,[†] so the solution is fairly straightforward.

A less rigorous approach to finding a (Q, R) solution would be to solve for Q and R separately. Note that $z = (R - \mu_{DLT})/\sigma_{DLT}$ gives a fractile of the distribution of demand over the lead time. Thus, we could set R to achieve a desired in-stock probability, along the lines of the newsvendor problem solution discussed above (i.e., to accumulate a given amount of probability under the lead-time demand distribution). In this setting, the in-stock probability is typically referred to as the *cycle service level* (CSL), or the expected in-stock probability in each replenishment cycle. With R fixed to achieve a desired CSL, we then use the first-order condition on Q in Equation 8.10,

$$\frac{\partial TAC(R,Q)}{\partial Q} = -\frac{AD}{Q^2} + \frac{h}{2} = 0 \tag{8.12}$$

to solve for

$$Q^*_{\text{EOQ}} = \sqrt{\frac{2AD}{h}} \tag{8.13}$$

the familiar *economic order quantity* (EOQ). Note also that if demand uncertainty is removed from our model altogether, then the safety stock term is dropped from Equation 8.10 and EOQ via Equation 8.13 provides the optimal order quantity.

Reorder point–order quantity (ROP-OQ) models are also relevant to periodic review systems, in which the system inventory is reviewed at some regular period, say every T days, and an order is placed if the inventory position is below a given level. Analysis of the periodic-review system is more complex and requires a dynamic programming solution (see, e.g., Clark and Scarf, 1960), with the result being an (s, S) policy, wherein an order of size $S - s$ is placed if the inventory position is below s at the current review. As pointed out by Silver et al. (1998), however, a reasonably good, approximate (s, S) policy can be generated from continuous-review-type parameters. Specifically, one could set s equal to an extension, of sorts, of the continuous-review reorder point R, set to cover demand over the lead time *plus* the review period and achieve a desired cycle service level. The order-up-to level S would then be given by $S = R + Q$, with Q given by EOQ (or a some slight adjustment—see Silver et al., 1998, pp. 331–336, for more details).

Multi-Item Models

Our treatment of ROP-OQ models up to this point has concerned only a single item at a single location. It would be the rare company, however, that managed only a single item at a single location. As one complication introduced with multiple items, consider a situation in

---

[*]Again, the backlogging assumption is important. If unmet demand is lost, then the replenishment quantity Q is never used to meet backlogged demand, and therefore, the total demand experienced in an average cycle is $Q + E[BO] = Q + S(R)$, yielding $\beta(R,Q) = 1 - \sigma_{DLT}L(z)/[Q + \sigma_{DLT}L(z)]$.

[†]See Hax and Candea (1984, p. 206) for a proof of the convexity of a related model.

which k items can be ordered jointly at a discount in the ordering cost—that is, where the cost of ordering k items independently is kA, but ordering these items jointly incurs only an incremental cost, $a_i \geq 0$, for each item i ordered beyond the single base cost of ordering, A, such that $A + \sum_{i=1}^{k} a_i < kA$.

A straightforward way to address this situation is to consider ordering all items jointly on a common replenishment cycle. In essence this changes the decision from "how much" to order—the individual order quantities, Q_i—to "how often" to order, which is given by n, defined as the number of joint orders placed per year. Then since $n = D_i/Q_i \Rightarrow Q_i = D_i/n$ for each item i, we can build a joint-ordering TAC model with

$$TAC(n) = \left(A + \sum_{i=1}^{k} a_i \right) n + \sum_{i=1}^{k} h_i \left(\frac{D_i}{2n} + SS_i \right) \tag{8.14}$$

where h_i is the annual cost of holding one unit of item i in inventory, $D_i/(2n) = Q_i/2$ is the cycle stock of item i, and SS_i is the safety stock of item i. For fixed values of SS_i (or if demand is certain and $SS_i = 0$, $i = 1, \ldots, k$), the optimal solution to Equation 8.14 is found by setting

$$\frac{\partial TAC(n)}{\partial n} = A + \sum_{i=1}^{k} a_i - \frac{1}{2n^2} \sum_{i=1}^{k} h_i D_i = 0 \tag{8.15}$$

yielding

$$n^* = \sqrt{\frac{\sum_{i=1}^{k} h_i D_i}{2 \left(A + \sum_{i=1}^{k} a_i \right)}} \tag{8.16}$$

Introducing safety stock and reorder points into the decision complicates the idea of joint replenishments. Clearly, it would negate the assumption that all items would appear in each replenishment since the inventory levels of the various items are unlikely to hit their respective reorder points at exactly the same point in time. Silver et al. (1998) discuss an interesting idea first proposed by Balintfy (1964), a "can-order" system. In this type of inventory control system, two reorder points are specified, a "can-order" point, at which an order *could* be placed, particularly if it would allow a joint-ordering discount, and a "must-order" point, at which an order *must* be placed to guard against a stockout. Although I will not discuss the details here, Silver et al. (1998: p. 435) provide a list of references for computing "must-order" and "can-order" levels in such a system. Thus, one could propose a review period for the joint-ordering system and set reorder points to cover a certain cumulative probability of the distribution of demand over the lead time plus the review period.

Let's consider an example that compares individual orders versus joint orders for two items. Floor-Mart—"We set the floor on prices" (and, apparently, on advertising copy)—is a large, (fictional) discount retailer. Floor-Mart stocks two models of 20-inch, LCD-panel TVs, Toshiba and LG, that it buys from two different distributors. Annual demand for the Toshiba 20-inch TV is $D_1 = 1600$ units, and the unit cost to Floor-Mart is \$400. Annual demand for the LG 20-inch TV is $D_2 = 2800$ units, and the unit cost to Floor-Mart is \$350. Assuming 365 sales days per year, this results in an expected daily demand of $\mu_{d,1} = 4.38$ units for the Toshiba TV and $\mu_{d,2} = 7.67$ units for the LG TV. Assume that Floor-Mart also has data on demand uncertainty such that $\sigma_{d,1} = 1.50$ and $\sigma_{d,2} = 2.50$. Annual holding costs at Floor-Mart are estimated to be 20% on each dollar held in inventory, and Floor-Mart's target fill rate on high-margin items like TVs is 99%. Floor-Mart uses the same contract carrier to ship TVs from each of the distributors. Let's put a slight twist on the problem formulation as it was stated above. Assume that Floor-Mart's only fixed cost of placing a replenishment order with its TV distributors is the cost the carrier charges to move the goods to Floor-Mart. The carrier charges \$600 for each shipment from the Toshiba distributor, and the mean and standard deviation of the replenishment lead time are $\mu_{L,1} = 5$ and $\sigma_{L,1} = 1.5$ days. The carrier charges \$500 for each shipment from the LG distributor, and the mean and standard deviation of the replenishment lead time are $\mu_{L,2} = 4$ and $\sigma_{L,2} = 1.2$ days. However, the carrier also offers a discounted "stop-off" charge to pick up TVs from *each* distributor on a *single* truck, resulting in a charge of \$700.

With the parameters above, we find that $\mu_{DLT,1} = 21.92$, $\sigma_{DLT,1} = 7.38$, $\mu_{DLT,2} = 30.68$, and $\sigma_{DLT,2} = 10.60$. Solving Equation 8.10 for each TV to find the optimal independent values of Q and R (via Excel Solver), we obtain $(Q_1, R_1) = (166, 25)$ and $(Q_2, R_2) = (207, 36)$. By contrast, independent EOQ solutions would be $Q_{\text{EOQ},1} = 155$ and $Q_{\text{EOQ},2} = 200$. Using Equation 8.16 for the joint solution, we obtain $n^* = \sqrt{(0.2)(400 \cdot 1600 + 350 \cdot 2800)/(2 \cdot 700)} = 15.21$, yielding $Q_1 = 105$ and $Q_2 = 184$. The joint solution—assuming that it does not exceed the truck capacity—saves approximately \$5100 in ordering and holding costs over either independent solution. The issue on the table, then, as we discuss above, is to set safety stock levels in this joint solution. As orders would be placed only every 3.4 weeks with $n = 15.21$, a review period in line with the order cycle would probably inflate safety stock too dramatically. Thus, the reader should be able to see the benefit of a "can-order" system in this case, perhaps with a one-week review period, a high CSL—that is, relatively large value of $z = (R - \mu_{DLT})/\sigma_{DLT}$—on the "can-order" reorder point, and a lower CSL on the "must-order" reorder point. In closing the example, we should point out that we have assumed a constant rate of demand over time, which may be reasonable for TVs, but clearly ignores the effects of intermittent advertising and sales promotions. Later in this section, we discuss distribution requirements planning, which addresses situations where demand is not stationary over time.

Finally, the more general case for joint ordering is to consider what Silver et al. (1998) call *coordinated replenishment*, where each item i is ordered on a cycle that is a multiple, m_i, of a base order cycle of T days. Thus, each item is ordered every $m_i T$ days. Some subset, possibly a single item, of the k items has $m_i = 1$, meaning that those items appear in every replenishment order. Silver et al. (1998, pp. 425–430) offer an algorithmic solution to compute m_i $(i = 1, \ldots, k)$ and T for this case. Jackson et al. (1985) consider the restrictive assumption that $m_i \in \{2^0, 2^1, 2^2, 2^3, \ldots\} = \{1, 2, 4, 8, \ldots\}$, which they call a "powers-of-two" policy, and show that this approach results in an easy-to-compute solution whose cost is no more than 6% above the cost of an optimal policy with unrestricted values of m_i. Another situation that could link item order quantities is an aggregate quantity constraint or a budget constraint. Hax and Candea (1984) present this problem and a Lagranian-relaxation approach to solving it; they refer to Holt et al. (1960) for several methods to solve for the Lagrange multiplier that provides near-optimal order quantities.

Multi-Echelon Models

In a *multi-echelon* supply chain setting, Clark and Scarf (1960) show that the supply-chain optimal solution at each echelon is an "echelon-stock policy" of (s, S) form. Clark and Scarf are able to solve their stochastic dynamic programming formulation of the problem

by redefining the manner in which inventory is accounted for at successive upstream echelons in the chain. Clark and Scarf's definition of *echelon stock* is "the stock at any given
installation plus stock in transit to or on hand at a lower installation" (p. 479). Thus, the
echelon stock at an upstream site includes stock at that site *plus* all stock downstream
that has not yet been sold to a customer. Using an echelon-stock approach coordinates the
ordering processes across the chain, or at least across the echelons that can be coordinated.
This last point refers to the complication that a supply-chain-wide, echelon-stock policy
would require all firms in the chain (except the one farthest upstream) to provide access to
their inventory data. Such supply-chain-wide transparency is perhaps unlikely to occur, but
coordination between company-owned production sites and company-owned distribution
sites is eminently reasonable. Silver et al. (1998, 477–480), for example, provide computational procedures to determine coordinated warehouse (upstream) and retailer (downstream) order quantities using an echelon-stock approach. In addition, I refer the reader to
Axsäter (2000) for an excellent overview of echelon-inventory methods and computational
procedures.

The astute reader should note that we have already discussed an approach to setting
safety stocks in a multi-echelon system, namely the non-linear optimization suggested by
Graves and Willems (2003). In the author's view, this approach is more intuitive and more
directly solved than an echelon inventory approach, but perhaps only slightly less problematic in terms of obtaining data about upstream and downstream partners in the supply
chain. An approach that uses the GW optimization (see the earlier section in this chapter
on "A Supply-Chain Wide Base-Stock Model") to set network-wide safety stock levels, coupled with simple EOQ computations as starting points for order quantities, and possibly
"tweaked" by simulation to find improved solutions, may be a promising approach to solving
network-wide inventory management problems.

8.2.3   Distribution Requirements Planning

In an environment in which independent-item demands vary over time, distribution requirements planning (DRP) provides a method for determining the time-phased inventory levels
that must be present in the distribution system to fulfill projected customer demands on
time. As Vollmann et al. (2005) point out, "DRP's role is to provide the necessary data for
matching customer demand with the supply of products being produced by manufacturing"
(p. 262). Note that in any make-to-stock product setting, the linchpin between product manufacturing and customer demand is the distribution system, and indeed, it is distribution
system inventory that coordinates customer demand with producer supply. While distribution system inventories could be set using the ROP-OQ control methods as described
above, the critical difference between DRP and ROP-OQ is that DRP plans *forward* in
time, whereas ROP-OQ systems react to actual demands whenever those demands drive
inventory to a level that warrants an order being placed. This is not to say that the two
approaches are inconsistent. Indeed, it is quite easy to show that, in a stationary-demand
environment with sufficiently small planning "time buckets," DRP and ROP-OQ produce
equivalent results. Where DRP provides a clear benefit, however, is in situations where
demand is *not* constant over time, particularly those situations in which surges and drops
in demand can be anticipated reasonably accurately.

The basic logic of DRP is to compute the time-phased replenishment schedule for each
stock-keeping unit (SKU) in distribution that keeps the inventory level of that SKU at or
above a specified safety stock level (which, of course, could be set to zero). Let the DRP
horizon span planning periods $t = 1, \ldots, T$, and assume that safety stock SS, order lot size
Q, and lead time L are given for the SKU in question. In addition, assume that we are given

a period-by-period forecast of demand for this SKU, D_t, $t = 1, \ldots, T$. Then, we can compute the *planned receipts* for each period t, namely the quantity—as an integer multiple of the lot size—required to keep inventory at or above SS.

To illustrate these concepts, let us consider an example, altered from Bozarth and Handfield (2006). Assume that MeltoMatic Company manufactures and distributes snow blowers. Sales of MeltoMatic snow blowers are concentrated in the Midwestern and Northeastern states of the United States. Thus, the company has distribution centers (DCs) located in Minneapolis and Buffalo, both of which are supplied by MeltoMatic's manufacturing plant in Cleveland. Table 8.4 shows the DRP records for the Minneapolis and Buffalo DCs for MeltoMatic's two SKUs, Model SB-15 and Model SBX-25. The first three lines in the table sections devoted to each SKU at each DC provide the forecasted demand (requirements), D_t; the scheduled receipts, SR_t, already expected to arrive in future periods; and the projected ending inventory in each period, I_t. This last quantity, I_t, is computed on a period-by-period basis as follows:

1. For period t, compute *net requirements* $NR_t = \max\{0, D_t + SS - (I_{t-1} + SR_t)\}$.
2. Compute planned receipts $PR_t = \lceil NR_t/Q \rceil \cdot Q$, where $\lceil x \rceil$ gives the smallest integer greater than or equal to x. (Note that this implies that if $NR_t = 0$, then $PR_t = 0$ as well.)
3. Compute $I_t = I_{t-1} + SR_t + PR_t - D_t$. Set $t \leftarrow t + 1$.
4. If $t \leq T$, go to step 1; else, done.

Then, planned receipts are offset by the lead time for that SKU, resulting in a stream of *planned orders*, PO_t, to be placed so that supply arrives in the DC in time to fulfill the projected demands (i.e., $PO_t = PR_{t+L}$ for $t = 1, \ldots, T - L$, and any planned receipts in periods $t = 1, \ldots, L$ result in orders that are, by definition, *past due*).

Note that an important aspect of DRP is that it *requires human intervention* to turn planned orders into *actual* orders, the last line in Table 8.4 for each SKU. To further illustrate this point, let's extend our example to consider a situation where the human being charged with the task of converting planned orders into actual orders might alter the planned order stream to achieve some other objectives. For example, let us assume that the carrier that ships snow blowers into MeltoMatic's warehouses in Minneapolis and Buffalo provides a discount if the shipment size exceeds a truckload quantity of 160 snow blowers, relevant to shipments of either SKU or to combined shipments containing a mix of both SKUs. Table 8.5 shows an altered actual order stream that generates orders across both SKUs of at least 160 units in all periods $t = 1, \ldots, T - L$ where $PO_t > 0$ for either SKU. In addition, Table 8.5 also shows the temporary overstock created by that shift in orders, with some units arriving in advance of the original plan. The question would be whether the projected savings in freight transportation exceeds the expense of temporarily carrying excess inventories. Interestingly, we should also note that the shift to larger shipment quantities, somewhat surprisingly, creates a smoother production schedule at the Cleveland plant (after an initial bump in Week 46).

Once again, therefore, we revisit the enormity of the problem of "optimizing the supply chain," noting the complexity of the above example for just two SKUs at a single firm. Granted, jointly minimizing inventory cost, transportation cost, and possibly also "production-change" costs might be a tractable problem at a single firm for a small set of SKUs, but the challenge of formulating and solving such problems clearly grows as the number of component portions of the objective function grows, as the number of SKUs being planned grows, and as the planning time scale shrinks (e.g., from weeks to days).

TABLE 8.4 DRP Records for MeltoMatic Snow Blowers Example with Actual Orders Equal to Planned Orders

Week		45	46	47	48	49	50	51	52	1	2	3	4
MINNEAPOLIS DC													
Model SB-15													
Forecasted requirements		60	60	70	70	80	80	80	80	90	90	95	95
Scheduled receipts		120											
Projected ending inventory	80	140	80	10	60	100	20	60	100	10	40	65	90
Net requirements		0	0	0	70	30	0	70	30	0	90	65	40
Planned receipts		0	0	0	120	120	0	120	120	0	120	120	120
Planned orders	0	0	**120**	**120**	0	**120**	**120**	0	**120**	**120**	**120**	0	0
Actual orders		0	**120**	**120**	0	**120**	**120**	0	**120**	**120**	**120**	0	0
LT (weeks): 2 Safety stock (units): 10 Lot size (units): 120													
Model SBX-25													
Forecasted requirements		40	40	50	60	80	90	100	100	110	110	100	100
Scheduled receipts		40											
Projected ending inventory	100	100	60	10	30	30	20	80	60	30	80	60	40
Net requirements		0	0	0	60	60	70	90	30	60	90	30	50
Planned receipts		0	0	0	80	80	80	160	80	80	160	80	80
Planned orders	0	0	**80**	**80**	**80**	**160**	**80**	**80**	**160**	**80**	**80**	**80**	**80**
Actual orders		0	**80**	**80**	**80**	**160**	**80**	**80**	**160**	**80**	**80**	**80**	**80**
LT (weeks): 2 Safety stock (units): 10 Lot size (units): 80													
BUFFALO DC													
Model SB-15													
Forecasted requirements		70	70	80	80	90	90	100	100	120	120	140	140
Scheduled receipts		100											
Projected ending inventory	30	60	110	30	70	100	10	30	50	50	50	30	10
Net requirements		0	20	0	60	30	0	100	80	80	80	100	120
Planned receipts		0	120	0	120	120	0	120	120	120	120	120	120
Planned orders	0	**120**	0	**120**	**120**	0	**120**	**120**	**120**	**120**	**120**	**120**	0
Actual orders		**120**	0	**120**	**120**	0	**120**	**120**	**120**	**120**	**120**	**120**	0
LT (weeks): 1 Safety stock (units): 10 Lot size (units): 120													
Model SBX-25													
Forecasted requirements		50	50	60	70	80	90	100	110	130	130	100	100
Scheduled receipts		60											
Projected ending inventory	30	40	70	90	20	20	20	80	45	80	65	90	70
Net requirements		40	0	5	0	75	75	95	75	45	65	5	25
Planned receipts		0	80	80	0	80	80	160	80	160	80	160	80
Planned orders	0	**80**	**80**	0	**80**	**80**	**160**	**80**	**160**	**80**	**160**	**80**	70
Actual orders		**80**	**80**	0	**80**	**80**	**160**	**80**	**160**	**80**	**160**	**80**	70
LT (weeks): 1 Safety stock (units): 15 Lot size (units): 80													
CLEVELAND PLANT													
Master production schedule (Gross requirements)	**SB-15**	120	120	240	120	120	240	120	240	240	240	90	70
	SBX-25	80	160	80	160	240	240	160	320	160	240	85	25
	Total	200	280	320	280	360	480	280	560	400	480		

TABLE 8.5 DRP Records for MeltoMatic Snow Blowers Example with Actual Orders Altered for Transportation Discount

Week			45	46	47	48	49	50	51	52	1	2	3	4
MINNEAPOLIS DC														
Model SB-15	80	0												
Forecasted requirements			60	60	70	70	80	80	80	80	90	90	95	95
Projected ending inventory			140	80	10	60	100	20	60	100	10	40	65	90
Planned orders			0	120	120	0	120	120	0	120	120	120	0	0
Actual orders			0	120	120	0	120	120	0	120	120	120	0	0
Model SBX-25	100	0												
Forecasted requirements			40	40	50	60	80	90	100	100	110	110	100	100
Projected ending inventory			100	60	10	30	30	20	80	60	30	80	60	40
Planned orders			0	80	80	80	160	80	80	160	80	80	0	0
Actual orders			0	80	80	160	80	80	160	80	80	80	0	0
Total for DC														
Planned orders			0	200	200	80	280	200	80	280	200	200	0	0
Actual orders			0	200	200	160	200	200	160	200	200	200	0	0
Temporary overstock			**0**	**0**	**0**	**80**	**0**	**0**	**80**	**0**	**0**	**0**	**0**	**0**
BUFFALO DC														
Model SB-15	30	0												
Forecasted requirements			70	70	80	80	90	90	100	100	120	120	140	140
Projected ending inventory			60	110	30	70	100	10	30	50	50	50	30	10
Planned orders			120	0	120	120	0	120	120	120	120	120	120	0
Actual orders			120	0	120	120	0	120	120	120	120	120	120	0
Model SBX-25	30	0												
Forecasted requirements			50	50	60	70	80	80	100	110	130	130	100	100
Projected ending inventory			40	70	90	20	20	20	80	50	80	30	90	70
Planned orders			80	80	0	80	80	160	80	160	80	160	80	0
Actual orders			80	160	80	80	160	80	80	80	80	80	80	0
Total for DC														
Planned orders			200	80	120	200	80	280	200	280	200	280	200	0
Actual orders			200	160	200	200	160	200	200	200	200	200	200	0
Temporary overstock			**0**	**80**	**160**	**160**	**240**	**160**	**160**	**80**	**80**	**0**	**0**	**0**
CLEVELAND PLANT														
Master production schedule (Gross requirements)	SB-15		120	120	240	120	120	240	120	240	240	240		
	SBX-25		80	240	160	240	240	160	240	160	160	160		
	Total		200	360	400	360	360	400	360	400	400	400		

The examples above, however, give us some insights into how inventory decisions could be affected by transportation considerations, a topic we consider further in the section that follows.

8.3   Managing Transportation in the Supply Chain

8.3.1   Transportation Basics

Ballou (2004) defines a *transport service* "as a set of transportation performance characteristics purchased at a given price" (p. 167). Thus, in general, we can describe and measure a freight transport mode by the combination of its cost and its service characteristics. Moreover, we can break down the service characteristics of a transport service into *speed* (average transit time), *reliability* (transit time variability), and risk of *loss/damage*. Freight transit modes are water, rail, truck, air, and pipeline; interestingly, Chopra and Meindl (2004) also include an *electronic* mode, which may play an increasing role in the future, and already has for some goods like business documents (e.g., e-mail and fax services) and recorded music (e.g., MP3 file services like Napster and Apple's iTunes). I will therefore distinguish *physical* modes of freight transit from this nascent electronic mode and focus our discussion exclusively on the physical modes. Ballou's characterization of a freight transport service leads nicely to a means of comparing physical modes—with respect to their relative cost, speed of delivery, reliability of delivery, and risk of loss and damage. Ballou (2004) builds a tabular comparison of the relative performance of the modes in terms of cost, speed, reliability, and risk of loss/damage, but I would suggest that there is some danger in trying to do this irrespective of the nature of the cargo or the transportation *lane* (i.e., origin–destination pair) on which the goods are traveling. Some comparisons are fairly obvious, however: Air is clearly the fastest and most expensive of the physical modes, whereas water is the slowest and perhaps the least expensive, although pipeline could probably rival water in terms of cost. In terms of "in-country" surface modes of travel, rail is much cheaper than truck, but also much slower—partially because of its lack of point-to-point flexibility—and also more likely to result in damage to or, loss of, the cargo due to the significant number of transfers of goods from train to train as a load makes its way to its destination via rail.

A Harvard Business School (Harvard Business School, 1998) note on freight transportation begins with a discussion of the deregulation of the rail and truck freight transport markets in the United States and states that "... an unanticipated impact of regulatory reform was a significant consolidation in the rail and LTL [less-than-truckload] trucking sectors" (p. 3). This stemmed from the carriers in these markets seeking what the HBS note calls "economies of flow," consolidating fixed assets by acquiring weaker carriers to gain more service volume, against which they could then more extensively allocate the costs of managing those fixed assets. This provides interesting insights to the cost structures that define essentially any transport service, *line-haul costs*—the variable costs of moving the goods—versus *terminal/accessorial costs*—the fixed costs of owning or accessing the physical assets and facilities where shipments can be staged, consolidated, and routed.

Even without the benefit of detailed economic analysis, one can infer that modes like pipeline, water, and rail have relatively large terminal/accessorial costs; that air and truck via LTL have less substantial terminal costs; and that truck via truckload (TL), as a point-to-point service, has essentially no terminal costs. Greater levels of fixed costs for the carrier create a stronger incentive for the carrier to offer per-shipment volume discounts on its services. Not only does this have a direct impact on the relative cost of these modes, but it also has implications for the relative lead times of freight transit modes, as shown in Figure 8.10, reproduced (approximately) from Piercy (1977). In this case, more substantial terminal operations—that is, rail yards to build trains or LTL terminals to build full trailer

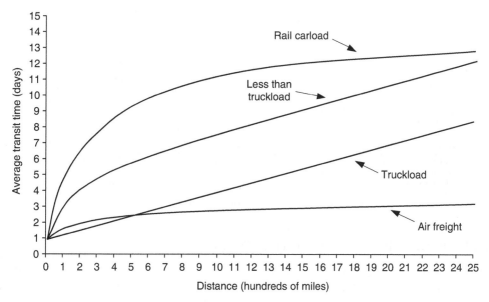

FIGURE 8.10   Comparison of shipping distance versus lead time for various modes. (Reproduced from Piercy, 1977.)

loads—result in more non-linearity in the distance-versus-transit time relationship. Note from the figure that for TL, this relationship is essentially linear as this mode does not require terminal operations to mix and consolidate loads.

In building a model of transportation cost as a function of the shipment quantity, the remainder of this section will focus on truck transit, as that is the mode for which the most data is available and for which the full-load versus less-than-full-load analysis is most interesting. The basic ideas of the results, however, apply to other transit modes as well. The basic tension for the decision maker is between using smaller shipments more frequently, but at a potentially higher per-unit shipping cost, versus making larger, but less expensive, shipments less frequently. Clearly, this has implications for transportation and inventory costs, and potentially for customer service as well. The analysis that follows, therefore, begins to round out the triangle of SCM strategy—inventory, transportation, locations, and customer service—that was presented earlier in this chapter.

In the trade press, Speigel (2002) picks up on exactly this tension in the TL versus LTL decision. As Speigel indicates, and as we draw out in an example later in this section, the answer to the TL–LTL question depends on a number of factors and explains why many shippers use transportation management system (TMS) software to attempt to generate minimum cost shipment decisions. Moreover, Speigel points out that it is getting increasingly difficult to fill truckloads in practice when one considers the many factors driving small shipment sizes: an increasing number of stock-keeping-units driven by customers' demands for greater customization, lean philosophies and the resulting push to more frequent shipments, and customer service considerations that are driving more decentralized distribution networks serving fewer customers per distribution center (this last theme returns in the succeeding section of this chapter on "Managing Locations"). As I point out after our example below, such trends probably lead to a need to consider mixed loads as a possible means of building larger, more economical shipments. This section closes with a general cost model that is well-positioned for carrying out such an analysis, even without the help of a costly TMS.

TABLE 8.6 Breakdown of United States Physical Distribution Costs, 1988 Versus 2002

	1988[*]		2002[†]	
	% of Sales	$/cwt[‡]	% of Sales	$/cwt
Transportation	3.02	12.09	3.34	26.52
Warehousing	1.90	11.99	2.02	18.06
Customer service/order entry	0.79	7.82	0.43	4.58
Administration	0.35	2.35	0.41	2.79
Inventory carrying cost @ 18%/year	1.76	14.02	1.72	22.25
Other	0.88	5.94	NR	NR
Total distribution cost	7.53	45.79	7.65	67.71

[*] From Davis, H. W., *Annual Conference Proceedings of the Council of Logistics Management*, 1988. With permission.
[†] From Davis, H. W. and W. H. Drumm, *Annual Conference Proceedings of the Council of Logistics Management*, 2002. With permission.
[‡] Weight measure "cwt" stands for "hundred-weight," or 100 lbs.

8.3.2 Estimating Transportation Costs

Table 8.6 shows a breakdown of total logistics cost as a percent of sales as given in Ballou (1992) and in Ballou (2004). From the table, one can note the slight growth in transportation cost as a fraction of sales and the slight decline in inventory cost over time. Thus, it would appear that effectively managing transportation costs is an important issue in controlling overall logistics costs, in addition to managing the impact of transportation decisions on warehousing and inventory carrying costs, the other large cost components in the table.

Nearly all operations management or logistics textbooks that incorporate quantitative methods (e.g., Nahmias, 1997; Silver et al., 1998; Chopra and Meindl, 2004; Ballou, 2004) present methods for computing reorder points and order quantities for various inventory models and, in some cases, also discuss accounting for transportation costs in managing inventory replenishment. To the author's knowledge, however, no textbook currently presents any models that effectively tie these interrelated issues together by incorporating a general model of transportation costs in the computation of optimal or near-optimal inventory policy parameters. Moreover, although many academics have previously proposed or studied joint inventory–transportation optimization models—extending back to the work of Baumol and Vinod (1970)—the author's experience is that the use of previously published approaches has not become commonplace among practitioners.

Given the recent estimate of Swenseth and Godfrey (2002) that transportation costs could account for as much as 50% of the total annual logistics cost of a product (consistent with Table 8.6, which shows aggregate data), it seems clear that failing to incorporate these costs in inventory management decisions could lead to significantly sub-optimal solutions—that is, those with much higher than minimal total logistics costs. Perhaps the most common model for transportation costs in inventory management studies is simply to assume a fixed cost to ship a truckload of goods (see, e.g., Waller et al., 1999, for use of such a model in studying vendor-managed inventory). This model, however, is insensitive to the effect of the shipment quantity on the per-shipment cost of transportation and seems unrealistic for situations in which goods are moved in smaller-sized, less-than-truckload shipments. In contrast, several authors have considered discrete rate-discount models, using quantity breaks to represent the shipment weights at which the unit cost changes. Tersine and Barman (1991) provide a review of some earlier modeling work. A recent example is the work of Çetinkaya and Lee (2002), who modify a cost model introduced by Lee (1986), which represents transportation costs as a stepwise function to model economies of scale by explicitly reflecting the lower per-unit rates for larger shipments.

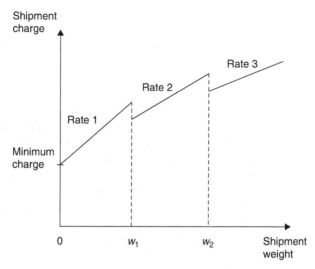

FIGURE 8.11   LTL rate structure as typically stated.

TABLE 8.7   Example Class 100 LTL
Rates for NYC Dallas Lane

Minimum Weight (lbs)	Rate ($/cwt)
1	107.75
500	92.26
1000	71.14
2000	64.14
5000	52.21
10,000	40.11
20,000	27.48
30,000	23.43
40,000	20.52

In contrast to explicit, discrete rate-discount models of freight rates, Swenseth and Godfrey (1996) review a series of models that specify the freight rate as a continuous, non-linear function of the shipment weight, building off of the work of Ballou (1991), who estimated the accuracy of a linear approximation of trucking rates. One advantage of continuous functions is that they do not require the explicit specification of rate breakpoints for varying shipment sizes nor do they require any embedded analysis to determine if it is economical to increase—or over-declare—the shipping weight on a given route. A second advantage is that a continuous function may be parsimonious with respect to specifying its parameters and can therefore be used in a wide variety of optimization models.

The idea that a shipper might benefit from "over-declaring" a shipment, or simply shipping more in an attempt to reduce the overall shipping cost, stems from the manner in which LTL rates are published by carriers. Figure 8.11 shows a pictorial view of the typical LTL rate structure as it is published to potential shippers. This rate structure uses a series of weight levels at which the per-unit freight rate—that is, the slope of the total shipment charge in Figure 8.11—decreases. A sample of such published rates is shown in Table 8.7, which contains rates for the New York City-to-Dallas lane for Class 100 freight,[*]

---

[*]Coyle et al. (2006) indicate that the factors that determine the freight class of a product are its density, its storability and handling, and its value. Class 100 freight, for example, would have a density between 6 and 12 lbs/ft^3 for LTL freight.

as given in Ballou (2004, table 6.5). Modeling per-unit rates for shipments of an arbitrary size requires one to account for *weight breaks*, which are used by carriers to determine the shipment weight at which the customer receives a lower rate, and which are typically *less* than the specific cutoff weight for the next lower rate to avoid creating incentives for a shipper to over-declare, as discussed above. For example, using the rate data in Table 8.7 for a Class 100 LTL shipment from New York to Dallas, if one were to ship 8000 pounds (80 cwt) on this lane, the charge without considering a possible weight break would be $80 \times 52.21 = \$4176.80$. If, however, one were to add 2000 pounds of "dummy weight," the resulting charge would be only $100 \times 40.11 = \$4011$. In Figure 8.11, this is the situation for a shipment weight just slightly less than w_2, for example. Thus, to discourage shippers from over-declaring or "over-shipping"—and also leaving space for more goods on the carrier's trailer and, therefore, providing the carrier with an opportunity to generate more revenue— the carrier actually bills any shipment at or above 7682 lbs (the weight break, or the weight at which the $52.21 rate results in a total charge of $4011) as if it were 10,000 lbs. Therefore, an 8000-lb shipment will actually result in carrier charges of only $4011. The *effective rate* for this 8000-lb shipment is $\$4011 \div 80 = \$50.14/\text{cwt}$. Figure 8.12 shows a pictorial view of the rates *as actually charged*, with all shipment weights between a given weight break at the next rate break point (i.e., w_1 or w_2) being charged the same amount.

Tyworth and Ruiz-Torres (2000) present a method to generate a continuous function from rate tables like those in Table 8.7. The functional form is $r(W) = C_w W^b$, where W is the shipment weight; $r(W)$ is the rate charged to ship this amount, expressed in dollars per hundred pounds (i.e., \$/cwt); and C_w and b are constants found using non-linear regression analysis. With knowledge of the weight of the item shipped, this rate function can easily be expressed as $r(Q) = CQ^b$ (in \$/unit), where Q is the number of units shipped and C is the "quantity-adjusted" equivalent of C_w. Appending the rate function $r(Q)$ to the annual cost equation for a continuous-review, ROP-OQ inventory system like the one presented earlier in this chapter results in a relatively straightforward non-linear optimization problem, as we will demonstrate below.

Thus, to generate the rate function parameters, we can, for a given origin–destination pair and a given set of LTL rates, compute effective rates for a series of shipment weights. The resulting paired data can be mapped to the continuous function $r(W)$ presented

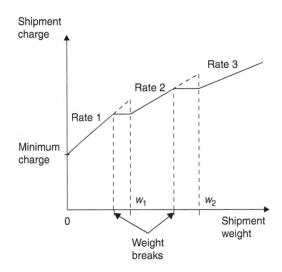

FIGURE 8.12   LTL rate structure as typically charged.

above. Although Tyworth and Ruiz-Torres (2000) present this mapping as a non-linear curve-fitting process, let us consider a computationally simpler linear regression approach. Since $r(W) = C_w W^b$, then $\ln r(W) = \ln[C_w W^b] = \ln C_w + b \ln W$, and therefore, the parameters of the continuous function that approximates the step-function rates can be found by performing simple linear regression analyses on logarithms of the effective rate–shipment weight pairs that result from computations like those described above to find the effective rate. Sample plots of the results of this process for the New York-Dallas rate data in Table 8.7 are shown in Figure 8.13, resulting in $C_w = 702.67$ and $b = -0.3189$. Thus, shipping Q units of a Class 100 item that weighs w lbs on this lane would result in a per-unit rate of $r(Qw) = 702.67(Qw)^{-0.3189}$ \$/cwt. Converting to \$/unit by multiplying this by

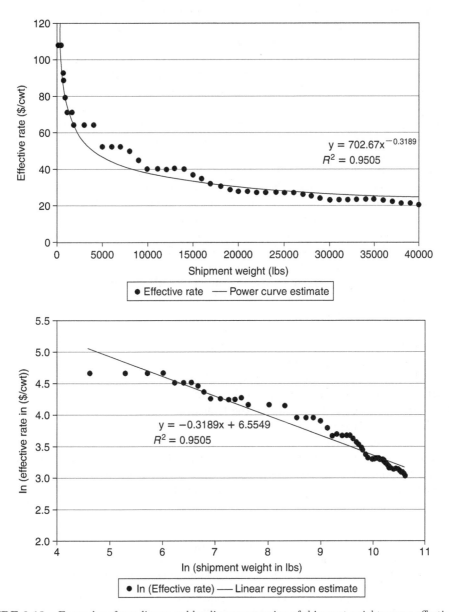

FIGURE 8.13 Examples of non-linear and log-linear regression of shipment weight versus effective rate.

$w/100$ cwt/unit, one obtains $r(Q) = 7.0267w^{0.6811}Q^{-0.3189}$ \$/unit. Depending on the source of the rates, it might also be necessary to assume a discount factor that would reflect what the shipper would actually contract to pay the LTL carrier for its services. For example, Tyworth and Ruiz-Torres (2000) apply a 50% discount, which they claim to be typical of contract-based discounts in practice.*

8.3.3 Transportation-Inclusive Inventory Decision Making

Let us return to the total annual cost equation developed earlier in the chapter, Equation 8.10. This equation accounts for the cost of placing orders and the costs of holding cycle stock and safety stock inventories. In the preceding subsections, we indicated that transportation costs tend to exhibit economies of scale, and we demonstrated, using LTL freight rates, a method for building a continuous function for transportation costs, $r(Q) = CQ^b$. Note also that, using this model of scale economies in transportation costs, the annual cost of freight transportation for shipments of size Q to satisfy demand of D units is $Q \cdot CQ^b \cdot D/Q = DCQ^b$. Although it seems immediately clear that this rate function *could* be incorporated into an annual cost equation, the question is, first, *whether* it should be incorporated into the TAC expression, and second, whether it implies any other inventory-related costs.

Marien (1996) provides a helpful overview of what are called *freight terms of sale* between a supplier shipping goods and the customer that ultimately will receive them from the carrier that is hired to move the goods. Two questions are answered by these negotiated freight terms, namely, "*Who owns the goods in transit?*" and "*Who pays the freight charges?*" Note that legally, the carrier that moves the goods is only the *consignee* and never takes ownership of the goods. The question of ownership of goods in transit is answered by the "F.O.B." terms,† indicating whether the transfer of ownership from supplier (shipper) to customer (receiver) occurs at the point of *origin* or the point of *destination*, hence the terms "F.O.B. origin" and "F.O.B. destination." The next issue is which party, shipper or receiver, pays the freight charges, and here there is a wider array of possibilities. Marien (1996), however, points out that three arrangements are common: (1) F.O.B. origin, freight collect—in which goods in-transit are owned by the receiver, who also pays the freight charges; (2) F.O.B. destination, freight prepaid—in which the shipper retains ownership of goods in transit and also pays the freight bill; and (3) F.O.B. origin, freight prepaid and charged back, similar to (1), except that the shipper pays the freight bill up-front and then charges it to the receiver upon delivery of the goods. These three sets of terms are described pictorially in Figure 8.14. The distinction of who pays the bill, and the interesting twist implied in (3), is important in that the party that pays the freight charges is the one who has the right, therefore, to hire the carrier and choose the transport service and routing.

Note that our TAC expression for inventory management is a decision model for ordering; thus, it takes the perspective of the *buyer*, the downstream party that will receive the goods being ordered. Therefore, the freight transportation terms that involve "freight collect" or "freight prepaid and charged" back imply that the decision maker should incorporate a transportation cost term in the TAC expression that is the objective function for the

---

*Using empirical LTL rate data from 2004, Kay and Warsing (2006) estimate that the average discount from published rates is approximately 46%.

†According to Coyle et al. (2006, pp, 473, 490), "F.O.B." stands for "free on board," shortened from "free on board ship." This distinguishes it from the water-borne transportation terms "free alongside ship" (FAS), under which the goods transfer ownership from shipper to receiver when the shipper delivers the goods to the port (i.e., "alongside ship"), leaving the receiver to pay the cost of lifting the goods onto the ship.

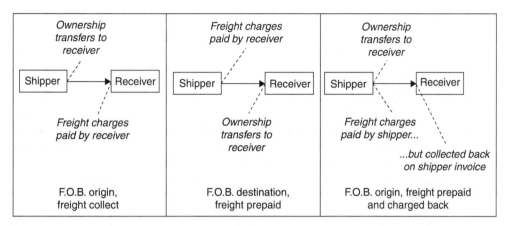

FIGURE 8.14 Common freight transportation terms.

order-placement decision. If the terms are freight prepaid, then the order-placing decision maker would not include the transportation cost in its objective function. At this point, one might object, pointing out that supply chain management must concern itself with a "supply-chain-wide" objective to minimize costs and maximize profits. My counterargument is that the Securities and Exchange Commission has not yet required, and likely never will require, any *supply chain* to report its revenues and costs; moreover, the Wall Street analysts who are so important in determining the market valuation of a firm care only about that firm's individual profits, not those of its supply chain. Thus, decision models that take the perspective of the decision-making firm and incorporate only those costs relevant to the firm are eminently reasonable. (An analysis of complementary, and to some extent competing, objective functions between a supplier and its customer is presented in Section 8.5.)

The next question we should address is whether the freight terms have any bearing on inventory holding cost. The answer is—as it often is—"It depends." Specifically, it depends on where the ownership of the goods transfers to the receiver. If the receiver is responsible for the goods in-transit, then it stands to reason that this liability should be reflected in annual inventory-related costs. From a conservative standpoint, those goods are "on the books" of the receiver, since, from the shipper's perspective, the goods were "delivered" as soon as the carrier took them under consignment. Although typical invoice terms dictate that money will not change hands between the supplier and customer for between 10 and 30 days, the goods are technically now part of the payable accounts of the customer. Thus, a conservative perspective would consider them to be "money tied up in inventory," in this case, inventory in-transit.

Using an approach suggested by Coyle et al. (2003, pp. 270–274), let us assume that the lead time L is composed of two parts, a random transit time T and a constant (and obviously non-negative) order processing time. If the expected transit time is μ_T, then every unit shipped spends a fraction μ_T/Y of a year in transit, where Y is the number of days per year. If h_p is the annual cost of holding inventory in transit, then annual *pipeline inventory* cost is given by $Dh_p\mu_T/Y$. The resulting total annual cost equation,[*] under F.O.B. origin

---

[*] It is straightforward to show that this cost equation is convex in Q and R. First, note that we are merely adding a convex term to an equation we already know to be convex. Moreover, one can modify the convexity proof of the transportation-exclusive equation from Hax and Candea (1984) to formalize the result, although the details are not included here.

terms that also require the receiver to pay the freight charges is

$$TAC(R,Q) = A\frac{D}{Q} + h\left(\frac{Q}{2} + R - \mu_{DLT}\right) + h_p D\frac{\mu_T}{Y} + DCQ^b \qquad (8.17)$$

Note that the pipeline inventory cost does not depend on the order quantity Q or the reorder point R. Thus, while it will not affect the optimal values of the decision variables for a given mode, it may affect the ultimate choice *between* modes through the effect of different values of μ_T on pipeline inventory. Note also that this approach to computing pipeline inventory costs ignores the time value of money over the transit time; one would assume, however, that such time value effects are insignificant in almost all cases.

Putting all of these pieces together, let's consider an example that compares cases with significant differences in item weight—affecting transportation costs—and item value—affecting holding costs. Assume that our company currently imports the item in question from a company based in Oakland, CA, and that our company's southeast regional distribution center is located in Atlanta, GA. We compare a reorder point–order quantity (Q, R) solution that excludes freight cost and pipeline inventory carrying costs—that is, from "F.O.B. destination, freight prepaid" terms—to solutions that incorporate these costs, both for LTL and TL shipments—using "F.O.B. origin, freight collect" terms—from Oakland to Atlanta. Assume that our target fill rate at the Atlanta DC is 99%. Our approach to computing the order quantity and reorder point for each of the cases—without transportation costs, using TL, and using LTL—is as follows:

- For "F.O.B. destination, freight prepaid" terms, compute (Q^*, R^*) using Equation 8.10 as the objective function, constrained by the desired fill rate.
- For "F.O.B. origin, freight collect" terms using LTL, compute (Q^*_{LTL}, R^*_{LTL}) using Equation 8.17 as the objective function, with the appropriate, empirically derived values of C and b, and with the solution constrained by the fill rate and by $Q_{LTL} \cdot w \leq 40{,}000$, where w is the item weight, which therefore assumes a truck trailer weight capacity of 40,000 lbs.*
- For "F.O.B. origin, freight collect" terms using TL, solve Equation 8.17 for (Q^*_{TL}, R^*_{TL}) as follows:

  1. $TAC_{\min} \leftarrow \infty$; $TAC_{TL} \leftarrow 0$; number of truckloads $= i = 1$
  2. Minimize Equation 8.17 with respect to R_{TL} with $Q_{TL} = i(40{,}000/w)$, C equal to the cost of i truckloads, $b = -1$, and constrained by the fill rate to obtain $TAC(Q_{TL}, R_{TL})$.
  3. $TAC_{TL} \leftarrow TAC(Q_{TL}, R_{TL})$
  4. If $TAC_{TL} < TAC_{\min}$,

     $$TAC_{\min} \leftarrow TAC_{TL}$$

     $$i \leftarrow i + 1; \text{go to step 2.}$$

  5. $(Q^*_{TL}, R^*_{TL}) = (Q_{TL}, R_{TL})$

---

*Actually, an LTL carrier would probably not take a shipment of 40,000 lbs, and the shipper would clearly get a better rate from a TL carrier for this full-TL weight. This is, however, the logical upper bound on the size of an LTL shipment. In addition, in using this value, we assume that a TL quantity will hit the trailer *weight* capacity before the trailer *space* (volume) capacity—i.e., that we "weigh-out" before we "cube-out."

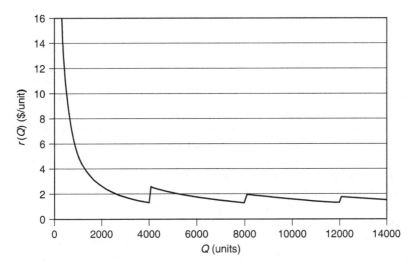

FIGURE 8.15 Example of truckload cost per unit versus number of units shipped.

TABLE 8.8 Example Class 100 LTL
Rates for Oakland-Atlanta Lane

Minimum Weight (lbs)	Rate ($/cwt)
1	136.26
500	109.87
1000	91.61
2000	79.45
5000	69.91
10,000	54.61
20,000	48.12
30,000	41.85
40,000	39.12

Before continuing with our example, note that the procedure laid out above to solve for (Q_{TL}^*, R_{TL}^*) is not necessarily guaranteed to generate a global optimum due to the non-smooth nature of the $r(Q)$ function for TL shipments. For example, Figure 8.15 shows a plot of $r(Q)$ versus Q for a scenario in which the cost per TL is $5180, the TL capacity is 40,000 lbs, and the item weight is $w = 10$ lbs. As Q increases, and therefore the number of truckloads increases, the freight rate per unit falls, but not in a smooth fashion. Thus, if $TAC_{TL,i}$ is the total cost of shipping i truckloads, depending on the problem parameters, even if $TAC_{TL,i} < TAC_{TL,i-1}$ and $TAC_{TL,i+1} > TAC_{TL,i}$, it does not immediately follow that $TAC_{TL,i+2} > TAC_{TL,i+1}$, leaving it unclear as to how $TAC_{TL,i}$ compares to $TAC_{TL,i+2}$. Thus, returning to our discussion at the outset of this section, one can see the possible value of a TMS in performing the analysis that takes these kinds of parametric complexities of the problem into account. Nonetheless, for our illustrative example, let us assume that our approach for computing (Q_{TL}^*, R_{TL}^*) is sufficient.

Continuing with our example, the distance from Oakland to Atlanta is 2471 miles.[*] Using a TL price of $0.105/ton-mi (adjusted upward from the TL revenue per mile of $0.0914/ton-mi from Wilson, 2003), the TL cost is $259/ton. For the LTL rates, we will assume that we are shipping Class 100 goods with a published rate table as shown in Table 8.8, resulting in $b = -0.2314$ and $C = 2.435w^{0.7686}$, after applying an assumed 50%

---

[*]The distances used in this example were obtained from MapQuest, www.mapquest.com.

discount from these published rates.* Further, we assume a cost of $80 to place a replenishment order, an on-site holding cost rate at the Atlanta DC of 15%, and an in-transit holding cost rate of 12%. Finally, we assume that the LTL shipment lead time is 5 days with a standard deviation of 1.5 days, that the TL lead time is 4 days with standard deviation of 0.8 days, that mean daily demand is 30 units with a standard deviation of 8 units, and that there are 365 selling days per year, such that $D = 10,950$. Table 8.9 displays the results, with item weights of $w \in \{10, 100\}$ (pounds) and item costs of $v \in \{\$20, \$500\}$. We report *TAC* only for the transportation-inclusive cases as this is the only relevant point of comparison. From the table, one can see that for this case, if our company pays transportation costs, TL is preferred in three of the four weight-cost combinations, with LTL being preferred only in the low-weight, high-item-cost case. This is reasonable, given the ability of LTL to allow small shipments sizes that keep inventory holding costs in check.

Let us now assume that we have the option to switch to an importer located in New York City (NYC), dramatically decreasing the transit distance for replenishment shipments to the Atlanta DC to 884 mi. Assume further that this importer charges a premium over the current importer located in Oakland so that the item cost will be 10% higher than in the Oakland-to-Atlanta case above. Given the distance from NYC to Atlanta, using the same approach as above, the TL cost is $92.82/ton. For an LTL shipment, the published rates are given in Table 8.10, resulting in $b = -0.3704$ and $C = 3.993w^{0.6296}$. Let us assume that the LTL lead time from NYC to Atlanta is 2 days, with a standard deviation of 0.6 days, and that the TL lead time is 1 day, with a standard deviation of 0.2 days. Therefore, although our annual materials cost will increase by 10% if we switch to the NYC importer, our transportation costs, where applicable, and inventory holding costs—safety stock and, where applicable, pipeline inventory—will decline. The comparative results are shown in Table 8.11. As the annual increase in materials cost is $21,900 in the low-item-cost case and $547,500 in the high-item-cost case, these values serve as the hurdles for shifting to the NYC source. For the high-item-cost cases, this increase in annual materials costs swamps the transportation and holding cost savings, even though the transportation costs are cut dramatically for the cases in which $w = 100$ lbs. The analysis shows, however, that the switch is viable in the case in which our company pays for freight transportation with a high-weight, low-cost item. In this case, TL is the preferred freight transit option, and the switch to the NYC source saves over $88,000 in annual transportation costs and over $91,000 overall.

8.3.4   A More General Transportation Cost Function

In most commercial transportation management and planning systems, LTL rates are determined using tariff tables, as in the examples above. The major limitations of this approach are that access to the tariff tables must be purchased and that it is necessary to know what discount to apply to the tariff rates in determining transportation costs. Kay and Warsing (2006), however, present a more general model of LTL rates, one that is not specific to a transportation lane or to the class of the items being shipped, with the item's density serving as a proxy for its class rating. This generality allows the rate model to be used in the early stages of logistics network design, when location decisions are being made and when the most appropriate shipment size for each lane in the network is being determined. Thus, the model developed by Kay and Warsing is important for the sort of analysis presented in Section 8.4.

In their model, Kay and Warsing (2006) use average, industry-wide LTL and TL rates empirically to develop an LTL rate model that is scaled to economic conditions, using the

---

*In general, $C = [\delta C_w w^{(b+1)}]/100$, in $/unit, where δ is the discount factor.

TABLE 8.9 Comparative Results from Inventory-Transportation Analysis

Oakland–Atlanta	Freight Terms	Mode	Q	R	Ordering Cost($)	Holding Cost($)	Pipeline Cost($)	Transp'n Cost($)	TAC ($)
$w = 10, v = \$20$	Origin–collect	TL	4000	120	219	6000	288	14,180	20,687
	Origin–collect	LTL	3750.99	150	234	5626	360	23,306	29,526
	Destination–prepaid	—	794.35	179.88	1103	1281	—	—	—
$w = 10, v = \$500$	Origin–collect	TL	4000	120	219	150,000	7200	14,180	171,599
	Origin–collect	LTL	361.52	201.19	2423	17,396	9000	40,047	68,866
	Destination–prepaid	—	176.38	217.94	4966	11,710	—	—	—
$w = 100, v = \$20$	Origin–collect	TL	800	127.86	1095	1224	288	141,803	144,409
	Origin–collect	LTL	400	198.65	2190	746	360	229,616	232,912
	Destination–prepaid	—	794.35	179.88	1103	1281	—	—	—
$w = 100, v = \$500$	Origin–collect	TL	400	140.70	2190	16,552	7200	141,803	167,745
	Origin–collect	LTL	400	198.65	2190	18,648	9000	229,616	259,454
	Destination–prepaid	—	176.38	217.94	4966	11,710	—	—	—

TABLE 8.10   Example Class 100 LTL
Rates for NYC-Atlanta Lane

Minimum Weight (lbs)	Rate ($/cwt)
1	81.96
500	74.94
1000	61.14
2000	49.65
5000	39.73
10,000	33.44
20,000	18.36
30,000	14.90
40,000	13.36

producer price index for LTL services provided by the U.S. Bureau of Labor Statistics. The LTL rate model requires inputs of W, the shipment weight (in tons); s, the density (in lbs/ft^3) of the item being shipped; and d, the distance of the move (in mi). From a non-linear regression analysis on CzarLite tariff rates* for 100 random O-D pairs (5-digit zip codes), with weights ranging from 150 to 10,000 lbs (corresponding to the midpoints between successive rate breaks in the LTL tariff), densities from approximately 0.5 to 50 lbs/ft^3 (serving as proxies for class ratings 500 to 50), and distances from 37 to 3354 mi, the resulting generalized LTL rate function is

$$r_{LTL}(W, s, d) = PPI_{LTL} \left[\frac{\dfrac{s^2}{8} + 14}{\left(W^{(1/7)} d^{(15/29)} - \dfrac{7}{2} \right) (s^2 + 2s + 14)} \right] \tag{8.18}$$

where, as suggested above, PPI_{LTL} is the producer price index for LTL transportation, reported by the U.S. Bureau of Labor Statistics as the index for "General freight trucking, long-distance, LTL."[†] Kay and Warsing (2006) report a weighted average residual error of approximately 11.9% for this functional estimate of LTL rates as compared to actual CzarLite tariff rates. Moreover, as industry-wide revenues are used to scale the functional estimate, the model reflects the average discount provided by LTL carriers to their customers, estimated by Kay and Warsing to be approximately 46%.

Thus, for a given lane, with d therefore fixed, the model given by Equation 8.18 provides estimates of LTL rates for various shipment weights and item densities. Kay and Warsing (2006) demonstrate the use of the model to compare LTL and TL shipment decisions under different item value conditions. Moreover, the functional estimate could easily be used to evaluate the comparative annual cost of sending truckloads of mixed goods with different weights and densities versus individual LTL shipments of these goods. On the other hand, for a given item, with a given density, Equation 8.18 provides a means of analyzing the effect of varying the origin point to serve a given destination, possibly considering different shipment weights as well. These two analytical perspectives are, as we suggest above, important in the early stages of supply chain network design, the subject of the section that follows.

---

*CzarLite is an LTL rating application developed by SMC^3 Company, and it is commonly used as the basis for carrier rates. The rates used in the study by Kay and Warsing (2006) were obtained in November–December 2005 from http://www.smc3.com/applications/webczarlite/entry.asp.

[†]See the Bureau of Labor Statistics website at http://www.bls.gov/ppi/home.htm, Series ID=PCU484122484122.

TABLE 8.11 Comparative Results from Extended Inventory-Transportation Analysis

NYC-Atlanta	Freight Terms	Mode	Q	R	Ordering Cost($)	Holding Cost($)	Pipeline Cost($)	Transp'n Cost($)	TAC($)	Change from OAK-ATL($)
$w = 10, v = \$22$	Origin–collect	TL	4000	30	219	6600	79	5082	11,980	-8707
	Origin–collect	LTL	2512.92	60	349	4146	158	10,253	14,906	-14,619
	Destination–prepaid	—	745.02	62.15	1176	1236	—	—	—	28
$w = 10, v = \$550$	Origin–collect	TL	4000	30	219	165,000	1980	5082	172,281	682
	Origin–collect	LTL	298.71	75.09	2933	13,567	3960	22,565	43,025	-25,841
	Destination–prepaid	—	157.04	82.59	5578	8341	—	—	—	-2757
$w = 100, v = \$22$	Origin–collect	TL	800	30.00	1095	1320	79	50,819	53,313	-91,096
	Origin–collect	LTL	400	71.31	2190	697	158	86,313	89,358	-143,554
	Destination–prepaid	—	745.02	62.15	1176	1236	—	—	—	28
$w = 100, v = \$550$	Origin–collect	TL	400	30.00	2190	16,500	1980	50,819	71,489	-96,256
	Origin–collect	LTL	400	71.31	2190	17,433	3960	86,313	109,895	-149,559
	Destination–prepaid	—	157.04	82.59	5578	8341	—	—	—	-2757

8.4   Managing Locations in the Supply Chain

At this point, we have discussed the inventory and transportation legs of Ballou's triangle, and along the way, had some brief discussion of customer service as given by α, the cycle service level, and β, the fill rate. Therefore, we now turn our attention to "location strategy," or designing and operating the supply chain network. Key questions to answer in this effort are as follows: How do we find a good (or perhaps the *best*) network? What are the key objectives for the network? What network alternatives should we consider? Once again, we need to consider the perspective of the decision maker charged with making the design decisions. Typically, the perspective will be that of a single firm. The objectives may be diverse. We clearly wish to minimize the cost of operating the network, and as product prices typically will not change as a function of our network design, this objective should be sufficient from a financial perspective. Customer service may also need to be incorporated in the decision, but this may pose a challenge in any explicit modeling efforts. Assuming an objective to minimize the costs of operating the network for a single firm, the key decision variables will be the number and locations of various types of facilities—spanning as far as supplier sites, manufacturing sites, and distribution sites—and the quantities shipped from upstream sites to downstream sites and ultimately out to customers.

To get a sense of the cost structure of these kinds of models, let us consider a single-echelon location-allocation problem. This problem is used to determine which sites to include in a network from a set of n production or distribution sites that serve a set of m customers. The fixed cost of including site i in the network and operating it across the planning period (say, a year) is given by f_i, and each unit of flow from site i to customer j results in variable cost c_{ij}. Site i has capacity K_i, and the total planning period (e.g., annual) demand at customer j is given by D_j. Therefore, our objective is to

$$\min \qquad \sum_{i=1}^{n} f_i y_i + \sum_{i=1}^{n}\sum_{j=1}^{m} c_{ij} x_{ij} \qquad\qquad (8.19)$$

$$\text{subject to} \qquad \sum_{i=1}^{n} x_{ij} = D_j, \qquad j = 1,\ldots,m \qquad\qquad (8.20)$$

$$\sum_{j=1}^{m} x_{ij} \leq K_i y_i, \qquad i = 1,\ldots,n \qquad\qquad (8.21)$$

$$y_i \in \{0,1\}, \qquad i = 1,\ldots,n \qquad\qquad (8.22)$$

$$x_{ij} \geq 0, \qquad\qquad i = 1,\ldots,n; \ j = 1,\ldots,m \qquad\qquad (8.23)$$

where decision variables y_i ($i = 1,\ldots,n$) determine whether site i is included in the network ($y_i = 1$) or not ($y_i = 0$) and x_{ij} ($i = 1,\ldots,n; \ j = 1,\ldots,m$) determine the number of units shipped from site i to customer j. As parameters f_i and c_{ij} are constants, this problem is a mixed-integer program (MIP). Depending on the values of n and m, solving it may be challenging, even by twenty-first-century computing standards, given the combinatorial nature of the problem.

Moreover, the problem represented by Equations 8.19 to 8.23 above is only a small slice of the overall network design formulation. If we do indeed wish to consider supplier location, production facility locations, *and* distribution facility locations, then we need location variables (y) and costs (f) at each of three echelons and flow variables (x) and costs (c) to represent the levels and costs of supplier-to-factory, factory-to-distribution, and distribution-to-customer flows. In addition, we need to add *conservation of flow* constraints that ensure that all units shipped *into* the factory and distribution echelons are also shipped back *out*—unless we consider the possibility of carrying inventories at the various echelons, another complication that would also possibly entail adding time periods to the model to reflect varying inventory levels over time. Further, we should keep in mind the data requirements and restrictions for formulating this problem. One must have a reasonably good estimate of the fixed cost of opening and operating each site, the capacity of each site (or at least a reasonable upper bound on this value), and the variable cost of producing or shipping goods from this site to other network locations or to customers. In this latter case, one must assume that these variable costs are linear in the quantities produced or shipped. Earlier in this chapter, however, we indicated that shipping costs typically exhibit economies of scale; production costs may do so as well. Moreover, the *discrete-location* problem formulated above assumes that a set of candidate locations has been precisely determined in advance of formulating the problem.

In this author's mind, the typical IE or operations management treatment of the facility location problem rushes to a description of the mathematical techniques without first considering *all* of the various inputs to the decision. As Ballou (2004, pp. 569–570) rightfully reminds his readers, "... It should be remembered that the optimum solution to the real-world location problem is no better than the model's description of the problem realities."

In addition, decisions regarding facility locations have significant *qualitative* aspects to them as well. As with customer service, these factors may be challenging to incorporate into a quantitative decision model. Indeed, in the author's opinion, the typical academic treatment of the facility location problem is one that could benefit from a serious consideration of what practitioners faced with facility location decisions actually *do*. This author's reading of the trade press indicates that those practitioners, rightly or wrongly, devote only a small portion—if *any*—of their decision-making process to solving traditional facility location models. Consider this list of the "ten most powerful factors in location decisions" from a 1999 survey of readers of *Transportation and Distribution* magazine (Schwartz, 1999a), in order of their appearance in the article: reasonable cost for property, roadway access for trucks, nearness to customers, cost of labor, low taxes, tax exemptions, tax credits, low union profile, ample room for expansion, community disposition to industry. In fact, only two, or perhaps three, of these ten—property and labor costs, and perhaps low taxes—could be directly incorporated into one of the classic facility location models. The others could be grouped into what another *T&D* article (Schwartz, 1999b) referred to as *service-oriented issues* (roadway access, nearness to customers) and *local factors* (tax exemptions, tax credits, union profile, expansion opportunity, and community disposition). That *T&D* article goes on to state, mostly from consulting sources, that service-oriented issues now "dominate," although this is not, in this author's mind, a "slam-dunk" conclusion. For example, 2004–2005 saw significant public debate in North Carolina as to the public benefit—or public detriment—of using state and local tax incentives to lure businesses to locate operations in a given community (Reuters, 2004; Cox, 2005; Craver, 2005). Given the coverage devoted to the effects of tax incentives on these location decisions, it would be hard to conclude as the *T&D* article did, that "local issues" are "least consequential" (p. 44). A more thoughtful perspective can be found in another trade publication discussion (Atkinson, 2002, p. S65), which states, "While everyone can identify the critical elements, not all of them agree on the order of priority."

From a broader perspective, Chopra and Meindl (2004) suggest that supply chain network design encompasses four phases: Phase I—SC Strategy, Phase II—Regional Facility Configuration, Phase III—Desirable Sites, Phase IV—Location Choices. They also suggest a number of factors that enter into these decisions. In our discussion, we will indeed start from supply chain strategy. The primary strategic choice in the treatment of supply chain network design that follows will be how *centralized* or *decentralized* the network design should be. Given the complexity of the network design process, it is reasonable to frame it as an iterative one, perhaps evaluating several network options that range from highly centralized to highly decentralized. Ultimately, a choice on this centralized-decentralized spectrum will lead to a sense of how many facilities there should be at each echelon of the network. Once the decision maker has an idea of the number of facilities and a rough regional configuration, he or she can answer the question of where good candidate locations would be using a *continuous-location* model, along the lines of that studied by Cooper (1963, 1964). This "ballpark" set of locations can then be subjected to qualitative analysis that considers service-oriented issues and local factors. Therefore, let us first turn our attention to understanding the differences between centralized and decentralized networks.

8.4.1 Centralized Versus Decentralized Supply Chains

Let us focus on the design of a distribution network and address the following questions: How centralized should the network be? What are the advantages of a more centralized or a more decentralized network? This issue can clearly be addressed in the context of the legs and center of Ballou's strategic triangle—inventory, locations, and transportation on the legs and customer service in the center. Consider the two DC networks laid out in Figure 8.16. Network 1 is more centralized, with only four distribution points, laid out essentially on the compass—west, central, northeast, and southeast. Network 2 is more decentralized, with a hub-spoke-type structure that has twenty additional DCs located near major population centers and ostensibly fed by the regional DCs that encompassed all of Network 1. Starting with transportation, we can break the tradeoffs down if we first consider transportation *outbound* from DCs to customers versus transportation *inbound* to the DCs, or essentially *within* the network. In the context of the facility location problem (Equations 8.19 to 8.23) laid out above, these would be costs c_{ij} for shipments from DC i to customer j and costs c_{0i} for shipments from the factory (location 0) to DC i.

As shipments to customers—often called the "last mile" of distribution—typically involve smaller quantities, perhaps only a single unit, it is often difficult to garner significant scale-driven discounts for these shipments. However, it is clear from our discussion above that distance, in addition to weight, is a significant factor in the cost of freight transportation. More weight allows greater economies in unit costs, but more distance increases the variable costs of the carrier and thereby increases the cost of the transportation service to the

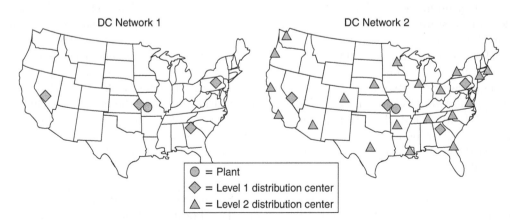

FIGURE 8.16 Comparative distribution networks.

shipper. Note that the centralized network will clearly involve longer "last-mile" shipments to customers than the decentralized network. Although it is possible that larger loads may be moved on vehicles outbound to a set of customers from a centralized site, it is quite possible that neither network affords much chance for significant bulk shipping activity outbound, particularly if the outbound shipments must be made to customers that are dispersed widely. Thus, one could argue persuasively that outbound distribution costs are larger in centralized networks.

On the inbound side of the DCs, however, the situation flips. With only a few DCs in the centralized network, the shipments from the supply points (e.g., plants, suppliers) into the final distribution points are more likely to garner significant economies of scale in transportation, and therefore, "within-network" freight costs should be lower in a centralized distribution network. In addition, Figure 8.16 clearly shows that there are fewer miles to travel in moving goods to the final distribution points in a centralized network, another source of lower within-network freight transportation costs in a centralized network. However, as shipments inbound will tend to be packed in bulk, more dense, and perhaps less time-sensitive, inbound transportation costs for a given network are likely to be much lower than outbound costs. Thus, it stands to reason that transportation costs will net out to be lower in a decentralized network, wherein relatively expensive "last-mile" transportation covers shorter distances. Moreover, it should be immediately clear that customers would prefer the decentralized network for its ability to reduce order fulfillment lead times.

From the standpoint of locations, there are obviously fewer locations in a centralized DC network, and therefore, this network will result in lower fixed and operating costs driven by facilities (e.g., amortization of loans, payment of leases, operating equipment, labor). In the context of our facility location problem since fewer DC locations (y_i) would be involved in the centralized problem (or more of them would be forced to be $y_i = 0$), the sum $\sum_{i=1}^{n} f_i y_i$ in the objective function would necessarily be smaller in a more centralized solution.

Thus, we have addressed two legs of the SC strategy triangle, transportation, and locations. We are left with inventory. The effects of centralizing, or not, on this leg are richer, and therefore, I will devote the section that follows to this topic.

8.4.2   Aggregating Inventories and Risk Pooling

Vollmann et al. (2005, p. 275) point out that ". . . if the uncertainty from several field locations could be aggregated, it should require less safety stock than having stock at each field location." Indeed, in one's mind's eye, it is relatively easy to see that the ups and downs of uncertain demand over time in several independent field locations are likely to cancel each other out, to some extent, when those demands are added together and served from an aggregated pool of inventory. This *risk pooling* result is the same as the one that forms the basis of financial portfolio theory, in which the risk—that is, the variance in the returns—of a portfolio of investments is less than the risk of any single investment in the portfolio. Thus, an investor can cushion the ups and downs, as it were, or reduce the variability in his investments by allocating his funds across a portfolio as opposed to concentrating all of his funds in a single investment.

Looking at this risk pooling effect from the standpoint of inventory management, since safety stock is given by $SS = z\sigma_{DLT}$, one would reasonably infer that the risk pooling effects of aggregating inventories must center on the standard deviation of demand over the replenishment lead time, σ_{DLT}. Indeed, this is the case. If we have, for example, descriptions of lead-time demand for N distribution regions, with a mean lead-time demand in region i of $\mu_{DLT,i}$ and standard deviation of lead-time demand in region i of $\sigma_{DLT,i}$, then aggregating demand across all regions results in a pooled lead-time demand with mean

$$\mu_{DLT,\text{pooled}} = \sum_{i=1}^{N} \mu_{DLT,i} \tag{8.24}$$

and standard deviation

$$\sigma_{DLT,\text{pooled}} = \sqrt{\sum_{i=1}^{N} \sigma_{DLT,i}^2 + 2 \sum_{i=1}^{N-1} \sum_{j=i+1}^{N} \rho_{ij} \sigma_{DLT,i} \sigma_{DLT,j}} \tag{8.25}$$

where ρ_{ij} is the correlation coefficient of the demand between regions i and j. If demand across all regions is independent, then $\rho_{ij} = 0$ for all $i, j = 1, \ldots, N$ ($i \neq j$), and Equation 8.25 reduces to

$$\sigma_{DLT,\text{pooled}} = \sqrt{\sum_{i=1}^{N} \sigma_{DLT,i}^2} \tag{8.26}$$

By the triangle inequality, $\sqrt{\sum_{i=1}^{N} \sigma_{DLT,i}^2} \leq \sum_{i=1}^{N} \sigma_{DLT,i}$, and therefore,

$$SS_{\text{pooled}} = z \sigma_{DLT,\text{pooled}} \leq z \sum_{i=1}^{N} \sigma_{DLT,i} = SS_{\text{unpooled}} \tag{8.27}$$

a result that holds for any values of ρ_{ij}, since perfect positive correlation ($\rho_{ij} = 1$ for all $i, j = 1, \ldots, N$) makes Equation 8.27 an equality ($SS_{\text{pooled}} = SS_{\text{unpooled}}$) and *negative* correlation actually *increases* the pooling-driven reduction in safety stock beyond the independent case.

Let's consider an example. Assume we are interested in designing a distribution network to serve 40 customers. An important decision is the number of distribution sites we need to serve those 40 customers. Our choices range from just a single distribution center (DC) up to, theoretically, 40 DCs, one dedicated to serving each customer. In our simplified system, let's assume that there is no fixed cost to place an order to replenish inventory at any distribution site and that demand over the replenishment lead time for any distribution site and any customer served is normally distributed with mean $\mu = 50$ and standard deviation $\sigma = 15$. Therefore, at any DC in the network, since there is no ordering cost, we can set the stocking level by using a base-stock policy, with a stocking level at DC i equal to $B_i = \mu_i + z_\alpha \sigma_i$, where μ_i is the mean demand over the replenishment lead time for the set of customers served by DC i, σ_i is the standard deviation of this demand, and z_α is the safety stock factor that yields a CSL of α. If DC i serves n_i ($1 \leq n_i \leq 40$) customers, then using Equations 8.24 and 8.25 above,

$$\mu_i = n_i \mu \tag{8.28}$$

and

$$\sigma_i = \sqrt{\sum_{j=1}^{n_i} \sigma^2 + 2 \sum_{j=1}^{n_i-1} \sum_{k=j+1}^{n_i} \rho_{jk} \sigma^2} \tag{8.29}$$

where ρ_{jk} is the correlation coefficient of the demand between customers j and k. If $\rho_{jk} = \rho$ for all $j, k = 1, \ldots, 40$ ($j \neq k$), then it is easy to show that Equation 8.29 reduces to

$$\sigma_i = \sqrt{[n_i(1 - \rho) + \rho n_i^2] \sigma^2} \tag{8.30}$$

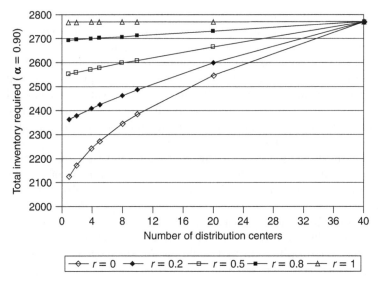

FIGURE 8.17 Example of risk-pooling-driven reduction in inventory.

TABLE 8.12 Risk Pooling-Based Reduction in Safety
Stock with $n = 40$ Customers

Number of Customers per DC	Number of DCs in Network	SS at $\rho = 0$	
		Units	% of $E[D] = n\mu$
1	40	768.9	38.45
2	20	543.7	27.19
4	10	384.5	19.22
5	8	343.9	17.19
8	5	271.9	13.59
10	4	243.2	12.16
20	2	171.9	8.60
40	1	121.6	6.08

Note that when $\rho = 0$ (i.e., the aggregated customer demands are independent), $\sigma_i = \sigma \sqrt{n_i}$, meaning that safety stock increases with the square root of the number of sites aggregated—that is, sub-linearly. In contrast, when $\rho = 1$ (i.e., the aggregated customer demands are perfectly correlated), $\sigma_i = n_i \sigma$, meaning that—as we pointed out in the more general discussion above—the standard deviation of the aggregated demands is simply the sum of the standard deviations of the disaggregated demands. Figure 8.17 shows the total system inventory required in the distribution system to achieve $\alpha = 0.90$ as n_i ranges from 1 to 40 at various levels of demand correlation (where ρ is expressed as r in the legend of the graph). Table 8.12 shows the reduction in system safety stock as a percentage of overall lead-time demand as the number of DCs in the network declines, assuming independent demands across customers.

The example above highlights the notion of postponement in the supply chain. In our network configuration, risk pooling through aggregation is a means of time (or place) postponement, where the forward movement of the goods to (or near) the final points of consumption is postponed until more is known about demand at those points of consumption. The opposite approach, in a decentralized network, is to speculate and move the goods to points closer to customers in advance of demand. As Chopra and Meindl (2004) point out, an approach that attempts to attain the best of both worlds—safety stock reduction through aggregation while still retaining the customer service benefits of a decentralized network—would be to virtually aggregate inventory in the network. The idea is to allow all DCs in the network to have visibility to and access to inventory at all other DCs in the network, such that when a particular DC's inventory approaches zero before a replenishment can arrive from the supplier, that DC can request a transshipment of inventory in excess at another DC. Chopra and Meindl (2004) point out that Wal-Mart has used this strategy effectively. With such an approach, demand at all locations is virtually pooled, with the reductions in safety stock and improvements in customer service coming at the expense of the transshipments that rebalance supply when necessary.

Rounding out this discussion, it is worthwhile to point out that another type of postponement is form postponement, whereby the final physical conversion of a product is delayed until more is known about demand. The two now-classic examples of form postponement in the operations management literature come from Hewlett-Packard (HP), the computing equipment manufacturer, and Benetton, a major apparel company. As described by Feitzinger and Lee (1997), the HP Deskjet printer was redesigned such that the power supply was converted to a drop-in item that could be added to a generic printer in distribution centers. The previous approach was to manufacture printers with country-specific power supplies—due to the different electrical power systems in various countries—in the manufacturing process. This change in product and packaging design allowed HP to aggregate all demand around the globe into demand for a single, generic printer and hold only the differentiating items, the power supply, and various cables, at in-country distribution centers. In the case of Benetton, it changed the order of manufacturing operations to re-sequence the dyeing and sewing processes for sweaters. The previous approach, time-honored in the apparel manufacturing industry, was first to dye bolts of "greige" (un-colored) fabric and then cut that dyed fabric into its final form to be sewn into apparel items. Benetton swapped this sequence to first cut and sew "greige" sweaters that would be dyed after more information could be collected about demand for color-specific items. By doing so, demand for all colors can be aggregated into demand for a single item until a point in time when the inherent uncertainty in demand might be reduced, closer to the point of consumption.

One final point regarding risk pooling effects on inventory is that these effects change significantly with changes in the coefficient of variation, $cv = \sigma/\mu$, of demand. This gets to the notion of classifying inventory items as slow movers versus fast movers. Fast movers have expected demands that are relatively large and predictable, thereby resulting in lower cv values. On the other hand, slow movers have expected demands that are relatively small and unpredictable, resulting in higher values of cv. Thus, the relative benefit of centralizing slow movers, in terms of safety stock reduction, will tend to be quite a bit larger than the benefit of centralizing fast movers. This effect is demonstrated in Figure 8.18, which uses our 40-customer example, still with mean lead-time demand of $\mu = 50$ for each customer, but now with standard deviations varying from $\sigma = 2.5$ ($cv = 0.05$) to $\sigma = 17.5$ ($cv = 0.35$—beyond which the normal distribution will exhibit more than a 1 in 1000 chance of a negative value, inappropriate for modeling customer demands). A simple rule of thumb that results from this data would be to centralize the slow movers to maximize the benefits of risk pooling-driven reductions in inventory and to stock the fast movers closer to points of consumption to maximize the benefits of rapid customer service for these high-demand items.

8.4.3 Continuous-Location Models

Note that the discrete-location MIP model for facility location does not, in the form we stated it above, include any inventory effects of the decision variables. Moreover, it is clear

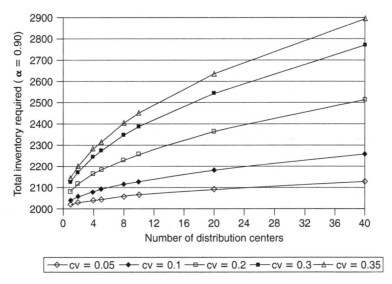

FIGURE 8.18 Effect of variability on risk-pooling impact on inventory levels.

from our discussion above that the effect of the centralizing decision on inventory levels in the network is non-linear. An MIP formulation could handle such non-linear effects via a piecewise linear approximation, but this would only add more integer variables to the formulation.* A better approach, however, might be to consider a facility location model that allows a non-linear objective function, as the most general formulation of the problem is likely to have a number of sources of non-linearity in costs: non-linear production costs driven by economies of scale in production, non-linear transportation costs driven by economies of scale in transportation, and non-linear inventory holding costs driven by the risk-pooling effects of centralizing inventory in the network. Moreover, in the opinion of the author, this approach is probably more consistent with the manner in which facility location decisions are made in practice. The resulting approach would be iterative and would not feed a specific set of candidate locations into the modeling effort a priori, but rather would use some preliminary modeling efforts to generate a rough geographic idea of what might be good candidate locations, perhaps guided by the customer service implications of fewer distribution locations farther from customers, on average, or vice versa. The resulting candidate locations from the cost-service tradeoffs could be subjected to a more qualitative analysis based on the types of "local factors" discussed earlier in this section.

The basis for this formulation of the facility location problem is commonly referred to as the "gravity model," typically formulated to minimize the costs of moving goods *through* the facility, in general, taking receipts from sources and ultimately satisfying demand at customers. In this formulation, the decision variables are x and y, the coordinates of the facility location on a Cartesian plane. The input parameters are D_i, the total volume—in

---

*Most introductory operations research textbooks include such formulations; see, for example, Winston (1994).

TABLE 8.13   Input Data for Gravity Location Example

	i		F_i ($/ton-mi)	D_i (Tons)	x_i	y_i
Source Locations	1	Buffalo	0.90	500	700	1200
	2	Memphis	0.95	300	250	600
	3	St. Louis	0.85	700	225	825
Customer Locations	4	Atlanta	1.50	225	600	500
	5	Boston	1.50	150	1050	1200
	6	Jacksonville	1.50	250	800	300
	7	Philadelphia	1.50	175	925	975
	8	New York	1.50	300	1000	1080

tons, for example—moved into (out of) the facility from source (to customer) $i \in \{1, \ldots, m\}$ and F_i, the cost—in $/ton-mi, for example—to move goods to (from) the facility from source (to customer) $i \in \{1, \ldots, m\}$. Then, the problem is to

$$\min \sum_{i=1}^{m} F_i D_i d_i \qquad (8.31)$$

where $d_i = \sqrt{(x_i - x)^2 + (y_i - y)^2}$, the Euclidean distance between the location (x, y) of the facility and the locations (x_i, y_i) $(i = 1, \ldots, m)$ of the sources and customers that interact with the facility being located. The "gravity" aspect of the problem stems from the fact that the combined cost and distance of the moves into and out of the facility drive the solution, with a high-volume or high-cost source or customer exerting significant "gravity" in pulling the ultimate location toward it to reduce the cost of moving goods between the facility and that source or customer.

Consider a small example, based on an example from Chopra and Meindl (2004), with three sources and five customers. The input data for this example is shown in Table 8.13. The optimal solution, found easily using the Solver tool embedded in Excel, is to locate the facility that interacts with these sources and customers at $(x, y) = (681.3, 882.0)$. A graphical view of the problem data and the solution (labeled as "initial solution") are shown in Figure 8.19.

To illustrate the "gravity" aspects of this model, consider an update to the original problem data. In this situation, assume that it is cheaper to move goods from the source in Buffalo ($F_1 = 0.75$) and more expensive to move them from the sources in Memphis ($F_2 = 1.00$) and St. Louis ($F_3 = 1.40$), and assume that it is also cheaper to move goods to the customers in Boston, New York, and Philadelphia ($F_5 = F_7 = F_8 = 1.30$), but that the costs of moving goods to Atlanta and Jacksonville stay the same. The combined effects of these changes in transportation costs is to exert some "gravity" on the optimal solution and shift it west and south, to attempt to reduce the mileage of the moves that are now more expensive. Indeed, the updated optimal solution is $(x, y) = (487.2, 796.8)$, shown as "updated solution" in Figure 8.19.

Of course, the example above is a simplified version of the larger-scale problem. A typical problem would involve a larger number of customers, and possibly also a larger number of sources. In addition, the cost of moving goods to/from the facility location in our example is a constant term, irrespective of the distance the goods are moved. In general, this cost could be some non-linear function of the distance between the source/customer and the facility, as discussed earlier in the chapter (e.g., the general function for LTL rates derived by Kay and Warsing, 2006). In addition, a more general formulation would allocate sources

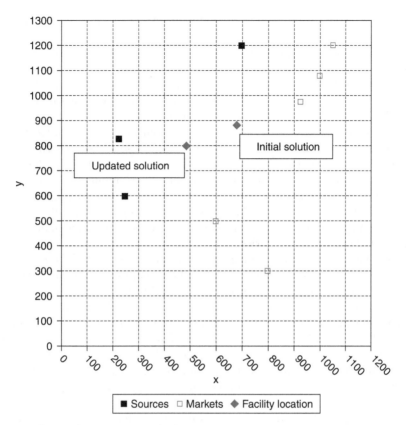

FIGURE 8.19 Gravity location data and solution.

and customers to n candidate locations. The general formulation would be to

$$\min \quad \sum_{i=1}^{n}\sum_{j=1}^{m_S}\delta_{ij}^{S}F_{ij}D_{j}d_{ij} + \sum_{i=1}^{n}\sum_{k=1}^{m_C}\delta_{ik}^{C}F_{ik}D_{k}d_{ik} \tag{8.32}$$

$$\text{subject to} \quad \sum_{j=1}^{m_S}\delta_{ij}^{S} = 1 \quad i=1,\ldots,n \tag{8.33}$$

$$\sum_{k=1}^{m_C}\delta_{ik}^{C} = 1 \quad i=1,\ldots,n \tag{8.34}$$

$$\delta_{ij}^{S} \in \{0,1\} \quad i=1,\ldots,n; \quad j=1,\ldots,m_S \tag{8.35}$$

$$\delta_{ik}^{C} \in \{0,1\} \quad i=1,\ldots,n; \quad k=1,\ldots,m_C \tag{8.36}$$

In addition to the generalized location decision variables (x_i, y_i) for each candidate facility location $i = 1,\ldots,n$, this formulation also introduces allocation decisions δ_{ij}^{S} and δ_{ik}^{C}, binary variables that allocate each source and each customer, respectively, to a candidate facility. Constraints (8.33) and (8.34) ensure that each source and each customer is allocated to exactly one candidate facility.

Given the nature of this problem—a constrained, non-linear optimization—it would appear that this is no easier to solve, and perhaps significantly more challenging to solve, than the MIP facility location problem we introduced at the outset of this section. Indeed, solving Equations 8.32 to 8.36 to optimality would be challenging. Note, however, that for a given allocation of sources and customers to candidate facilities—that is, if we *fix* the values of δ_{ij}^S and δ_{ik}^C—the problem decomposes into n continuous-location problems of the form of Equation 8.31. In fact, this approach has been suggested by Cooper (1964), in what he described as an "alternate location-allocation" (ALA) algorithm. Starting from some initial allocation of sources and customers to sites, one solves the n separate location problems exactly. Then, if some source or customer in this solution could be allocated to another candidate location at a net reduction in total cost, that source or customer is reallocated and the overall solution is updated. The process continues until no further cost-reducing reallocations are found. This algorithm results in a local minimum to the problem, but not necessarily a global minimum. Cooper shows, however, that the algorithm is fast and generates reasonably good solutions. More recent work (Houck et al., 1996) has shown that the ALA procedure is actually quite efficient in reaching an optimal solution in networks with fewer than 25 candidate facilities.

In addition, the ALA approach appears to be promising in allowing for a more generalized functional form of the cost to move goods and fulfill demands in the network. Bucci and Kay (2006) present a formulation similar to Equations 8.32 to 8.36 above for a large-scale distribution problem—limited to customer allocations only—applied to the continental United States. They use U.S. Census Bureau data to generate 877 representative "retail locations" from 3-digit ZIP codes, with the relative demand values based on the populations of those ZIP codes. They consider locating six fulfillment centers to serve these 877 customers. The fulfillment centers exhibit economies of scale in production such that the cost of producing a unit of output decreases by an exponential rate with increasing size of the facility. Facility size is determined by the amount of demand allocated to the facility, with the relative costs given by $c_2/c_1 = (S_2/S_1)^b$, where c_i is the production cost for facility i, S_i is the size of facility i, and b is a constant value provided as an input parameter. Bucci and Kay (2006) cite Rummelt (2001), who in turn cites past studies to estimate that $b \approx -0.35$ for U.S. manufacturers. In addition, Bucci and Kay consider the ratio of production and transportation costs as an input parameter to their problem formulation. Thus, their formulation minimizes the sum of the production and transportation costs that result from the allocation of candidate facilities to customers. Figure 8.20 shows a series of solutions that result from their analysis, using $b = -0.4$ and production costs at 16% of transportation costs. The initial solution at the top of the figure is a static solution that does not incorporate production economies; the middle solution is intermediate in the ALA process; and the bottom solution is the final ALA solution. This solution process provides a graphical sense of the manner in which Bucci and Kay's modified ALA procedure balances production and transportation costs, with the final solution aggregating the demands of the densely populated eastern United States to be served by a single large-scale facility and consolidating what initially were three western facilities into two, ultimately resulting in a five-facility solution.

8.5   Managing Dyads in the Supply Chain

Much of the discussion in previous sections in this chapter, particularly in the immediately preceding section, considers issues related to the *design* of the supply chain. Our emphasis up to this point has been to carefully consider and understand the interaction of inventory, transportation, and location decisions on SC design. In our inventory discussion, we got some sense of how *operating policies* can affect the performance of the SC. The perspective of the entire chapter thus far has really been from the standpoint of a single decision maker.

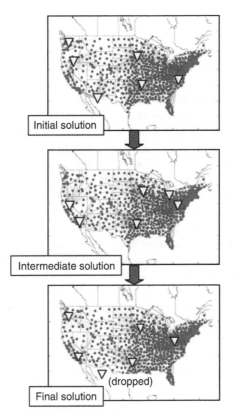

FIGURE 8.20   Series of ALA solutions (from Bucci and Kay, 2006).

An important issue in managing the supply chain, however, is how firms in the supply chain interact with one another. From a supply and demand perspective, the firms in any supply chain interact through orders placed upstream to replenish supply and shipments sent downstream to fulfill the demand represented by those orders. Vollmann et al. (2005), however, cogently point out that understanding how to manage the supply chain really requires a focus on firm-to-firm interactions in *dyads*. While it is indeed the complex network of these dyads that truly comprises the supply chain, one cannot possibly hope to understand all of the complexity of those many interactions without first stepping back to try and understand the factors that drive each dyad relationship individually.

In the past decade, a number of practitioners and supply chain management researchers have lent a significant attention to the level of variability in the stream of orders placed over time. Empirical evidence (Blinder, 1982, 1986; Lee et al., 1997b; Chen et al., 2000; Callioni and Billington, 2001) shows that replenishment orders demonstrate increasing variability at successive upstream echelons of a supply chain. Lee et al. (1997b) report that Procter & Gamble noted this phenomenon, called the *bullwhip effect*, in its supply chain for Pampers diapers. Underlying demand for diapers over a relatively short period of time, say a few months, tends to be quite stable, as one would predict—babies clearly go through them at a pretty steady rate, as any parent knows first-hand! However, P&G noticed that, although underlying consumer demand for Pampers was stable, the orders from retailers to their wholesalers were more volatile than consumer demand, while orders from wholesalers to P&G were more volatile still, and orders from P&G to its suppliers were more volatile still.

Indeed, the bullwhip effect is formally defined by this phenomenon, an increasing level of variability in orders at successive upstream echelons of the supply chain.

Although the effect has been the subject of much discussion in the supply chain management literature since, roughly, the mid-1990s, the effect has been studied quite a bit longer than this. As we indicated above, economists began to study the effect and its origins in earnest in the mid-1980s (Blinder, 1982, 1986; Caplin, 1985; Khan, 1987). Moreover, the famous systems dynamics studies of Forrester (1961) ultimately have had a tremendous impact on supply chain management education, with scores of students each academic year playing the famous "Beer Game" that resulted from follow-on work to Forrester's by Sterman (1989). In playing the Beer Game, students run a four-echelon supply chain for "beer" (typically, only coins that represent cases of beer—to the great relief of many university administrators, and to the great dismay of many students), ultimately trying to fulfill consumer demand at the retailer by placing orders with upstream partners, eventually culminating in the brewing of "beer" by a factory at the farthest upstream echelon in the chain. Indeed, the author's experience in playing the game with business students at all levels—undergraduates, MBAs, and seasoned managers—has never failed to result in bullwhip effects in the chains playing the game, sometimes in a quite pronounced fashion.

A natural question, and one that researchers continue to debate, is what causes this effect. If the underlying demand is relatively stable, and orders are meant ultimately to replenish this underlying demand, why should orders not also be relatively stable? The astute reader may quickly identify an issue from our discussion of the inventory above, namely that the economics of the ordering process might dictate that orders be batched in order to, reduce the administrative costs of placing orders or the scale-driven costs of transporting order fulfillment quantities from the supplier to the customer. Indeed, Lee et al. (1997a,b) identify order batching as one of four specific causes of the bullwhip effect, and the economic studies of the 1980s (Blinder, 1982, 1986; Caplin, 1985) clearly note this cause. The other causes as described by Lee et al. are demand forecast updating, price fluctuations, and supply rationing and shortage gaming. The last two effects are fairly straightforward to understand, namely that a customer might respond to volume discounts or short-term "special" prices by ordering out-of-sync with demand, or that a customer might respond to allocations of scarce supply by placing artificially-inflated orders to attempt to secure more supply than other customers. For the purposes of understanding reasonably simple dyad relationships, we will focus only on order batching and demand forecast updating, leaving the complexities of pricing and shortage gaming outside the scope of our discussion.

Indeed, the method used to generate the demand forecast at each echelon in the supply chain can have a significant impact on the stream of orders placed upstream. Vollmann et al. (2005) provide a simple example to demonstrate the existence of the bullwhip effect; this example is repeated and extended in Table 8.14 and Figure 8.21. We extend what is a 10-period example from Vollmann et al. to 20 periods that cover two repeating cycles of a demand stream with slight seasonality. In addition, we institute a few policy rules: (1) that orders must be non-negative, (2) that shortages at the manufacturer result in lost consumer demand, (3) that shortages at the supplier are met by the manufacturer with a supplemental source of perfectly reliable, immediately available supply. The example of Vollman et al. employs a relatively simple, and essentially *ad hoc*, forecast updating and ordering policy at each level in the chain: in any period t, order $2D_t - I_t$, where D_t is actual demand in period t and I_t is the ending inventory in period t, computed after fulfilling as much of D_t as possible from the beginning inventory. Thus, the current demand is used as the forecast of future demand, and as the replenishment lead time is one period (i.e., an order placed at the end of the current period arrives at the beginning of the next period—one could obviously

TABLE 8.14 Bullwhip Example from Vollmann et al. (2005) Altered for Minimum Order Size of 0 and Horizon of 20 Periods

Period	Consumer Sales	Manufacturer				Supplier			
		Beg Inv	End Inv	Lost Dmd	Order	Beg Inv	End Inv	Lost Dmd	Order
1	50	100	50	0	50	100	50	0	50
2	55	100	45	0	65	100	35	0	95
3	61	110	49	0	73	130	57	0	89
4	67	122	55	0	79	146	67	0	91
5	74	134	60	0	88	158	70	0	106
6	67	148	81	0	53	176	123	0	0
7	60	134	74	0	46	123	77	0	15
8	54	120	66	0	42	92	50	0	34
9	49	108	59	0	39	84	45	0	33
10	44	98	54	0	34	78	44	0	24
11	50	88	38	0	62	68	6	0	118
12	55	100	45	0	65	124	59	0	71
13	61	110	49	0	73	130	57	0	89
14	67	122	55	0	79	146	67	0	91
15	74	134	60	0	88	158	70	0	106
16	67	148	81	0	53	176	123	0	0
17	60	134	74	0	46	123	77	0	15
18	54	120	66	0	42	92	50	0	34
19	49	108	59	0	39	84	45	0	33
20	44	98	54	0	34	78	44	0	24
Min	44		38		34		6		0
Max	74		81		88		123		118
Avg	58.1		58.7		57.5		60.8		55.9
Fill rate				100.0%				100.0%	
Range	30				54				118
Std Dev	9.16				17.81				39.04
				Ratio	1.94			Ratio	4.26

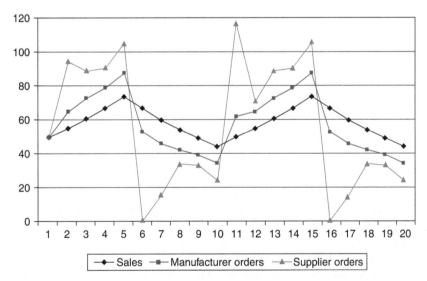

FIGURE 8.21 Bullwhip example from Vollmann et al. (2005) with horizon of 20 periods.

debate whether this a lead time of zero), the policy amounts to carrying a period's worth of safety stock.

We measure the severity of the bullwhip effect by taking the ratio of the standard deviation of the orders placed at each echelon to the standard deviation of consumer sales. As one can see, the example in Table 8.14 and Figure 8.21 demonstrates a clear bullwhip effect, with manufacturer orders exhibiting about twice the volatility of consumer sales and supplier orders exhibiting more than four times the volatility. One conclusion that we could

TABLE 8.15   Bullwhip Example with Two-Period Lead Time

Period	Consumer Sales	Manufacturer Beg Inv	End Inv	Lost Dmd	Order	Supplier Beg Inv	End Inv	Lost Dmd	Order
1	50	100	50	0	50	100	100	–	0
2	55	50	0	5	110	100	50	0	170
3	61	50	0	11	122	50	0	60	244
4	67	110	43	0	91	170	48	0	134
5	74	165	91	0	57	292	201	0	0
6	67	182	115	0	19	335	278	0	0
7	60	172	112	0	8	278	259	0	0
8	54	131	77	0	31	259	251	0	0
9	49	85	36	0	62	251	220	0	0
10	44	67	23	0	65	220	158	0	0
11	50	85	35	0	65	158	93	0	37
12	55	100	45	0	65	93	28	0	102
13	61	110	49	0	73	65	0	0	146
14	67	114	47	0	87	102	29	0	145
15	74	120	46	0	102	175	88	0	116
16	67	133	66	0	68	233	131	0	5
17	60	168	108	0	12	247	179	0	0
18	54	176	122	0	0	184	172	0	0
19	49	134	85	0	13	172	172	0	0
20	44	85	41	0	47	172	159	0	0
Min	44		0		0		0		0
Max	74		122		122		278		244
Avg	58.1		59.55		57.4		130.8		55.0
Fill rate				98.6%				94.8%	
Range	30				122				244
Std Dev	9.16				35.23				77.32
				Ratio	3.85			Ratio	8.44

draw from this example is that an ad hoc ordering policy can cause a bullwhip effect. Fair enough—but an important aspect of this conclusion is that many supply chains are, in practice, driven by similarly ad hoc ordering policies, an observation noted by the author in conversations with practicing managers and also supported by Chen et al. (2000).

Thus, even without the type of complexities in the demand stream such as the serial correlation in demand studied by Khan (1987), Lee et al. (1997a), and Chen et al. (2000), our simple example shows the bullwhip effect in action. Chen et al. (2000), however, demonstrate a key point—that the magnitude of the bullwhip effect can be shown to be a function, indeed in their case a *super-linear* function, of the replenishment lead time. Thus, let us alter the Vollmann et al. (2005) example to consider a situation in which the replenishment lead time is *two* periods—that is, where an order placed at the end of period t arrives at the beginning of period $t+2$. Using the same ad hoc ordering policy as above, the results are shown in Table 8.15 and Figure 8.22. Clearly, the bullwhip effect gets dramatically worse in this case, essentially doubling the variability in orders at both echelons with respect to consumer demand.

Now let's try some simple things to mitigate the bullwhip effect in this example. First, we will use a base-stock ordering policy at both the manufacturer and the supplier. As the original example did not specify an ordering cost, we will assume that this cost is zero or negligible. Thus, for *stationary* demand, a stationary base-stock policy would be warranted; however, in our example, demand is not stationary. Let us therefore use an "updated base-stock" policy, where the base-stock level will be updated periodically, in this case every five periods. We will assume that the manufacturer bases its base stock policy on a forecast of demand, with a forecasted mean and standard deviation of per-period demand of $\mu_1^M = 65$ and $\sigma_1^M = 10$ in the periods 1–5 and $\mu_2^M = 55$ and $\sigma_2^M = 10$ in the periods 6–10. These parameters are also used in the second demand cycle, with the higher mean being used as the demand rises in periods 11–15 and the lower mean as demand falls in periods 16–20. We will assume that the supplier uses a forecast of the manufacturer's orders to compute

FIGURE 8.22   Bullwhip example with two-period lead time.

TABLE 8.16   Bullwhip Example with Updated Base-Stock Orders

Period	Consumer Sales	Manufacturer Beg Inv	End Inv	Lost Dmd	B	Order	Supplier Beg Inv	End Inv	Lost Dmd	B	Order
1	50	100	50	0	83	33	100	100	–	96	0
2	55	50	0	5	83	83	100	67	0	96	29
3	61	33	0	28	83	83	67	0	16	96	96
4	67	83	16	0	83	67	29	0	54	96	96
5	74	99	25	0	83	58	96	29	0	96	67
6	67	92	25	0	73	48	125	67	0	102	35
7	60	83	23	0	73	50	134	86	0	102	16
8	54	71	17	0	73	56	121	71	0	102	31
9	49	67	18	0	73	55	87	31	0	102	71
10	44	74	30	0	73	43	62	7	0	102	95
11	50	85	35	0	83	48	78	35	0	60	25
12	55	78	23	0	83	60	130	82	0	60	0
13	61	71	10	0	83	73	107	47	0	60	13
14	67	70	3	0	83	80	47	0	26	60	60
15	74	76	2	0	83	81	13	0	67	60	60
16	67	82	15	0	73	58	60	0	21	94	94
17	60	96	36	0	73	37	60	2	0	94	92
18	54	94	40	0	73	33	96	59	0	94	35
19	49	77	28	0	73	45	151	118	0	94	0
20	44	61	17	0	73	56	153	108	0	94	0
Min	44		0			33		0			0
Max	74		50			83		118			96
Avg	58.1		20.65			57.35		45.45			45.8
Fill rate				97.2%					84.0%		
Range	30					50					96
Std Dev	9.16					16.11					36.22
				Ratio		1.76			Ratio		3.95

its base-stock level in the first five periods, with $\mu_1^S = 60$ and $\sigma_1^S = 20$. In the next three five-period blocks, the supplier uses actual orders from the manufacturer to compute the mean and standard deviation of demand, so that $(\mu_i^S, \sigma_i^S) = \{(64.8, 20.8), (50.4, 5.3), (68.4, 14.2)\}$, $i = 2, 3, 4$. At both echelons, therefore, the base-stock policy is computed as $B_i = \mu_i + z\sigma_i\sqrt{2}$ $(i = 2, 3, 4)$, with $z = 1.282$, yielding a target cycle service level of 90%, and $\sigma_{DLT,i} = \sigma_i\sqrt{2}$. As one can see from Table 8.16 and Figure 8.23, the base-stock policies work quite well in reducing the bullwhip effect, but at the expense of lower fill rates at both the manufacturer

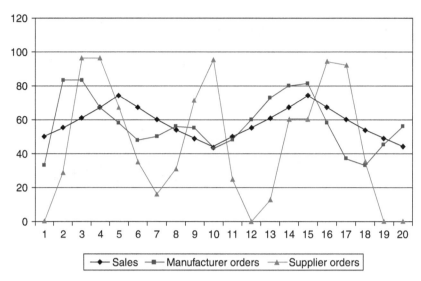

FIGURE 8.23   Bullwhip example with updated base-stock orders.

and the supplier. (Note that a higher CSL target does not have much impact on this outcome.)

One key prescription of the literature on the bullwhip effect is to centralize demand information so that demand forecasts are not inferred from the orders of the echelon immediately downstream—the essence of the "demand forecast updating" (Lee et al., 1997b) cause of the bullwhip effect—but instead come directly from consumer sales data. Therefore, let us update our example to account for this. We will assume the same base-stock policy at the manufacturer as above, based on its forecasts of rising and falling consumer demand. At the supplier, however, we will assume that the initial base stock level is computed directly from the manufacturer's forecast so that $\mu_1^S = 65$ and $\sigma_1^S = 10$. In the succeeding five-period blocks, the supplier updates its base-stock levels using actual consumer sales data, such that $(\mu_i^S, \sigma_i^S) = \{(61.4, 9.5), (54.8, 9.0), (61.4, 9.5)\}$, $i = 2, 3, 4$. As proved in the literature (Chen et al., 2000), this will mitigate the bullwhip effect at the supplier, as our results in Table 8.17 and Figure 8.24 demonstrate, but only slightly, and it certainly does not eliminate the effect. Also, the supplier fill rate in this latter case improves from the previous base-stock case, but only slightly.

Finally, let us consider the effect of batch ordering at the manufacturer. Let us assume that the manufacturer has occasion to order in consistent lots, with $Q = 150$. We further assume that the manufacturer follows a (Q, R) policy with R set to the base-stock levels from the prior two examples. In contrast, we assume that the supplier has no need to batch orders and follows a base-stock policy, with $\mu_1^S = 100$ and $\sigma_1^S = 30$, set in anticipation of the large orders from the manufacturer. The results are shown in Table 8.18 and Figure 8.25, in which the supplier's base-stock levels in the remaining five-period blocks are updated using the manufacturer orders, resulting in $(\mu_i^S, \sigma_i^S) = (60.0, 82.2)$, $i = 2, 3, 4$, as each block contains exactly two orders from the manufacturer. (Note that the assumption of a normal distribution for lead-time demand probably breaks down with such a large standard deviation, due to the high probability of negative values, but we retain this assumption nonetheless.) From the table and figure, one can see that the variability in the order stream increases dramatically at the manufacturer. Interestingly, however, the significant increase in variability is not further amplified at the supplier when it uses a base-stock policy.

TABLE 8.17 Bullwhip Example with Updated Base-Stock Orders Using Consumer Demand

Period	Consumer Sales	Manufacturer					Supplier				
		Beg Inv	End Inv	Lost Dmd	B	Order	Beg Inv	End Inv	Lost Dmd	B	Order
1	50	100	50	0	83	33	100	100	–	83	0
2	55	50	0	5	83	83	100	67	0	83	16
3	61	33	0	28	83	83	67	0	16	83	83
4	67	83	16	0	83	67	16	0	67	83	83
5	74	99	25	0	83	58	83	16	0	83	67
6	67	92	25	0	73	48	99	41	0	79	38
7	60	83	23	0	73	50	108	60	0	79	19
8	54	71	17	0	73	56	98	48	0	79	31
9	49	67	18	0	73	55	67	11	0	79	68
10	44	74	30	0	73	43	42	0	13	79	79
11	50	85	35	0	83	48	68	25	0	71	46
12	55	78	23	0	83	60	104	56	0	71	15
13	61	71	10	0	83	73	102	42	0	71	29
14	67	70	3	0	83	80	57	0	16	71	71
15	74	76	2	0	83	81	29	0	51	71	71
16	67	82	15	0	73	58	71	0	10	79	79
17	60	96	36	0	73	37	71	13	0	79	66
18	54	94	40	0	73	33	92	55	0	79	24
19	49	77	28	0	73	45	121	88	0	79	0
20	44	61	17	0	73	56	112	67	0	79	12
Min	44		0			33		0			0
Max	74		50			83		100			83
Avg	58.1		20.65			57.35		34.45			44.85
Fill rate				97.2%					84.9%		
Range	30					50					83
Std Dev	9.16					16.11					29.47
				Ratio		1.76			Ratio		3.22

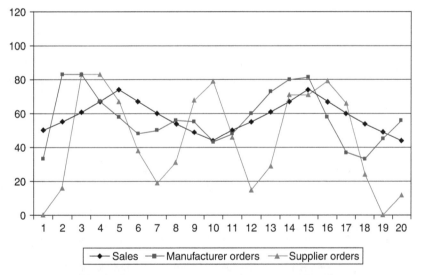

FIGURE 8.24 Bullwhip example with updated base-stock orders using consumer demand.

These examples are not intended to prove any general results, per se, but merely to demonstrate the effect of ordering policies on supply chain performance and to emphasize the challenge of identifying policy prescriptions to mitigate these effects. Note that Lee et al. (1997a,b) lay out a more extensive set of prescriptions that may be used in practice to attempt to address the full set of causes for the bullwhip effect—order batching, forecast updating, price fluctuations, and shortage gaming. From our examples, interestingly, the fill rate performance is actually best with the ad hoc policy from the original example

TABLE 8.18   Bullwhip Example with (Q, R) Orders at Manufacturer and Base-Stock Orders at Supplier

	Consumer	Manufacturer					Supplier				
Period	Sales	Beg Inv	End Inv	Lost Dmd	R	Order	Beg Inv	End Inv	Lost Dmd	B	Order
1	50	100	50	0	83	150	100	100	–	154	54
2	55	50	0	5	83	150	100	0	50	154	154
3	61	150	89	0	83	0	54	0	96	154	154
4	67	239	172	0	83	0	154	154	0	154	0
5	74	172	98	0	83	0	308	308	0	154	0
6	67	98	31	0	73	150	308	308	0	209	0
7	60	31	0	29	73	150	308	158	0	209	51
8	54	150	96	0	73	0	158	8	0	209	201
9	49	246	197	0	73	0	59	59	0	209	150
10	44	197	153	0	73	0	260	260	0	209	0
11	50	153	103	0	83	0	410	410	0	209	0
12	55	103	48	0	83	150	410	410	0	209	0
13	61	48	0	13	83	150	410	260	0	209	0
14	67	150	83	0	83	0	260	110	0	209	99
15	74	233	159	0	83	0	110	110	0	209	99
16	67	159	92	0	73	0	209	209	0	209	0
17	60	92	32	0	73	150	308	308	0	209	0
18	54	32	0	22	73	150	308	158	0	209	51
19	49	150	101	0	73	0	158	8	0	209	201
20	44	251	207	0	73	0	59	59	0	209	150
Min	44		0			0		0			0
Max	74		207			150		410			201
Avg	58.1		85.55			60.0		169.85			68.2
Fill rate				94.1%					87.8%		
Range	30					150					201
Std Dev	9.16					75.39					75.37
				Ratio		8.23			Ratio		8.23

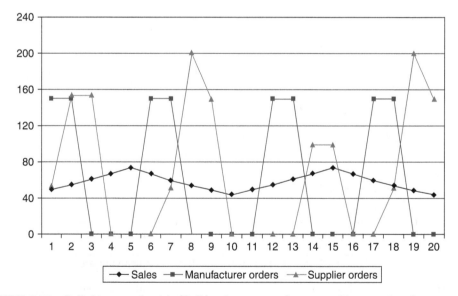

FIGURE 8.25   Bullwhip example with (Q, R) orders at manufacturer and base-stock orders at supplier.

in Vollmann et al. (2005). This is important, in that lost demand at the supplier in our example is a proxy for additional cost at the manufacturer to supplement supply. (A more complex assumption would be to allow backorders at the supplier, thus deflating the service performance at the manufacturer further and likely exacerbating the bullwhip effect at the supplier.) Our example might really speak to a need for a richer model of the interaction between customer and supplier, especially in a case where there is an underlying pattern to demand, and particularly in the case where the customer has access to more information about consumer demand than the supplier could infer on its own simply from the customer's stream of orders. Indeed, in practice, this has led to burgeoning efforts such as collaborative planning, forecasting, and replenishment (CPFR).*

Lest one conclude that sharing forecast information *always* improves the outcome for all parties in the supply chain, let us briefly consider some other research in progress that identifies cases when forecast sharing is helpful, and when it might actually make things worse. Thomas et al. (2005) offer a richer model of a specific dyad relationship than the simple ordering relationships we consider above. Their model considers a contract manufacturer (CM) that builds a product for an original equipment manufacturer (OEM). The product is composed of two sets of components, a long lead time set and a shorter lead time set, such that orders for the long lead time components must be committed by the CM to its supplier some time before orders are placed for the short lead time components. In the intervening period between these lead times, the OEM may be able to provide an updated forecast of the single-period consumer demand to the CM. In the base model, the only cost to the OEM is the opportunity cost of lost sales if the CM does not order sufficient quantities of both component sets to assemble complete products and fulfill consumer demand for the product. The CM, however, bears both a lost-sales cost, based on its share of the product profit margin, *and* the overage loss related to having components left over in excess of demand. Since the CM bears the costs of both underage and overage, it may tend to be risk averse in its ordering, particularly when its profit margin is relatively small, meaning that the OEM could benefit by offering contract terms to share a portion of the overage cost with the CM.

With D denoting demand, ξ the demand-related information update, and γ_1 and γ_2 the fraction of overage loss the OEM bears on the long lead time and short lead time components, respectively, the OEM's cost function in Thomas et al. (2005) is

$$TC_{\text{fs}}^{\text{OEM}}(\gamma_1, \gamma_2) = E_\xi[E_{D|\xi}[m_o(D - Q_2^*)^+ + \gamma_1 l_1(Q_1^* - \min\{D, Q_2^*\})^+ + \gamma_2 l_2(Q_2^* - D)^+]] \tag{8.37}$$

where m_o is the OEM's share of the product margin, l_1 and l_2 are the overage losses (unit cost minus salvage value) on the long lead time and short lead time components, respectively, and Q_1^* and Q_2^* are the CM's optimal ordering decisions (for long and short lead time components, respectively), given the forecast sharing and risk sharing decisions of the OEM. The CM's optimal ordering decisions come from solving

$$TC_{\text{fs}}^{\text{CM}}(Q_1) = E_\xi[E_{D|\xi}[m_c(D - Q_2^*)^+ + (1 - \gamma_1)l_1(Q_1 - \min\{D, Q_2^*\})^+ + (1 - \gamma_2)l_2(Q_2^* - D)^+]] \tag{8.38}$$

where m_c is the CM's share of the product margin and

$$Q_2^* = \min\{F_{D|\xi}^{-1}(\alpha_2), Q_1\} \tag{8.39}$$

with $F_{D|\xi}$ being the cdf of D given ξ and with α_2 given by

$$\alpha_2 = \frac{m_c + (1 - \gamma_1)l_1}{m_c + (1 - \gamma_1)l + (1 - \gamma_1)l_1} \tag{8.40}$$

If the demand forecast update is not shared, Equations 8.37 and 8.38 are simplified, with the expectation taken over the random variable D alone. Using a model that employs nested

---

*See the Web site of the Voluntary Interindustry Commerce Standards (VICS) organization at http://www.vics.org/committees/cpfr/ for more details on this set of practices.

TABLE 8.19 OEM-CM Cost Reductions from Forecast and Risk Sharing, Thomas et al. (2005)

			No Risk Sharing (%)	Limited Risk Sharing (%)	Complete Risk Sharing (%)
Reduction in TC^{OEM}	No forecast sharing	Mean	–	4.60	6.87
		Max	–	55.23	68.90
	Forecast sharing	Mean	8.69	12.83	15.80
		Max	43.75	50.17	72.97
Reduction in TC^{CM}	No forecast sharing	Mean	–	6.63	10.90
		Max	–	58.97	78.30
	Forecast sharing	Mean	3.16	9.19	14.28
		Max	17.49	49.05	79.87

uniform distributions for demand and demand-related information, Thomas et al. are able to express solutions for Q_1^* and Q_2^* in closed-form. They also demonstrate that this nested-uniform demand model is a very good approximation to the case where the demand forecast and the related updating information form a bivariate normal pair.

Thomas et al. consider different contracts for sharing overage risk, with the extremes being no risk sharing ($\gamma_1 = \gamma_2 \equiv 0$), limited risk sharing (only on the long lead time components—i.e., $\gamma_2 \equiv 0$), and complete risk sharing (on both long and short lead time components). The interaction of forecast sharing and risk sharing creates some interesting results in this relatively complex situation involving components with a complementary relationship—noting that the CM's order for short lead time components is bounded above by its order for long-lead time components, per Equation 8.39 above. Table 8.19 shows the comparative reductions in OEM and CM costs, on average, of sharing forecast information and sharing overage risk from the base case of "no forecast sharing—no risk sharing" across 18,225 computational experiments with m_o and m_c ranging from 0.1 to 1.5 and l_1 and l_2 ranging from 10% to 90% of unit cost. Interestingly, 12.5% of the computational experiments under limited risk sharing exhibit *increased* cost for the OEM—by an average of about 4% and as much as 12%—when the forecast update is shared with the CM. This effect is completely mitigated, however, when the OEM shares the overage risk on both sets of components.

8.6 Discussion and Conclusions

Let us now step back and tie together the themes and ideas presented in this chapter. An important theme, and one that appeared throughout the chapter, is that supply chain management strategy is framed by three important decision areas—inventory, transportation, and locations—and their joint impact on customer service. Moreover, we can now turn that statement around and state that it is the *interaction* of the inventory, transportation, and location decisions—and their joint impact on customer service—that frame the design of the supply chain itself. Indeed, it is the author's view that SC design initiatives that are undertaken without a clear understanding and appreciation of those interactions will generate poor results, at best, and may be doomed to spectacular failure, at worst.

Another theme, albeit one that appeared late in the chapter and to which a much smaller number of pages has been devoted, is the effect of SCM *policies* on the operation of the supply chain. Policies in SCM take many forms, and many IE and operations management researchers have devoted significant time and effort to the study of inventory management policies. Indeed, a relatively wide-ranging chapter on SCM such as this one could not do justice to the study of inventory management, about which many books have been written. However, my hope in this chapter was to take at least a small step beyond inventory policies and also consider coordinating mechanisms like the sharing of overage risk that help to better align the incentives of the parties in a supply chain. This theme focuses on the *dyad relationships* in the supply chain, the point at which any two parties in the chain interact, and, in the author's mind, the *only* point at which one can hope to study supply chain management—as opposed to making the bold and rather far-fetched claim that one could, for example, "optimize the management of the supply chain," whatever that might mean. Similar to the manner in which the decisions and interactions of the "triangle-leg decisions" (inventory, transportation, and location) frame SC design, it should be clear that an understanding and appreciation of the interaction of dyad partners frames the design of SCM policies.

So, we return to the "guy walks into a bar" scenario—or at least the "software salesperson walks into the IE's office" scenario—that opened the chapter. Could one "optimize the supply chain" with software? Certainly not with a single piece of software solving a single problem. After all of the discussion presented in this chapter, my sincere hope is that the reader has an appreciation for the significant scope, complexity, and subtlety of the types of problems underlying the relatively large domain that is supply chain management. Could one, however, optimize portions of the supply chain? ... individually optimize the various decision problems that together frame SCM? ... find policies that minimize or jointly improve the independent objective functions of dyad partners? Certainly, all of those things can be done. Will they, together, represent a "supply chain optimal" solution? That seems like a loaded question, and one that I'll avoid answering.

Perhaps the discussion at the outset of the chapter was a bit too critical, however. Softening that criticism somewhat, I will close by noting the relatively recent advent of advanced planning systems (APS) software, which seems to indicate that the marketplace for SCM software recognizes that "optimizing" the supply chain in one fell swoop would be a Herculean, if not impossible, task. Indeed, referring back to Figure 8.2 in the Introduction to this chapter, one is reminded that at least one company that does business by attempting to create and implement supply chain (re)design initiatives does so in modular fashion, employing tools that solve a variety of discrete problems in the process of building an overall SC design solution. An excellent reference on this subject is the book chapter by Fleischmann and Meyr (2003), which includes a mapping of their conceptual model of APS activities to the APS software offerings of the major software vendors in this market—namely Baan, i2 Technologies, J.D. Edwards, Manugistics, and SAP. Each company offers its software in a series of modules that focus on discrete problem areas—for example, forecasting and demand management, transportation management, inventory management—very much along the lines of the discussion in this chapter.

Indeed, supply chain management is a wide-ranging and challenging subject. Solving the decision problems that frame SCM requires a similarly wide-ranging knowledge and an appreciation of the manner in which the various aspects of SC design solutions and SCM policy solutions interact to impact company objectives like minimizing operating costs at acceptably high levels of customer service. However, armed by this chapter with an appreciation of the challenges implied by designing and managing—let alone *optimizing*— the supply chain, my sincere hope is that you, the reader, won't be tempted to rush into buying that bridge in Brooklyn.

References

1. Atkinson, W. (2002), DC siting—What makes most sense?, *Logistics Management and Distribution Report*, May issue, 41(5), S63–S65.
2. Axsäter, S. (2000), *Inventory Control*, Boston: Kluwer Academic.

3. Balintfy, J. (1964), On a basic class of multi-item inventory problems, *Management Science*, 10(2), 287–297.

4. Ballou, R.H. (1991), The accuracy in estimating truck class rates for logistical planning, *Transportation Research—Part A*, 25A(6), 327–337.

5. Ballou, R. H. (1992), *Business Logistics Management*, 3rd ed., Englewood Cliffs, NJ: Prentice Hall.

6. Ballou, R. H. (2004), *Business Logistics/Supply Chain Management*, 5th ed., Upper Saddle River, NJ: Pearson Prentice Hall.

7. Baumol, W. J. and H. D. Vinod (1970), An inventory theoretic model of freight transport demand, *Management Science*, 16(7), 413–421.

8. Blinder, A. S. (1982), Inventories and sticky prices: More on the microfoundations of macroeconomics, *American Economic Review*, 72(3), 334–348.

9. Blinder, A. S. (1986), Can the production smoothing model of inventory behavior be saved?, *Quarterly Journal of Economics*, 101(3), 431–453.

10. Bozarth C. C. and D. P. Warsing (2006), *The Evolving Role of Manufacturing Planning and Control Systems in Supply Chains*, College of Management working paper, North Carolina State University, Raleigh, NC.

11. Bozarth, C. C. and R. B. Handfield (2006), *Introduction to Operations and Supply Chain Management*, Upper Saddle River, NJ: Pearson Prentice Hall.

12. Bozarth. C. C. (2005), Executive education session notes on *Inventory Operations*, North Carolina State University, Raleigh, NC.

13. Bryksina, E. A. (2005), *Assessing the impact of strategic safety stock placement in a multi-echelon supply chain*, M.S. thesis, North Carolina State University, Raleigh, NC.

14. Bucci, M. J. and M. G. Kay (2006), A modified ALA procedure for logistic network designs with scale economies, *Proceedings of the Industrial Engineering Research Conference*, Orlando, FL.

15. Cachon, G. and C. Terwiesch (2006), *Matching Supply with Demand: An Introduction to Operations Management*, New York: McGraw-Hill/Irwin.

16. Callioni, G. and C. Billington (2001), Effective collaboration, *OR/MS Today*, October issue, 28(5), 4–39.

17. Caplin, A. S. (1985), The variability of aggregate demand with (S,s) inventory policies, *Econometrica*, 53(6), 1395–1410.

18. Çentinkaya, S. and C. Y. Lee (2002), Optimal outbound dispatch policies: Modeling inventory and cargo capacity, *Naval Research Logistics*, 49(6), 531–556.

19. Chen, F., Z. Drezner, J. K. Ryan, and D. Simchi-Levi (2000), Quantifying the bullwhip effect in a simple supply chain: The impact of forecasting, lead times, and information, *Management Science*, 46(3), 436–443.

20. Chopra, S. and P. Meindl (2001),*Supply Chain Management: Strategy, Planning, and Operation*, Upper Saddle River, NJ: Pearson Prentice Hall.

21. Chopra, S. and P. Meindl (2004), *Supply Chain Management: Strategy, Planning, and Operation*, 2nd ed., Upper Saddle River, NJ: Pearson Prentice Hall.

22. Clark, A. J. and H. Scarf (1960), Optimal policies for a multi-echelon inventory problem, *Management Science*, 6(4), 475–490.

23. Cooper, L. (1963), Location-allocation problems, *Operations Research*, 11(3), 331–343.

24. Cooper, L. (1964), Heuristic methods for location-allocation problems, *SIAM Review*, 6(1), 37–53.

25. Cox, J. B. (2005), They got Dell. Now what?, *The News & Observer*, October 9 edition, Raleigh, NC.

26. Coyle J. J., E. J. Bardi, and C. J. Langley, Jr. (2003), *The Management of Business Logistics*, 7th ed., Mason, OH: South-Western.

27. Coyle, J. J., E. J. Bardi, and R. A. Novack (2006), *Transportation*, 6th ed., Mason, OH: Thomson–South-Western.

28. Craver, R. (2005), Dell computer-assembly plant opens in Forsyth County, N.C., *Winston-Salem Journal*, October 5 edition, Winston-Salem, NC.

29. Ettl, M., G. E. Feigin, G. Y. Lin, and D. D. Yao (1997), *A supply network model with base-stock control and service requirements*, IBM Corporation technical report RC 20473, Armonk, NY.

30. Ettl, M., G. E. Feigin, G. Y. Lin, and D. D. Yao (2000), A supply network model with base-stock control and service requirements, *Operations Research*, 48(2), 216–232.

31. Feigin, G. E. (1999), Inventory planning in large assembly supply chains, in *Quantitative Models for Supply Chain Management*, S. Tayur, R. Ganeshan, and M. Magazine, Eds., Boston: Kluwer Academic Publishers.

32. Feitzinger, E. and H. L. Lee (1997), Mass customization at Hewlett-Packard: The power of postponement, *Harvard Business Review*, 75(1), 116–121.

33. Fisher, M. L. (1997), What is the right supply chain for your product?, *Harvard Business Review*, 75(2), 105–116.

34. Fleischmann, B. and H. Meyr (2003), Planning hierarchy, modeling and advanced planning systems, in *Handbooks in OR & MS, Vol. 11*, A. de Kok and S. Graves, Eds., Amsterdam: Elsevier.

35. Forrester, J. W. (1961), *Industrial Dynamics*, Cambridge, MA: MIT Press.

36. Graves, S. and S. Willems (2003), Supply chain design: Safety stock placement and supply chain configuration, in *Handbooks in OR & MS, Vol. 11*, A. de Kok and S. Graves, Eds., Amsterdam: Elsevier.

37. Harvard Business School (1998), *Note on the U.S. Freight Transportation Industry*, # 9-688-080, Boston, MA: Harvard Business School Publishing.

38. Hax, A. C. and D. Candea (1984), *Production and Inventory Management*, Englewood Cliffs, NJ: Prentice-Hall.

39. Holt, C. C., F. Modigliani, J. F. Muth, and H. A. Simon (1960), *Planning Production, Inventories, and Work Force*, Englewood Cliffs, NJ: Prentice-Hall.

40. Hopp, W. J. and M. L. Spearman (2004), To pull or not to pull: What is the question?, *Manufacturing & Service Operations Management*, 6(2), 133–148.

41. Houck, C. R., J. A. Joines, and M. G. Kay (1996), Comparison of genetic algorithms, random restart, and two-opt switching for solving large location-allocation problems, *Computers & Operations Research*, 23(6): 587–596.

42. Jackson, P., W. Maxwell, and J. Muckstadt (1985), The joint replenishment problem with a powers-of-two restriction, *IIE Transactions*, 17(1), 25–32.

43. Kay, M. G. and D. P. Warsing (2006), *Modeling Truck Rates Using Publicly Available Empirical Data*, North Carolina State University working paper, Raleigh, NC.

44. Khan, J. A. (1987), Inventories and the volatility of production, *American Economic Review*, 77(4), 667–679.

45. Kimball, G. E. (1988), General principles of inventory control, *Journal of Manufacturing and Operations Management*, 1(1), 119–130.

46. Koch, C. (2004), Nike rebounds: How (and why) Nike recovered from its supply chain disaster, *CIO*, 17 (Jun 15 issue).

47. Kulkarni, V. G. (1995), *Modeling and Analysis of Stochastic Systems*, Boca Raton, FL: Chapman & Hall/CRC.

48. Lee, C. Y. (1986), The economic order quantity for freight discount costs, *IIE Transactions*, 18(3), 318–320.

49. Lee, H. L. and C. Billington (1993), Material management in decentralized supply chains, *Operations Research*, 41(5), 835–847.

50. Lee, H. L., V. Padmanabhan, and S. Whang (1997a), Information distortion in a supply chain: The bullwhip effect, *Management Science*, 43(4), 546–558.

51. Lee, H. L., V. Padmanabhan, and S. Whang (1997b), The bullwhip effect in supply chains, *Sloan Management Review*, 38(3), 93–102.

52. Lockamy, A. and K. McCormack (2004), Linking SCOR planning practices to supply chain performance, *International Journal of Operations and Production Management*, 24(11/12), 1192–1218.

53. Marien, E. J. (1996), Making sense of freight terms of sale, *Transportation & Distribution*, 37(9), 84–86.

54. Marien, E. J. (2000), The four supply chain enablers, *Supply Chain Management Review*, 4(1), 60–68.

55. Nahmias, S. (1997), *Production and Operations Analysis*, 3rd ed., Chicago: Irwin.

56. Piercy, J. E. (1977), A Performance Profile of Several Transportation Freight Services, unpublished Ph.D. dissertation, Case Western Reserve University, Cleveland, OH.

57. Reuters (2004), N. Carolina lures new Merck plant with incentives, on-line *Update*, April 6 edition.

58. Rumelt, R. P. (2001), *Note on Strategic Cost Dynamics*, POL 2001-1.2, Anderson School of Management, UCLA, Los Angeles, CA.

59. Schwartz, B. M. (1999a), Map out a site route, *Transportation & Distribution*, 40(11), 67–69.

60. Schwartz, B. M. (1999b), New rules for hot spots, *Transportation & Distribution*, 40(11), 43–45.

61. Silver, E. A., D. F. Pyke, and R. Peterson (1998), *Inventory Management and Production Planning and Scheduling*, 3rd ed., New York: John Wiley & Sons.

62. Speigel, R. (2002), Truckload vs. LTL, *Logistics Management*, 47(7), 54–57.

63. Stedman, C. (1999), Failed ERP gamble haunts Hershey, *Computerworld*, 33(44), 1–2.

64. Sterman, J. D. (1989), Modeling managerial behavior misperceptions of feedback in a dynamic decision making experiment, *Management Science*, 35(3), 321–329.

65. Supply-Chain Council (2006), *Supply-Chain Operations Reference-Model—SCOR Version 8.0 Overview*, Washington, D.C. (available via http://www.supply-chain.org/).

66. Swenseth, S. R. and M. R. Godfrey (1996), Estimating freight rates for logistics decisions, *Journal of Business Logistics*, 17(1), 213–231.

67. Swenseth, S. R. and M. R. Godfrey (2002), Incorporating transportation costs into inventory replenishment decisions, *International Journal of Production Economics*, 77(2), 113–130.

68. Tersine, R. J. and S. Barman (1991), Lot size optimization with quantity and freight rate discounts, *Logistics and Transportation Review*, 27(4), 319–332.

69. Thomas, D. J., D. P. Warsing, and X. Zhang (2005), *Forecast Updating and Supplier Coordination for Complementary Component Purchases*, Smeal College of Business working paper, Pennsylvania State University, University Park, PA.

70. Tyworth, J. E. and A. Ruiz-Torres (2000), Transportation's role in the sole- versus dual-sourcing decision, *International Journal of Physical Distribution and Logistics Management*, 30(2), 128–144.

71. Vollmann, T., W. Berry, D. C. Whybark, and R. Jacobs (2005), *Manufacturing Planning & Control for Supply Chain Management*, 5th ed., New York: McGraw-Hill/Irwin.

72. Waller, M., M. E. Johnson, and T. Davis (1999), Vendor-managed inventory in the retail supply chain, *Journal of Business Logistics*, 20(1), 183–203.

73. Warsing, D. P., E. A. Helmer, and J. V. Blackhurst (2006), Strategic Safety Stock Placement in Production Networks with Supply Risk, *Proceedings of the Industrial Engineering Research Conference*, Orlando, FL.

74. Wilson, R. A. (2003), *Transportation in America*, Washington, D.C.: Eno Transportation Foundation.

75. Winston, W. L. (1994), *Operations Research: Applications and Algorithms*, 3rd ed., Belmont, CA: Duxbury Press.

9

E-Commerce

Sowmyanarayanan Sadagopan
Indian Institute of Information Technology

9.1   Introduction

Commerce is the exchange of goods and services between a buyer and a seller for an agreed price. The commerce ecosystem starts with a customer who interacts with a *salesperson*, typically at a *storefront*; the *store* gets the supply from a *distributor*; the distribution is typically from a *warehouse* or a *factory*; if the customer decides to buy the goods, the seller *invoices* the buyer for the *price* of goods, including taxes and shipping charges, if any; and finally the invoiced amount is *collected* either immediately or through an intermediary such as a bank or credit company. This commerce ecosystem has evolved over centuries. With the explosion of the Internet since 1995, every aspect of commerce could be supplemented by electronic support leading to an e-commerce ecosystem similar to the commerce eco-system. This includes systems to support customer acquisition, price negotiation, order processing, contracting, order fulfillment, logistics and delivery, bill presentation and payment, and post-sales support [1–3].

> E-commerce or electronic commerce is commerce conducted electronically, typically over the Internet. E-commerce includes support to all stages of commerce.

E-commerce initially was limited to "digital goods"—software, e-books, and information. Thanks to "digitization," today digital goods encompass a large set of items including

- *Computer and telecom software*;
- *Music* with analog recording on tapes evolving to digital music—first as CD audio and currently digital audio in the form of MP3 and other forms like RealAudio;

- *Video* with the evolution of digital cinematography and MPEG compression scheme;
- *Images* including still images captured by digital camera, video images captured by camcorders, document images and medical images like ultrasound image, digital X-ray image, CAT scan, and MRI image;
- *Documents* including books, journals, tech reports, magazines, handbooks, manuals, catalogs, standards documentation, brochures, newspapers, phone books, and a multitude of other "forms" often created using DTP; and
- *Drawings* including building plans, circuit diagrams, and flow diagrams typically created using CAD tools.

Digital goods could be completely delivered electronically, and formed the first focus of e-commerce. Soon it was clear that a whole range of information services lend themselves to electronic delivery. This includes

- *Financial services* that include banking, insurance, stock trading, and advisory services;
- *E-governance* including electronic delivery of justice and legal services, electronic support to legislation and law making, issue of identities like passport and social service numbers, visa processing, licensing for vehicles, toll collection, and tax collection;
- *Education* that includes classroom support, instructional support, library, laboratories, textbooks, conducting examinations, evaluation, and issue of transcripts; and
- *Healthcare delivery* including diagnostics using imaging, record keeping, and prescription handling.

Thanks to the success of digital goods delivery and electronic delivery of information services, e-commerce could address the support issues for buying and selling of physical goods as well. For example,

- *Customer acquisition* through search engine-based advertisements, online promotion, and e-marketing;
- *E-ticketing* for airline seats, hotel rooms, and movie tickets;
- *E-procurement* including e-tendering, e-auction, reverse auction, and electronic order processing;
- *Electronic payment* that includes electronic shopping cart and online payments; and
- *Electronic post-sales support* through status monitoring, tracking, health-checking, and online support.

To support digital goods delivery, delivery of information services, and support commerce related to physical goods, a huge Internet infrastructure had to be built (and continues to get upgraded). Optimizing this infrastructure that converged data communications with voice communications on the one hand and wire-line with wireless communications on the other hand is leading to several interesting challenges that are being addressed today. Behind those challenges are the key techniques and algorithms perfected by the OR/MS community over the past five decades [4–6]. In this chapter, we will address the key OR/MS applications relating to the distribution of digital goods, electronic delivery of information services, and optimizing the supporting Internet infrastructure.

9.2 Evolution of E-Commerce

Electronic commerce predates the widespread use of the Internet. Electronic data interchange (EDI) used from the early 1960s did facilitate transfer of commercial data between seller and buyer. The focus, however, of EDI was mostly to clear congestion of goods movement at seaports and airports due to rapid globalization of manufacturing, improved transportation, and the emergence of global logistics hubs such as Rotterdam and Singapore. EDI helped many governments in the process of monitoring of export and import of goods between countries and also control the movement of undesired goods such as drugs and narcotics across borders [1,2,7–15].

Less widely known e-commerce applications in the pre-Internet era include

- *Online databases* like *MEDLARS* for medical professionals and *Lexis-Nexis* for legal professionals;
- *Electronic funds transfer* such as SWIFT network; and
- *Online airline tickets reservations* such as SABRE from American Airlines.

In all these cases electronic data transfer was used to facilitate business between *partners*:

- Pre-determined intermediaries such as port authorities, customs inspectors, and export-import agencies in the case of EDI;
- Librarians and publishing houses in the case of online databases;
- Banks and banking networks in the case of electronic funds transfer; and
- Airlines and travel agents in the case of online reservations systems.

The ordering of books online on Amazon.com in 1994 brought e-commerce to the end users. One could order books electronically, and by the year 2000 make payments also electronically through a secure payment gateway that connects the buyer to a bank or a credit company. All this could be achieved from the desktop computer of an end user, a key differentiator compared to the earlier generation of electronic commerce. Such a development that permitted individual buyers to complete end-to-end buying (and later selling through eBay) made e-commerce such a revolutionary idea in the past decade.

The rise of digital goods (software, music, and e-books), the growth of the Internet, and the emergence of the "Internet-economy" companies—Yahoo and eBay—led to the *second stage* of e-commerce. Software was the next commodity to embrace e-commerce, thanks to iconic companies like Microsoft, Adobe, and Macromedia. Selling software over the Internet was far more efficient than any other means, particularly for many small software companies that were writing software to run on Wintel (Windows + Intel) platform (as PCs powered by Intel microprocessors were a dominant hardware platform and Windows dominated the desktop software platform). The outsourcing of software services to many countries including India gave further impetus to e-commerce.

The spectacular growth of the Internet and World Wide Web (WWW) fueled the demand for widespread access and very large network-bandwidth, leading to phenomenal investment in wire-line (fiber optic networks and undersea cables) and wireless networks in the past decade (the trend is continuing even today). With consumers demanding broadband Internet access both at home and in the office, developed countries (USA, Western Europe, Japan, and Australia) saw near ubiquitous access to Internet at least in urban areas. Widespread Internet reach became a natural choice to extend e-commerce to support the sale of physical goods as well.

- E-procurement (e.g., CommerceOne)
- Electronic catalogs (e.g., Library of Congress)

- Electronic shopping cart (e.g., Amazon)
- Payment gateway (e.g., PayPal), and
- On-line status tracking (e.g., FedEx)

are tools used by ordinary citizens today. Practically every industry dealing with physical goods—steel, automotive, chemicals, retail stores, logistics companies, apparel stores, and transportation companies—could benefit from this *third stage* of e-commerce.

With the widespread adoption of e-commerce by thousands of corporations and millions of consumers by the turn of the century, the equivalent of Yellow Pages emerged in the form of online catalogs like Yahoo. The very size of the Yellow Pages and the need for speed in exploration of the desired items by the buyer from online catalogs led to the arrival of the search engines. The algorithmic approach to search (in place of database search) perfected by Google was a watershed in the history of e-commerce, leading to a *fourth stage* in e-commerce.

The next stage of growth in e-commerce was the end-to-end delivery of information services. Banks and financial institutions moved to online banking and online delivery of insurance booking. NASDAQ in USA and National Stock Exchange (NSE) in India moved to all-electronic stock trading. Airlines moved beyond online reservations to e-ticketing. Hotels moved to online reservations, even linking to online reservations of airline tickets to offer seamless travel services. Several movie theaters, auditoriums, and games/stadiums moved to online ticketing. Even governments in Australia and Canada started delivering citizen services online. Many libraries and professional societies such as IEEE shifted to electronic libraries and started serving their global customers electronically from the year 2000, representing the *fifth stage* of e-commerce.

The *sixth stage* is the maturing of the Internet into a highly reliable, available, and scaleable infrastructure that can be depended upon for 24×7 delivery of services worldwide. Services like Napster selling music to millions of users (before its unfortunate closure) and the recent success of Apple iTunes puts enormous pressure on network bandwidth and availability. In turn, this demand led to significant improvements in content delivery mechanisms, pioneered by companies like Akamai.

A whole range of issues including next generation network infrastructure, such as

- Internet II
- IPv6
- Security issues
- Privacy issues
- Legal issues, and
- Social issues

are getting addressed today to take e-commerce to the next stage.

During the dotcom boom, e-commerce created a lot of hype. It was predicted that all commerce will become e-commerce and eventually "brick and mortar" stores will be engulfed by "click and conquer" electronic storefront. Commerce in its conventional form is far from dead (it may never happen); yet e-commerce has made considerable progress [1,16].

- By December 2005 the value of goods and services transacted electronically exceeded $1 trillion, as per Forrester Research.
- Stock trading, ticketing (travel, cinema, and hospitality), and music are the "top 3" items traded electronically today.
- Online trading accounts for 90% of all stock trading.

- Online ticketing accounts for 40% of the total ticketing business.
- Online music accounts for 30% of all "label" music sales is increasing dramatically, thanks to iPod and iTunes from Apple.
- By the year 2010, it is expected that nearly 50% of all commercial transactions globally will happen electronically.

9.3   OR/MS and E-Commerce

"Operations Research (OR) is the discipline of applying analytical methods to help make better decisions using the key techniques of Simulation, Optimization and the principles of Probability and Statistics," according to Institute of Management Science and Operations Research (INFORMS) [17]. Starting with the innovative application of sophisticated mathematical tools to solve logistics support issues for the Allied Forces during the Second World War, OR has moved into every key industry segment. For example, OR has been successfully used in the

- Airline industry to improve "yield" (percentage of "paid" seats);
- Automotive industry to optimize production planning;
- Consumer goods industry for supply chain optimization;
- Oil industry for optimizing product mix;
- Financial services for portfolio optimization; and
- Transportation for traffic simulation.

Practically all Fortune 500 companies use one or the other of the well-developed tools of OR, namely,

- Linear, integer, nonlinear, and dynamic programming
- Network optimization
- Decision analysis and multi-criteria decision making
- Stochastic processes and queuing theory
- Inventory control
- Simulation, and
- Heuristics, AI, genetic algorithms, and neural networks.

OR tools have been extensively used in several aspects of e-commerce in the past decade. The success behind many iconic companies of the past decade can be traced to OR embedded within their patented algorithms—Google's Page Ranking algorithm, Amazon's Affiliate program, and i2's Supply Chain Execution Engine, for example [4–6,18–20].

During the dotcom boom a number of e-commerce companies started using OR tools extensively to offer a range of services and support a variety of business models, though many of them collapsed during the dotcom bust. As early as the year 1998 Geoffrion went to the extent of stating "OR professionals should prepare for a future in which most businesses will be e-business" [5,6] and enumerated numerous OR applications to e-commerce including

- *OptiBid* using integer programming for Internet-based auction;
- *Trajectra* using stochastic optimization for optimizing credit card portfolio planning;
- *SmartSettle* using integer programming for automated settling of claims;

- Optimal Retirement Planner that used linear programming to provide advisory services;
- Home Depot using integer programming-based bidding for truck transportation planning;

indicating the importance of OR to e-commerce. Over the past 8 years the applications have increased manyfold and the area has matured.

In this chapter, we will outline a number of OR applications that are widely used in the different stages of e-commerce. This being a handbook, we will limit ourselves to outlining the interesting applications, indicating the possible use of OR tools to address the selected applications, and the benefits generated. The actual formulation and the algorithmic implementation details of specific applications are beyond the scope of this chapter, though interested readers can refer to literature for further details. There are dozens of OR tools, hundreds of algorithms, and thousands of e-commerce companies, often with their own patented implementation of algorithms. Instead of making a "shopping list" of the algorithms and companies, we will limit our scope to nine key applications that provide a broad sense of the applications and their richness.

We limit our scope to Internet-based e-commerce in this chapter. With the spectacular growth of wireless networks and mobile phones—particularly in India and China with lower PC penetration and land-line phones—e-commerce might take different forms including

- Mobile commerce (m-commerce)
- Smart cards, and
- Internet kiosks.

This is an evolving area with not-so-clear patterns of e-commerce and best captured after the market matures.

In the next section, we will outline the OR tools that are useful in the different stages of e-commerce, namely,

- Customer acquisition
- Order processing
- Order fulfillment
- Payment, and
- Post-sales support.

Section 9.4.1 will be devoted to the case of e-commerce related to digital goods. Section 9.4.2 will address the issues relating to e-commerce for physical goods. Section 9.4.3 will be devoted to the case of electronic delivery of services. Section 9.4.4 will focus on the optimization of the infrastructure to support e-commerce.

A very large number of e-commerce applications and business models have emerged over the past decade. For brevity, we have identified a set of nine key tools/applications that form the core set of tools to support e-commerce today. These nine candidates have been so chosen that they are representative of the dozens of issues involved and *comprehensive* enough to capture the essence of contemporary e-commerce.

We will discuss the chosen nine core tools/applications under one of the four categories mentioned earlier, where the specific tool/application is most relevant to the particular category. Accordingly,

i. Internet search
ii. Recommendation algorithms

   iii. Internet affiliate programs

will be discussed in Section 9.4.1.
   Section 9.4.2 will address

   iv. Supply chain optimization
   v. Auction and reverse auction engines

   Section 9.4.3 will be devoted to

   vi. Pricing engines

And finally Section 9.4.4 will focus on

   vii. Content delivery network (CDN)
  viii. Web-site performance tuning
   ix. Web analytics.

9.4 OR Applications in E-Commerce

OR applications predate Internet e-commerce, as demonstrated by the airline industry that used seat reservation information to optimize route planning; in the process, innovative concepts like "planned overbooking" were developed that dramatically improved "yield." In the past decade (1995–2005) a lot more has been achieved in terms of the use of algorithms to significantly improve user experience, decrease transactions cost, reach a larger customer base, or improve service quality. We will outline the major applications of OR to the key aspects of e-commerce in this section.

9.4.1 E-Commerce Relating to Digital Goods

Goods that are inherently "soft" are referred to as "digital goods" as they are likely to be created by digital technology. Packaged software such as Microsoft Office and Adobe Acrobat that are used by millions of users globally are among the first recognizable digital goods. Until the mid-1990s shrink-wrapped software was packaged in boxes; with the widespread availability of the Internet and high-speed access at least in schools and offices, if not homes, software could be distributed electronically. It is common these days even for lay people, let alone IT professionals, to routinely "download" and "install" software on to their desktop or notebook computers. With the arrival of MP3 standard digital music and freely downloadable digital music players for PCs, such as Real Networks' RealAudio player, and Microsoft Media Player, digital music became a widely traded digital good, particularly during the Napster days. In recent years, with the phenomenal success of the Apple iPod MP3 player and the digital music stores Apple iTunes, music has become the fastest growing digital goods. With the world's largest professional society IEEE launching its Digital Library in the year 2000, online books from Safari Online have become popular; and with many global stock exchanges going online, e-books, digital libraries, and stock trading became popular digital goods in the past 5 years.

i. Internet Search

The first stage in commerce is the identification and acquisition of customers by the seller. In traditional commerce this is accomplished by advertising, promotion, catalog printing and distribution, and creation of intermediaries like agents. In e-commerce, particularly in the case of digital goods, the customer acquisition process itself often happens on the Internet. Typically, a customer looking for a digital good is likely to "Google" (a term that has become synonymous with "searching" thanks to the overwhelming success of Google over the past 7 years) for the item over the Internet.

Search engines like Google and portals like Yahoo, MSN, eBay, and AOL have become the focal points of customer identification today and many corporations, big and small, are fine-tuning their strategies to reach potential customers' eyeballs over the portal sites. Internet advertising is only next to TV advertising today. The portal sites Yahoo and MSN and online bookstore Amazon have their own search engines today; eBay has started using Yahoo's search engine recently.

Search engines help the customer acquisition process by way of helping potential customers to explore alternatives. The reason for the phenomenal success of Google is its speed and relevance, as well as the fact that it is free. Very few users would use Google if every search took 5 minutes, however good the search is; similarly, Google usage would be miniscule if users have to wade through a large file with millions of "hits" to find the item they are looking for, even if the million-long list is delivered within seconds.

The secret behind the speed–relevance combination of Google (and other search engines today) is the sophisticated use of algorithms, many of them being OR algorithms.

Search engines broke the tradition followed by librarians for centuries; instead of using databases (catalogs in the language of librarians), search engines use indexes to store the links to the information (often contained in Web pages); they use automated tools to index, update the index, and most importantly, use algorithms to rank the content. For the ranking of the content another principle perfected by information scientists in the form of "citation index" is used; it is quite natural that important scientific papers will be cited by many scientific authors and the count of citation can be used as a surrogate for the quality of the paper. Scientific literature goes through peer review and fake citations are exceptions. In the "free-for-all" world of WWW publishing, a simple count of citation will not do; also the sheer size of the content on the WWW and the instantaneous response expected by Internet users (unlike the scientific community that is used to waiting for a year for the citations to be "compiled") makes it imperative to use fast algorithms to aid search. Google manages speed through its innovative use of server farms with nearly 1 million inexpensive servers across multiple locations and distributed algorithms to provide availability. Search engine vendors like Google also use some of the optimization techniques discussed in Section 9.4.4 (multicommodity flows over networks, graph algorithms, scheduling and sequencing algorithms) to optimize Internet infrastructure. The success behind the "relevance" of search engines is contributed by another set of algorithms, the famous one being Google's patented "page ranking" algorithm [21–23].

A typical user looking for an item on the Internet would enter a couple of keywords on the search window (another successful idea by Google of a simple text-box on a browser screen as user interface). Google uses the keywords to match; instead of presenting all the items that would match (often several millions), Google presents a list of items that are likely to be the most relevant items for the keywords input by the user. Often the search is further refined using local information (location of the computer and the network from where the search is originating and user preferences typically stored in cookies). Google is able to order the relevant items using a rank (patented Page Rank); the rank is computed using some simple graph theoretic algorithm.

A typical Web page has several pieces of content (text, graphics, and multimedia); often the contents are hyperlinked to other elements in the same Web page or to other Web pages. Naturally, a Web page can be reduced to a graph with Web pages as nodes and hyperlinks as edges. There are outward edges that point to other Web pages and inward edges that link other pages. Inward edges are equivalent to citations and outward edges are similar to references. By counting the number of edges one can compute a surrogate for citation; this can be a starting point for computing the rank of a Web page; the more the citations the better the rank. What is special about Google's page ranking algorithm is that it recognizes that there are important Web pages that must carry a lot of weight compared to a stray Web page. While a page with several inward links must get a higher importance, a page with a single inward link from an important page must get its due share of importance. However, the importance of a page would be known only after all the pages are ranked. Using a recursive approach, Google's page ranking algorithm starts with the assumption that the sum of ranks of all Web pages will be a constant (normalized to unity). If page A has inward links (citations) T1, T2, ... , Tn and C(A) is the number of outward links of page A, the page rank (PR) could be defined as

$$PR(A) = (1 - d) + d\{PR(T1)/C(T1) + PR(T2)/C(T2) + \cdots + PR(Tn)/C(Tn)\} \quad (9.1)$$

In a sense, instead of adding the citations linearly (as in the case of citation index), page rank algorithm takes the weighted rank approach with citation from higher-ranked page getting higher weight. "d" is used as a "dampening" constant with an initial value of 0.85 in the original Google proposal. In a sense, page rank is used as a probability distribution over the collection of Web pages. Defining page rank for the millions of Web pages leads to a very large system of linear equations; fortunately, OR professionals have the experience of solving such large system of linear equations using linear programming for years. The special structure of the set of equations admits even a faster algorithm. According to the original paper by Google co-founders Sergey and Page, their collection had 150 million pages (nodes) and 1.7 billion links (edges) way back in 1998. Their paper reports the following performance: 24 million pages had links to 75 million unique URLs; with an average 11 links per page and 50 Web pages per second it took 5 days to update the content. Distributing the load to several workstations they could get convergence to the set of equations (Equation 9.1) within 52 iterations; the total work distributed across several workstations took 5 hours (about 6 min/workstation). Based on this empirical performance the algorithm is extremely efficient (with complexity of O(log n) and can scale to Web scale [21]. That explains why Google search performs so well even today, though Web page collection has increased severalfold over the period 1998 to 2006 and Google is still serving results real fast. There are a number of special issues such as dangling links that have been addressed well; in the year 2005 Google introduced the no-follow link to address spam control, too.

To constantly keep pace with the changing content of the Web pages, search engines use a crawler to collect new information from the sites all over the world. The background infrastructure constantly updates the content and keeps an index of links to documents for millions of possible keyword combinations as a sorted list. After matching the user-provided keywords, Google collects the links to the documents, formats them as HTML page, and gives it to the browser that renders it over the user screen.

Google is not the only search engine; there are many other search engines including specialized search engines used by eBay for searching price lists. Many of them use specialized algorithms to improve ranking [20]; there are others like Vivisimo [24] and Kartoo [25] that use clustering techniques and visual display to improve user experience with search; yet Google remains the supreme search engine today.

The key OR algorithms used by search engines include linear programming to solve large-scale linear equations, neural networks for updating new information (learning), and heuristics to provide a good starting solution for the rank.

ii. Recommendation Algorithms

When a customer visits an online store to order an item there is a huge opportunity to up sell and cross sell similar items. Recommending such potential buy items to prospective buyers is a highly researched area. While such direct selling is not new, online direct selling is new. Common sense suggests that the online store recommends items similar to those bought by the customer during the current interaction or based on prior purchases (using database of prior purchases); a more refined common sense is to use the customer profile (based on cookies or customer-declared profile information) and match items that suit the customer profile. A more sophisticated approach is to use similarity between customers with preferences that match the current customer. Many users are familiar with Amazon Recommends [26,27] that talks of "people who ordered *this* book also ordered *these* books." Knowing customer profiles and customer interests it makes sense to suggest items that are likely to appeal "most" to the customer (taking care not to hurt any customer sensitivities). All such approaches form the core of recommendation algorithms [28].

Recommendation needs a lot of computation; both associative prediction and statistical prediction have been widely used by market researchers for ages, but at the back-end; tons of junk promotional materials and free credit card offers are the results of such recommendation. Recently recommendation has invaded e-mail too. But recommendation in e-commerce space needs superfast computation, as online visitors will not wait beyond seconds; this poses special challenges. Considering the fact that online stores like eBay or Amazon stock tens of thousands of items, millions of customers, and billions of past sales transaction data, the computational challenge is immense even with super computers and high-speed networks. Brute force method will *not* work! There are other constraints too. A typical user browsing the Internet will typically see the results on a PC screen that at best accommodates a handful of products' information; even if the screen size is large the customer is unlikely to view hundreds of products. The challenge is to present just a handful of the most relevant products out of hundreds or thousands.

Starting with crude technology that simply recommended items others have purchased, recommendation technology has evolved to use complex algorithms, use large volume of data, incorporate tastes, demographics, and user profiles to improve customer satisfaction and increased sales for the e-commerce site. The Group Lens project of the University of Minnesota was an early pioneer in recommendation algorithms. E-commerce sites today have much larger data sets leading to additional challenges. For example, Amazon had 29 million customers in 2003 and a million catalog items. While the earlier generation of OR analysts had the problem of powerful methods and scanty data, today's recommendation algorithms have to contend with excessive data!

Common approaches to recommendation include [27,28]

- *Simple "similarity" search* that treats recommendation as a search problem over the user's purchased item and related items stocked by the e-commerce seller.
- *Collaborative filtering* uses a representation of a typical customer as an N-dimensional vector of items where N is the number of distinct catalog items with positive vector for items purchased or rated positively and negative vector for negatively rated products. The vector is scaled by the number of customers who purchased the item or recommended the item. Typically the vector is quite sparse as most customers purchase just a few items. Collaborative filtering algorithm measures the similarity between two customers (cosine of the angle between the two vectors) and uses this similarity measure to recommend items.
- *Item-to-item collaborative filtering* goes beyond matching the purchased item to each of the user's purchased items and items rated by the user to similar

items and combines them into a recommendation list. The Amazon site "Your recommendations" leads to the customer's earlier recommendations categorized by product lines and provides feedback as to why an item was recommended.

- *Cluster methods* that divide the customer base into many segments using a classification scheme. A prospective user is classified into one of the groups and a recommendation list is based on the buying pattern of the users in that segment.

Recommendation algorithms have been fairly successful; with unplanned online purchases amounting to just 25%, it is important to recommend the right items—those the customer likes but did not plan to buy, a sure way to increase sales.

Major recommendation technology vendors include *AgentArts, ChoiceStream, Expert-Maker,* and *Movice* in addition to Google.

The tools of OR that recommendation algorithms use include graphs—to capture similarity and identify clusters—and in turn implement collaborative filtering. Simulation and heuristics are used to cut down computation (dimensionality reduction).

iii. Internet Affiliate Programs

The Internet search techniques help in reaching the eyeballs of potential customers; the recommendation algorithms address the issue of recommending the right products to the customers. Obviously the customer might purchase some or all of the items recommended. The next stage is to retain the customer and attempt to convert the recommendation into an actual sale. As the Internet is a platform available to billions of people and a great equalizer that makes the platform accessible even to a small vendor in a remote area, it is not possible for all vendors to be able to reach potential customers through search engines or portals; an alternative is to ride on the major e-commerce sites like Amazon; the small vendors act as intermediaries or referrals that bring customers to the major sites and get a small commission in return. These are generally known as affiliate programs in marketing literature. The Internet gave a new meaning to the entire area of affiliate programs [9,29].

Intermediation is not new. For decades the print advertisement media have been engaging agents who design and get the advertisement for their agents displayed in the newspapers and get paid a small commission from the publishers for bringing advertisement revenue and a nominal fee from clients toward the labor of designing the advertisement. The Internet, however, is more complex in the sense that the intermediary has no precise control on the real estate, being a personalizeable medium rather than a mass medium, the advertisement—typically "banner ad"—gets displayed differently on different user screens.

Amazon is credited to be a pioneer in affiliate programs. Starting in 1994, Amazon let small booksellers link their site to Amazon site; if a prospective buyer who visits the affiliate site is directed to the Amazon site and buys an item from Amazon, the small bookstore vendor is entitled to receive a commission (a small percentage of the sale value). At a later date, it was extended to any associate that brings the buyer to the Amazon site. Today Amazon has more than 600,000 affiliates (one of the largest Internet affiliate programs). Google started its affiliate program in 2003; initially it was *AdWords* meant for large Web sites and extended later as *AdSense* in May 2005 for all sites. Google and others took the program to further sophistication through cost per lead, cost per acquisition, and cost per action and also extended it to cost per thousand that is more like the print advertisements. Dealing with such large intermediaries (affiliates) calls for significant management resources (track earnings, accounting, reporting, quality of service monitoring, and filter sensitive content) leading to Internet-based solutions to manage affiliate programs from third party companies. LinkShare [29] is one of the most successful agencies to manage Internet-based affiliate programs.

Yahoo Search Marketing and Microsoft *adCenter* are other large-size affiliate programs to watch.

Managing the banner ad that is typically 460 pixels wide $\times$ 60 pixels high is itself a sophisticated space optimization; the banner ad must be placed within the small real estate of the space earmarked for banner ad within the display page of the browser window on the user screen—an interesting integer programming problem (Knapsack problem).

Internet affiliate programs use optimization techniques to optimize banner ads and estimate initial cost per click; they use genetic algorithms and neural networks to adaptively optimize and improve estimates; and graph theory to capture associations.

9.4.2   E-Commerce Relating to Physical Goods

The success of e-commerce in digital goods sales prompted the brick and mortar industry to utilize the power of the Internet. One of the key challenges of many physical goods companies embracing e-commerce is the speed mismatch between electronic information delivery and the physical movement of goods that must be shipped to the customer who has ordered electronically. It is one thing to be able to order groceries over the Internet and yet another thing to receive it at your home by the time you reach home.

Obviously, logistics and supply chain optimization are two key issues that would make or mar the success of e-commerce for physical goods.

iv. Supply Chain Optimization

Traditional OR models focused on the optimization of inventory, typically in a factory. Typically one would forecast sales as accurately as possible, including variations, build optimal level of safety stocks suiting a particular replenishment policy. Supply chain optimization refers to the optimization of inventory across the entire supply chain—factory to warehouse to distributors to final stores. Supply chain optimization must ensure that at every stage of the supply chain there is the *right material* in the *right quantity* and *quality* at the *right time* and supplied at the *right price*. Often there will be uncontrollability in predicting the demand; this must be protected against through the cushion of safety stocks at different levels. One must optimize the safety stocks so that they are neither too small leading to nonavailability when needed, nor too large leading to unnecessary locking up of capital. Naturally, supply chain optimization would involve multistage optimization and the forecasting of inventory must evolve from a simple probability distribution to a sophisticated stochastic process.

A key success behind today's supply chain optimization is the power of online databases to forecast at low granularity involving hundreds, if not thousands of items and to aggregate the forecasts; today's distributed computing power permits statistical forecasting to arrive at the best fit using past demand pattern, even if it involves thousands of items.

Today's supply chain optimization engines [30,31] use academically credible algorithms and commercially effective implementations. Supply chain optimization leads to

- Lower inventories,
- Increased manufacturing throughput,
- Better return on investment,
- Reduction in overall supply chain costs.

Pioneering companies like i2 and Manugistics took many of the well-studied algorithms in multistage optimization, inventory control, network optimization, and mixed-integer programming, and combining them with recent concepts like third party logistics and comanaged inventory along with exceptional computing power available to corporations, they created supply chain optimization techniques that could significantly save costs. The simultaneous development of enterprise resource planning (ERP) software—SAP R/3 and PeopleSoft (now part of Oracle)—helped supply chain optimization to be integrated with the core enterprise functions; together it was possible to integrate

- Supply chain network design
- Product families
- Suppliers/customers integration
- Multiple planning processes
- Demand planning
- Sales and operations planning
- Inventory planning
- Manufacturing planning
- Shipment planning

with

- Execution systems

paving the way for optimizing across the supply chain (compared to single-stage optimization earlier).

Today's supply chain optimizing engines use established OR tools like

- Mixed integer programming
- Network optimization
- Simulation
- Theory of constraints
- Simulated annealing
- Genetic algorithms, and
- Heuristics

routinely; and solvers like iLog are part of such supply chain optimization engines.

Dell and Cisco are two vendors who used supply chain optimization extensively and built highly successful extranets that enabled suppliers and customers to conduct end-to-end business online—ordering, status monitoring, delivery, and payment—leading to unprecedented improvements in managing the supply chain. For example, Dell reported the following metrics for its online business [30]

- User configurable ordering with maximum of 3 days for feasibility checking
- Online order routing to the plant with a maximum of 8 h
- Assembly within 8 h
- Shipping within 5 days.

With the estimated value of 0.5–2.0% value depletion per week in the PC industry, optimizing supply chain makes a lot of sense; the leadership position that Dell enjoyed for several years is attributed to its success with supply chain optimization—a classic success story in e-commerce for physical goods.

Supply chain uses multistage optimization and inventory control theory to reduce the overall inventory and hence the costs; it uses networks to model multistage material flows; queuing theory and stochastic processes to model demand uncertainty; simulation to model scenarios; and heuristics to reduce computation.

v. Internet Auction

In e-commerce relating to physical goods, supply chain optimization dramatically improves the delivery side of many manufacturing enterprises. Material inputs (raw materials, sub-assemblies) account for a significant part of the overall business of most manufacturing enterprises and procuring the inputs, if optimized, will lead to significant savings in costs. Another area where e-commerce made a big difference to physical goods sales is that of auction over the Internet pioneered by eBay. Another interesting twist pioneered by Freemarket Online is that of reverse auction used by many in the procurement space. Internet-based reverse auction brings unprecedented efficiency and transparency in addition to billions of dollars of savings to large corporations like GE (though not always sustainable over a longer period).

Auction is not new; antiques, second-hand goods, flower auctions, or tea auctions are centuries old. The technology industry is familiar with spectrum auction. Electricity auction is an idea that is being experimented with in different parts of the world, though with limited success [32].

Auctioning is an idea that has been well researched. Simply stated, a seller (generally a single seller) puts out an item for sale; many buyers bid for it by announcing a bid price; the process is transparent with all bidders knowing the bid price of other bidders; the bidders can quote a higher price to annul the earlier bid; generally, an auction is for a limited time and at the end of the time limit, the seller sells the item to the highest bidder; the seller reserves the right to fix a floor price or reserve price below which bids are not accepted. If there are no bids at a price higher than the reserve price, the sale does not go through, though the seller might bring down the reserve price.

The Internet brings a number of efficiencies to the process of auctioning; unlike physical auction electronic auction can happen across time zones globally and on a 24×7 basis; with geographic constraints removed electronic auction (e-auction) may bring larger numbers of bidders leading to network economics and improving the efficiency of the auction process [33].

To conduct e-auction an auction engine has to be built either by the service provider (eBay) or the seller can use third-party e-auction engines from a number of vendors Over-Stock, TenderSystem, and TradeMe, for example.

The auction engine has modules to publish the tender, process queries, accept quotations from the bidders, present the highest bid at any time (call out) electronically to any of the bidders over the Net, provide time for the next higher bidder to bid, close the bid at the end of the time, collect commission charges for conducting the auction, and declare the results—all in a transparent manner. Though the process is fairly simple, sophisticated algorithms have to work behind the screen as tens of thousands of auctions would be taking place simultaneously; even the New Zealand-based TradeMe.com reports 565,000 auctions running at any day, while eBay talks of millions of auctions running simultaneously [34,35].

A more interesting Internet auction is reverse auction (also called procurement auction) pioneered by an ex-GE employee, who founded Freemarkets Online (later acquired by Ariba). In reverse auction, instead of the seller putting out the auction for the buyers to bid, the buyer puts out the auction for all the sellers to bid; a typical application is the procurement manager wanting to buy items from multiple suppliers.

In the conventional purchase, the specifications for an item, that the buyer is interested to buy, is put out as tender. Typically, sellers put out a sealed bid; after ensuring that all the bids have quoted for the same item the sealed bids are opened and the seller who quotes the minimum price is awarded the contract. In reverse auction, there is no sealed bid; everyone would know the bid price, leading to higher levels of transparency and efficiency. Such buying is a huge opportunity globally—*Fortune* magazine estimated that industrial goods buying alone was $50 billion in year 2000—and even a single percentage improvement could translate to millions of dollars of savings.

A typical reverse auction engine facilitates the process of auctioning [33–35] by way of

- Preparation (help in preparing standardized quotations that makes supply/ payment schedule, payment terms, and inventory arrangements common across suppliers to get everyone a "level playing field"). In the process, many industrial items with complex specifications that are vaguely spelt out get so standardized that they become almost a "commodity" that can be "auctioned";
- Finding new suppliers (being an intermediary and a "market maker," reverse auction vendors bring in a larger range of suppliers to choose from);
- Training the suppliers (reverse auction engine vendors train the suppliers so that they take part more effectively in the process, leading to efficiency and time savings);
- Organizing the auction (including the necessary electronic support, server capacity, network capacity, interface software, special display to indicate the various stages of bidding);
- Providing auction data (analysis for future auctions).

Several large buyers including General Electric and General Motors have found reverse auctions extremely valuable. There are also consortiums like GM, Ford, and Daimler Chrysler who run their own "specialized" reverse auction sites. A whole range of industry-specific portals such as *MetalJunction* also offer reverse auction as a standard option today.

Internet auctions use economic theory (often modeled through simultaneous equations that are solved through LP Solver), multicriteria decision making to model conflicting requirements between bidder and buyer, and neural networks to model learning.

9.4.3 Electronic Delivery of Services

Internet search helps in the identification of potential customers; recommendation algorithms help in suggesting the most relevant items for possible buy by a visitor to the specific site or portal; affiliate programs bring in intermediaries who help in converting the potential customer visit on the Internet to a possible sale; supply chain optimization helps in supporting the sales of physical good; auction and reverse auction help in bringing efficiency of online markets to business-to-business commerce, particularly in the case of physical goods [36].

There are a whole range of services that are delivered electronically; these include information services—hotel room booking, airline booking, e-trading, banking, online stores, e-governance services, and e-learning. The delivery is completely online in the case of digital goods and information services while all but the physical shipment (seeking, price negotiation, ordering, order tracking, and payment) happen over the Net. For example [37],

- Hotel rooms are physical that the visitor physically occupies during the visit, but all supporting services are delivered online.
- University classes are held in the classrooms of a University, but registration for the course, payment of fees, reading materials, assignment submission, and award of the grades happen electronically.
- In electronic trading if the stock exchange is online (like NASDAQ) the entire process happens electronically.

- Many banks provide full service-banking electronically, with the customer using the "physical world" (the branch of a bank or ATM) only for cash withdrawal/ payment.

- Many governments provide electronic access to their services (marriage registration, passport, tax payment, drivers license) where all but physical documents (actual passports, driver license, license plates) are delivered electronically.

- Many stores (bookstores, computers, electronic gadgets, video stores, merchandise, and even groceries) support electronic access for all but except the actual delivery.

vi. Price Engines

A key advantage of e-commerce is a far superior "comparison shopping" that the electronic stores can offer. Normal paper-based comparison shopping is limited to a few stores, a few items, and a few merchants/agents, typically in the same area. Thanks to the global footprint of the Internet and the 24×7 availability, "pricing engines" [38] can offer a far richer information source, electronic support to compare "real" prices (calculating the tax, freight charges, and accounting for minor feature variations all within seconds) and displaying the results in a far more user-friendly manner. Special discounts that are in place in specific time zones or geographies for specific customer segments (age groups, profession, and sex) can be centrally served from a common database with the user not having to "scout" for all possible promotions for which he/she is eligible on that day in that geography.

A number of price engines are available today. Google offers "Froogle"; the early pioneer *NexTag* started with a focus on electronics and computer products and recently expanded to cars, mortgages, and interestingly online education programs! PriceGrabber is an innovator in comparative shopping with 22 channels and brings the entire ecosystem together that includes merchants, sellers, and other intermediaries; *BestBuy*, Office Depot, and Wal-Mart are *PriceGrabber* customers.

Comparison shopping can be viewed as a specialized search engine, specialized e-marketplace, or as a shopping advisory service firm. Yahoo introduced *YahooShopping*, another comparative shopping service.

Comparative shopping sites organize structured data—price lists, product catalog, feature lists—unlike ordinary search engines like Google that deal with unstructured Web pages. Specialized techniques are needed (table comparison) as well as "intelligence" to check which accessory goes with what main item. Often many of these higher level processes are done manually by experts and combined with automated search, though attempts to automate structured data are underway by many comparative shopping engines. Specialized comparative shopping engines focusing on travel would address issues like connectivity of different segments, multimodal availability checking like room availability, transportation facilities, along with airline booking.

Comparative shopping engines use expert systems, ideas from AI, heuristics, and specialized algorithms like assignment and matching to get the speed and richness of the recommendations (in that sense a comparative shopping engine can be viewed as another "recommendation engine").

Price engines use simulation to study millions of alternatives to generate initial alternatives, linear and nonlinear programming to optimize costs, and multiattribute utility theory to balance the conflicting requirements of buyers and sellers.

9.4.4　Support for E-Commerce Infrastructure

The sale of digital goods, electronic support for sale of physical goods, or electronic delivery of other services need a reliable, highly available, and scaleable Internet infrastructure with a range of challenges. We address the challenges relating to the distribution of content by building a specialized infrastructure for mission critical sites; we address the issues relating to content distribution of very demanding sites like the World Cup games site or Olympic games site during the peak period using Web-caching.

Finally, e-commerce should address post-sales support also. While there are many issues, we address the specific technology of "Web analytics" that provides insights into customer behavior in the e-commerce context.

vii. Content Delivery Network

Mission critical servers that distribute large content (software downloads, music, video) must plan for sustained high performance. Downloading of software, if it takes too much time due to network congestion, would lead to "lost customers" as the customer would "give up" after some time; dynamic information like Web-casting (for Webinar support) would need guaranteed steady network availability to sustain delivery of audio or video content; any significant delay would lead to poor quality sessions that the users would give up. Companies like Akamai [39] have pioneered a way of distributing content by equipping the "edge" with sufficient content delivery for high availability, replicating the content at multiple locations for faster delivery, and running an independent managed network for sustained high performance, in addition to tools to manage the whole process without human intervention.

A content delivery network (CDN) consists of servers, storage, and network working in a cooperative manner to deliver the right content to the right user as fast as possible, typically serving the user from the nearest possible server location, and if possible using the cheapest network. Obviously, it is a continuously adaptive optimization problem for which multicommodity flow algorithms have been extensively applied.

The key issues that must be optimized are the location of the servers and the choice of multiple backbones that improves performance (user gets the content delivered real fast) and decreases cost (minimal replication, least number of servers/least expensive servers, minimal usage of bandwidth/least cost bandwidth). Naturally the objective is to provide consistent experience to the users without their having to direct their search to specific locations; working behind the infrastructure must be a smart system that constantly watches the network, copies content at multiple locations, replicates servers at multiple networks/locations, and negotiates bandwidth.

Typically, content delivery networks are built as "overlay network" over the public Internet and use DNS re-direct to re-direct traffic to the overlay network. The key elements of CDN include

- High-performance servers
- Large storage
- Backup/storage equipment and "smart" policies to manage backup/storage
- Automated content distribution/replication
- Content-specific (audio/video, file transfer) delivery
- Content routing
- Performance measurement, and
- Accounting for charging the clients.

A content delivery network can be viewed as "higher layer" routing [39]; "Ethernet routing" happens at Layer 2 (as per OSI protocol); IP routing happens at Layer 3; port-based routing takes place at Layer 4; and content-based routing can be viewed as Layer 4 to Layer 7 routing. New protocols that permit query of cache have led to the maintenance of "cache farms" to improve content delivery.

Akamai, Speedera (now part of Akamai), *Nexus, Clearway*, and *iBeam* Broadcasting (that pioneered the use of satellite for replication) are the pioneers in CDN technology development. This is an evolving area of research currently.

The importance of CDN can be gauged by the fact that the leader in CDN, Akamai, has Google, Microsoft, AOL, Symantec, American Express, and FedEx as its customers today.

Google uses commodity servers and high availability algorithms that use a distributed computing model to optimize search; content delivery is optimized using sophisticated caching servers; as search results are less demanding (compared to video streaming, Webinars, and Webcasts) Google is able to provide high performance at lower costs; with Google getting into video, some of the strategies used by other CDN vendors might be pursued by Google also.

CDN uses a distributed computing model to improve availability, networks to model the underlying infrastructure, linear and dynamic programming to optimize flows, stochastic processes to model user behavior, and heuristics to decide routing at high-speed for real-time high performance.

viii. Web Site Performance Tuning

With Internet penetration on the rise, end users depending on Google, Yahoo, MSN, AOL, and eBay for their day-to-day operations (search, mail, buy), the demand on the Internet infrastructure is increasing both quantitatively and qualitatively. Users expect better performance and ISPs want more customers. Users want screen refresh to happen within a couple of seconds irrespective of network congestion or the number of users online at a time. The users also expect the same response irrespective of the nature of the content—text, graphics, audio, or video. Several researchers have addressed this issue. Google uses several server farms with tens of thousands of servers to make its site available all the time and the site contents refresh fairly fast. A whole area of "load-balancing" content delivery has emerged over the past decade with several novel attempts [40,41].

One attempt to improve Web-site performance by load balancing is to use content distribution strategies that use differentiated content delivery strategies for different content [41]. In this scheme, the differential content is distributed to servers of varying capability:

- High-end mainframe database servers for critical content
- Medium-end Unix box running low-end DBMS for less-critical content
- Low-end commodity server for storing other information as text files.

For example, in an airline reservation system,

- Fast-changing and critical content like "seat matrix" could be stored on a mainframe/high-performance server,
- Less frequently changing content like "fare data" that is less critical for access performance could be stored on a mid-range database server, and
- "Time table" information that does not change for a year can be kept as a flat file on a commodity server.

By storing differentiated content on servers of different capability, criticality, and costs, it is possible to cut down costs without compromising on quality of service; of course, sophisticated replication schemes must be in place to ensure that the data across the three levels of servers are consistent. The queries would arrive at an intermediate server and be directed to the right server based on the content, yet another content delivery distribution scheme.

With the rising popularity of some sites, hit rates are reaching staggering proportions—access rates of up to million hits per minute on some sites are not uncommon, posing great challenges to servers, software, storage, and network. Content-based networking is a solution for some premier sites (Microsoft and FedEx), but ordinary sites too need enhanced performance.

One approach to improve Web-site performance is to create concurrency through

- Process-based server (e.g., Apache server)
- Thread-based server (e.g., Sun Java server)
- Event-driven server (e.g., Zeus server), or
- In-kernel servers (very fast though not portable).

OR algorithms have been used to combine the various schemes; in addition, the "popularity index" of the content is used in a prioritization algorithm to improve Web site performance of critical application infrastructure.

Such strategies have been used successfully to optimize Web servers for critical applications like the Olympic Games site during the days of the Olympic Games with telling performance. For example, the Guinness Book of World Records announced on July 14, 1998, has two notable records:

- Most popular Internet record of 634.7 million requests over the 16 days of the Olympic Games.
- Most hits on an Internet site in a minute—110,114 hits—occurred at the time of women's freestyle figure skating.

Using special techniques that use embedded O/S to improve router performance by optimizing TCP stack, such strategies helped IBM achieve superior performance—serving 5000 pages/second on a single PowerPC-processor-based PC class machines running at 200 MHz [41].

Yet another technique widely used for Web-caching uses "greedy algorithms" to improve server throughput and client response; typical strategies include

1. Least recently used (LRV) caching and
2. Greedy dual size that uses cost of object cashing (how expensive it is to fetch the object) and size of the object (Knapsack problem)

to fill the cache.

Web site performance tuning uses simulation to get broad patterns of usage, optimization, and decision analysis for arriving at strategies, stochastic processes and queuing theory to model network delays, and inventory control to decide on network storage needs.

ix. Web Analytics

The final stage of e-commerce is post-sales support. In the context of the Web it is important to analyze the consumer behavior in terms of the number of visits, how much time it took for the user to find the information that was being searched, the quality of the recommendations, the time the user spent on the various activities of the buyer–seller interaction that takes place on the Internet. In turn, such insights would help the seller to offer better buying experience and offer more focused products in the next buying cycle. Web analytics is the technology that addresses this issue [42].

From the early days, Web site administrators have used logs to study the visit pattern of the visitors visiting the site—number of visitors, distribution of visitors over time, popular Web pages of the site, most frequently visited Web page, least frequently visited Web page, most downloaded Web page, and so on. With the proliferation of the Web and the sheer size of the Web content, many Web masters could not cope up with the large numbers thrown up by the logs to get any key insights. It is the marketing and marketing communications professionals along with tool vendors that have helped the ideas of Web analytics mature over the past decade.

The technique of page tagging that permitted a client to write an entry on an external server paved the way for third-party Web analytics vendors who could use the "logs" and "page tag databases" to provide a comprehensive analysis of the way a particular e-commerce Web site was being used. JavaScript technology permitted a script file to be copied (say, for example, a "counter" application) that would let third party vendors remotely "manage" the site and provide analytics.

Google started providing "Google Analytics" free of charge on an invitation basis since 2005. SAS Web Analytics, *Webtrends, WebsideStory, Coremetrics*, Site Catalyst, and *HitBix* are other Web analytics vendors who pioneered many ideas in this emerging area [42]. A comprehensive Web analytics report goes beyond simple page counts, unique visitors, or page impressions to provide insights into usability, segmentation, campaigns to reach specific customers, and identification of potential customers, in the process becoming a powerful marketing tool.

Web analytics uses data mining including optimization and simulation for pattern generation and pattern matching, stochastic processes, and simulation for traffic analysis and neural networks, and genetic algorithms for postulating patterns and hypotheses.

9.5 Tools–Applications Matrix

We have outlined a range of OR applications for e-commerce in this section. Instead of listing the tools and techniques we have discussed the different OR algorithms within the life-cycle of typical e-commerce—customer acquisition, promotion, price negotiation, order processing, delivery, payment, and after-sales support. We have indicated at the end of every subsection describing the nine tools and applications (Internet search to Web analytics), the key OR algorithms that the researchers have found useful. To provide a better perspective of the interplay of e-commerce applications and tools of OR, the following tools–applications matrix indicates the applicability of all relevant tools for each of the nine key e-commerce tools and applications discussed in this chapter. The chart is more indicative and includes the tools most likely to be used; after all, most tools can be used on every occasion.

OR Techniques used in	1	2	3	4	5	6	7
Internet search	☑						☑
Recommendation algorithms	☑	☑				☑	☑
Internet affiliate programs	☑						☑
Supply chain optimization	☑	☑		☑	☑	☑	☑
Internet auction	☑		☑				☑
Price engines	☑		☑			☑	☑
Content delivery network	☑	☑		☑		☑	☑
Web performance tuning	☑	☑	☑	☑	☑	☑	☑
Web analytics	☑	☑		☑		☑	☑

1. Linear, integer, nonlinear, and dynamic programming
2. Network optimization
3. Decision analysis and multi-criteria decision making
4. Stochastic processes and queuing theory
5. Inventory control

6. Simulation, and

7. Heuristics, AI, genetic algorithms, and neural networks.

9.6   Way Forward

E-commerce is fundamentally transforming every business—corporations, universities, NGOs, media, and governments. With Internet still reaching only 2 billion of the 6+ billion people in the world, e-commerce has the potential to grow; the existing users of e-commerce will see further improvements in the services they see today and a whole new set of services that have not even been touched. We are still in the growth phase of global e-commerce. OR algorithms have influenced significantly the various business models and services offered by e-commerce today. Several of the e-commerce applications pose special challenges; OR professionals would find new algorithms and techniques to solve the challenges arising out of e-commerce powered by WWW 2.0 and Internet II. It will be exciting to see OR influencing the future of e-commerce and OR getting influenced by e-commerce applications.

9.7   Summary

In this chapter, we have outlined a number of OR applications in the context of e-commerce. With e-commerce growing phenomenally, it is impossible to document all the e-commerce applications and the OR tools used in those applications. We have taken a more pragmatic view of outlining the key OR tools used in the different stages of the life-cycle of a typical e-commerce transaction, namely, customer acquisition, promotion, order processing, logistics, payment, and after-sales support. A set of nine key e-commerce applications, namely,

1. Internet search
2. Recommendation algorithms
3. Internet affiliate programs
4. Supply chain optimization
5. Internet auction
6. Price engines
7. Content delivery network
8. Web-site performance tuning
9. Web analytics

have been discussed as a way to provide insight into the role of OR models in e-commerce. It is not possible to list the thousands of e-commerce applications and the use of hundreds of OR tools; what is indicated in this chapter is the immense potential of OR applications to e-commerce through a set of nine core tools/applications within the world of e-commerce. A tools–applications matrix provides the e-commerce applications versus OR tools match. While a lot has happened in the past decade, one expects a much larger application of OR to the next stage of growth in global e-commerce.

References

1. E-commerce-times www.ecommercetimes.com/.
2. Kalakota, Ravi and Andrew Whinston, *Electronic Commerce: A Manager's Guide*, Addison-Wesley (1997).
3. Wikipedia (www.wikipedia.com).

4. e-Optimization site http://www.e-optimization.com/.

5. Geoffrion, A.M., "A new horizon for OR/MS," *INFORMS Journal of Computing*, 10(4) (1998).

6. Geoffrion, A.M. and Ramayya Krishnan, "Prospects of Operation Research in the E-Business era," *Interfaces*, 31(2): 6–36 (March–April 2001).

7. Chaudhury, Abhijit and Jean-Pierre Kuilboer, *E-Business and E-Commerce Infrastructure: Technologies Supporting the E-Business Initiative*, McGraw-Hill College Publishing (January 2002).

8. Kalakota, Ravi and Andrew B. Whinston, *Frontiers of Electronic Commerce* (1st edition), Addison-Wesley (January 1996).

9. Krishnamurthy, S., *E-Commerce Management: Text and Case*, South Western Thomson Learning (2002).

10. Laudon, Kenneth C. and Carol Guercio Traver, *E-Commerce: Business, Technology, Society* (2nd edition), Prentice Hall (2004).

11. Robinson, Marcia, Don Tapscott, Ravi Kalakota, *e-Business 2.0: Roadmap for Success* (2nd edition), Addison-Wesley Professional (December 2000).

12. MIT E-commerce resource center, http://ebusiness.mit.edu/.

13. Penn State e-biz center, Smeal.psu.edu/ebrc.

14. Smith, Michael et al., "Understanding digital markets," in *Understanding the Digital Economy*, Erik Brynjolfsson (ed.) MIT Press (2000).

15. University of Texas at Austin Center for Research in Electronic Commerce, http://crec.mccombs.utexas.edu/.

16. e-marketer site http://www.emarketer.com/.

17. INFORMS (The Institute of Operations Research and Management Science) www.informs.org.

18. Bhargava, H. et al., "WWW and implications for OR/MS," *INFORMS Journal of Computing*, 10(4): 359–382 (1998).

19. Bhargava, H. et al., "Beyond spreadsheets," *IEEE Computer*, 32(3): 31–39 (1999).

20. Willinger, W. et al., "Where mathematics meets the Internet," *American Math Society*, 45(8): 961–970 (1998).

21. Brin, Sergey and Larry Page, "The anatomy of a large-scale hyper-textual Web search engine," Working Paper, CS Department, Stanford University (1998).

22. Google Page Rank Algorithm, "Method for node ranking in a linked database," U.S. Patent 6,285,999, dated September 4, 2001.

23. Page, L., S. Brin, R. Motwani, and T. Winogard, "Page rank citation ranking—bringing order to the Web," Technical Report, Stanford University (1999).

24. Vivisimo search engine (www.vivisimo.com).

25. Kartoo search engine (www.kartoo.com).

26. Amazon one-click patent, "Method and system for placing a purchase order via a communications network," U.S. Patent 5,960,411 dated Sep 28, 1999.

27. Linden, Greg et al., "Amazon.com recommendation," *IEEE Internet Computing*, pp. 76–80 (Jan–Feb 2003).

28. Leavitt, Neal, "Recommendation technology: will it boost e-commerce?," *IEEE Computer*, May 2006, pp. 13–16.

29. Linkshare patent, "System method and article of manufacture for internet based affiliate pooling, U.S. Patent 6.804,660 dated October 12, 2004.

30. Kapuscinski, R. et al., "Inventory decisions in Dell's supply chain," *Interfaces*, 34(3): 191–205 (May–June 2004).

31. Lapide, Larry, "Supply chain planning optimization: Just the facts," *AMR Research* (1998).

32. Beam, C., Segev, A., and Shantikumar, G. "Electronic Negotiation Through Internet-Based Auctions," Working Paper 96-WP-1019, University of California, Berkeley.

33. *Fortune* magazine, "Web auctions," March 20, 2000.

34. Nissanoff, Daniel, *Future Shop: How the New Auction Culture Will Revolutionize the Way We Buy, Sell and Get the Things We Really Want*, The Penguin Press (2006) ISBN 1-594-20077-7.

35. Sawhney, Monhanbir, "Reverse auctions," *CIO Magazine*, June 1, 2003.

36. Alvarez, Edurado et al., "Beyond shared services—E-enabled service delivery," Booze Allen & Hamilton Report (1994).

37. Seybold, P., *Customers.com: How to Create a Profitable Business Strategy for the Internet and Beyond*, Time Books (1998).

38. Priceline patent "Method and apparatus for a cryptographically assisted commercial network system designed to facilitate buyer-driven conditional purchase offers," U.S. Patent 5,794,207 dated Aug 11, 1998.

39. Akamai patent, "Method and apparatus for testing request-response service using live connection traffic, U.S. Patent 6,981,180 dated December 27, 2005.

40. Chen, H. and A. Iyengar, "A tiered system for serving differentiated content," *World-Wide Web: Internet and Web Information Systems Journal*, no. 6, Kluwer Academic Publishers, 2003.

41. Iyengar, A., Nahum, E., Shaikh, A., and Tewari, R., "Improving web-site performance," in *Practical Handbook of Internet Computing*, Munindar, P. Singh (ed.) CRC Press (2004).

42. Peterson, Eric, "Web analytics demystified," Celilo Group Media (2004), ISBN 0974358428.

Further Reading

1. Awad, E., *Electronic Commerce* (2nd edition), Prentice Hall (2003).

2. Chan, H. et al., *E-Commerce: Fundamentals and Applications*, 0-471-49303-1, Wiley (January 2002).

3. Deitel, Harvey M. and Paul J. Deitel, *e-Business & e-Commerce for Managers* (1st edition), Prentice Hall (December 2000).

4. Hillier, Frederick S. and Gerald J. Lieberman, *Introduction to Operations Research*, (8th edition), McGraw Hill, ISBN 0-07-732-1114-1 (2005).

5. Jeffrey F. Rayport et al., *Introduction to e-Commerce*, McGraw-Hill/Irwin (August 2001).

6. Carroll, Jim, Rick Broadhead, *Selling Online: How to Become a Successful E-Commerce Merchant*, Kaplan Business (2001).

7. Murty, Katta, *Operations Research: Deterministic Optimization Models*, Prentice Hall (1994).

8. Ravindran, A., D. T. Phillips, and J. Solberg, *Operations Research: Principles and Practice* (2nd edition), John Wiley and Sons, Inc., New York, 1987.

9. Schneider, G., *Electronic Commerce* (4th edition), Canada: Thompson Course Technology (2003).

10. Schneiderjans, Marc J. and Qing Cao, *E-commerce Operations Management*, Singapore: FuIsland Offset Printing (S) Pte. Ltd. (2002).

11. Taha, Hamdy, *Operations Research—An Introduction* (4th edition), Macmillan (1987).

12. Turban, E. and David King, *E-Commerce Essentials*, Prentice Hall (2002).

13. Wagner, H. M., *Principles of Operations Research*, Prentice Hall (1969).

14. Winston, Wayne, *Operations Research: Applications & Algorithms* (4th edition), Duxbury (2004).

10

Water Resources

G.V. Loganathan
Virginia Tech

10.1 Introduction

10.1.1 Sustainability

There has been an increased awareness and concern in sustaining precious natural resources. Water resources planning is rising to the top of the international agenda (Berry, 1996). In the following, we present some of the problems that are affecting water resources. Chiras et al. (2002) list the following water-related problems in the United States: increasing demand in states such as Florida, Colorado, Utah, and Arizona; water used for food has tripled in the last 30 years to about $24(10^6)$ hectares straining one of the largest aquifers, Ogallala aquifer, ranging from Nebraska to Texas with a water table decline of 1.5 m per year in some regions; high industrial and personal use; unequal distribution; and water pollution. In the United States, the estimated water use is about 408 billion $[408(10^9)]$ gallons per day with a variation of less than 3% since 1985. The water use by categories is as follows: 48% for thermoelectric power, 34% for irrigation, 11% for public supply, and 5% for other industrial use (Hutson et al., 2004).

Sustaining this resource is attracting significant attention (Smith and Lant, 2004). Sustainability is defined as "meeting current needs without compromising the opportunities of future generations to meet their need" (World Commission, 1987). According to the American Society of Civil Engineers (1998) and UNESCO, "Sustainable water systems are designed and managed to fully contribute to the objective of the society now and in the future, while maintaining their ecological, environmental and hydrologic integrity." Heintz (2004) points to economics to provide a framework for sustainability as not spending the "principal." He also elucidates the use of technology for resource substitution especially for nonrenewable resources. Dellapenna (2004) argues that if the replenishment time is too long, such as for iron ore or oil cycles, the resource should not be considered sustainable. Replenishing groundwater in general is a long-term process. Kranz et al. (2004), Smith (2004), Loucks (2002), and Loucks and Gladwell (1999) provide indicators for sustaining water resources. An indicator used for assessing the receiving water quality is the total maximum daily load (TMDL), which is the maximum mass of contaminant per time that a waterway can receive from a watershed without violating water quality standards; an allocation of that mass to pollutant's sources is also required. In 2004, about 11,384 miles of Virginia's 49,220 miles of rivers and streams were monitored; of these, 6948 miles have been designated as impaired under Virginia's water quality standards (Younos, 2005). Reinstating water resources to acceptable standards requires a holistic engineering approach. Delleur (2003) emphasizes alternative technological solutions that should be assessed for (i) scientific and technical reliability, (ii) economic effectiveness, (iii) environmental impact, and (iv) social equity.

10.1.2   Minor Systems

In the following, we draw attention to a minor system, namely the home plumbing system. Domestic copper plumbing pipes (for drinking water use) are experiencing pinhole leaks. Figure 10.1 shows the distribution of locations reporting pinhole leaks around the country. We define the major system as the public utility water distribution system that brings drinking water to houses. While the major system is readily recognized as a vast infrastructure system of nearly 1,409,800 km of piping within the United States (Material

FIGURE 10.1   Plumbing pipe pinhole leak distribution (2000–2004) (small triangles show number of leaks less than 10). (From Lee and Loganathan, *Virginia Water Central*, Virginia Water Resources Research Center, 2006. With permission.)

Performance, 2002), the minor system that is at least 5–10 times larger is generally not well addressed. When a pipe has a leak, the homeowner is faced with the following issues: water damage cost, repair cost, service disruption, possible lowering of home value, home insurance/premium increase/nonrenewal, and health consequences, resulting from brown mold growth and mental stress. The interaction between hydraulics, water quality parameters, and pipe material has to be understood. There is a need for a decision model recommending whether to continue to repair or replace the system.

Another example is related to urbanization, which increases imperviousness along with quick draining channelization for runoff. These modifications result in increases in runoff volume and peak flow and decreases in time to peak and recharge to groundwater storage. It is required by many local governments that the post development peak flows of certain designated frequencies must not exceed those of the predevelopment conditions and proper flood controlling detention ponds must be installed to contain the increased post development peak flows. An associated problem is that the falling rain can capture airborne pollutants with it as well as dislodge particulate matter settled on the ground. The surface runoff can dissolve as well as carry the adsorbed pollutants downstream. The water quality impacts of stormwater discharges on receiving lakes and rivers are quite significant, as indicated by the Environmental Protection Agency's (EPA) guidelines for stormwater discharge permits in the National Pollutant Discharge Elimination System (NPDES) program. Civil action may be brought by the EPA against large municipalities that do not comply with the permit requirements.

The above description addresses the most frequent drainage problem within an urban area. The control measures put in place together to alleviate the problem constitute the *minor drainage system* consisting of street gutters, inlets, culverts, roadside ditches, swales, small channels, and pipes. A less frequent but significant flood damage problem arises when an urban area is located along a water course draining a large basin. The issue here is controlling the flood plain use and providing for proper drainage of flood waters. The Federal Emergency Management Agency (FEMA) issues guidelines with regard to flood plain management under the National Flood Insurance Program (NFIP). The components involved in this latter problem are the natural waterways, large man-made conduits including the trunk-line sewer system that receives the stormwater from the minor drainage system, large water impoundments, and other flood protection devices. These components together are designated as the *major drainage system*. Distributing, dimensioning, maintaining, and operating these devices for maximum benefit is part of efficient water resources management.

10.1.3 Chapter Organization

Even though the foregoing discussion seems to delineate large and small-scale systems, in reality they all are intertwined. For example, phosphate is typically added as a corrosion inhibitor at the treatment plant but may have to be removed at the wastewater plant before it reaches the receiving water body. It has become clearer that water systems cannot be engineered in isolation. This introduction emphasizes the need for a holistic approach. In the following, we present a set of water resources problem formulations with the understanding that by no means are these exhaustive. All real life water resources problems require an examination of alternatives. These problems typically require a combined simulation-optimization approach highly suited to evolutionary optimization techniques.

There are also specific structures that can be exploited for many of the problems. They seem to break into the following categories, namely (1) project planning including sequencing of multiple projects, (2) design, (3) operation, and (4) replacement. The goals and data needs under these categories differ. Project planning involves national and regional economic development, damage reduction, protection of ecosystems, and improvement of quality of life. *Planning* process requires a reconnaissance study followed by feasibility assessment on a system-wide basis (National Research Council, 1999). Typically, a design life is used. It includes factors such as period of cost recovery, meeting the intended function, technological obsolescence, and the life of the weakest component. The *design* aspect mainly focuses on dimensioning of the components subject to regulatory, physical, societal, and technical constraints. The *operational* phase addresses system behavior of an already built system and its coexistence under changing needs. McMahon and Farmer (2004) provide a lucid account of the need for reallocation of federal multipurpose reservoirs because of changing social preferences and purposes that were originally intended. The *replacement* facet assumes that the facility's continued service is essential and it fails periodically. Here the analysis focuses on failure patterns and alternative components to provide improved service.

In this chapter, we cover two major systems, namely, reservoirs and water distribution networks. The minor system, namely, the home plumbing system, is actually much larger than the public water distribution system and is addressed next. We present outcomes of a focus study group on consumer preferences toward different attributes of plumbing pipes. Urban stormwater management related to nonpoint source pollution is a major problem. In addition to regulations, financial considerations due to limited space and price of land steer the engineering design. Groundwater contamination is widespread and remains a significant cause of concern. All these problems require a process simulation-optimization approach. In this chapter, mathematical formulations for the efficient utilization of water resources are presented.

10.2   Optimal Operating Policy for Reservoir Systems

10.2.1   Introduction

In this section, we consider one of the largest and most complex water resource systems, namely, the reservoir. Labadie (2004) provides a comprehensive updated review of optimization of reservoir systems. He also includes a discussion on available software. He points out that with construction of no new large-scale projects there is a need for methods that can guide in efficient operation of these systems. It is not clear whether the increased demands on fossil fuels will result in a re-examination of policy toward building dams. A comprehensive analysis of reallocation of reservoir storage for changing uses and social preferences is given in McMahon and Farmer (2004) (also see National Research Council, 1999). A reservoir system captures, stores, and distributes water in sufficient quantities at sufficient head when and where needed. Such a system is typically subjected to three kinds of analysis. At preconstruction planning stage, determine the required capacity using the lowest inflows called the critical duration sequence that shifts burden to storage in meeting the demand; or compute the maximum yield that is obtainable for a given capacity. At post-construction planning stage, reevaluate performance with multiple inflow sequences in meeting the demand.

The United States Army Corps of Engineers (USACE) is the primary agency responsible for flood control for major reservoir systems. In the following, USACE's storage classification (Feldman, 1981; Wurbs, 1991) is adopted. In broad terms, reservoir storage is divided into three zones. The bottom zone, called *inactive pool*, is located beneath the low-level outlet so that no release is possible. This storage is meant to trap sediment and may provide for fish habitat. The top zone, called *flood control storage*, is empty space and is used only during periods of high flows so that releases will not exceed downstream channel capacity. When the top of the flood control zone is exceeded, the reservoir is out of control and an emergency spillway is used to divert the bulk of the flow. Major dams are designed for large flood events called *probable maximum floods* based on meteorological factors. The moisture maximization procedure employed in computing *probable maximum precipitation* provides the upperbound flood and it is highly unlikely this design event can ever be exceeded (Wurbs, 1991). The middle zone, called *conservation storage*, is utilized to satisfy day to day demands.

The USACE employs an *index level* associated with the top of each storage zone from the inactive pool, conservation pool to the flood pool. For example, a reservoir of capacity 700,000 ac-ft may have an index level of 1 assigned to a storage of 50,000 ac-ft at the top of the inactive pool; index 2 may correspond to 200,000 ac-ft at the top of the conservation pool; and index 3 corresponds to 700,000 ac-ft at the top of the flood control pool. The number of index levels can be increased by further subdividing the storage zones, mainly in the conservation pool. When the index levels at all reservoirs coincide, the entire system is considered to be in equilibrium and deviations from the equilibrium should be minimized. The equilibrium maintenance of index levels sets up a framework for a mathematical formulation in determining the reservoir operations schedule (Loganathan, 1996).

In real-time operations, forecasted inflows are used. The problem is solved with a rolling horizon of a few days with the results being implemented for the current day only. For the next day, the problem is re-solved with the updated forecast information from the current day observations. To aid the operator in day to day operations a *rule curve* is used. It specifies where the water level (storage) should be as a function of a particular month or week. These rule curves are based on the river flow sequence and the storage capacity that are available to smooth out operations. Because they are typically based on long-term simulations subjecting the system to a variety of inflow sequences, inherent in a rule curve is the anticipatory nature with regard to the stream flow. Hedging is the ability to shift storage to later times by providing only a portion of the demand when in fact it is possible to meet the entire demand. It is done to minimize drastic deficits over prolonged low flow periods.

The maximum flow that can be guaranteed during the most adverse stream flow period, also known as *critical period*, is the *firm yield*. The energy that can be produced under the most adverse flow conditions is called *firm energy* (Mays and Tung, 1992). The firm yield and firm energy set the lower bounds on flow and energy that are available for all times from a reservoir. A *flow duration curve*, which is a plot of flow versus percentage of times flow that is equaled or exceeded, is utilized to select the firm yield and firm energy at the near 100% exceedance level. This curve is important because it shows the inflow potential for a reservoir. The larger the reservoir, the higher the firm yield and energy that can be sustained for a prolonged time period. Preliminary power plant capacity is determined based on the average flow between selected threshold percentage exceedance values (American Society of Civil Engineers, 1989; Warnick et al., 1984). These limits help to impose a realistic demand pattern over the system.

The demand requirements are as follows. The water supply demands require specific amounts of flow at various locations and time; hydropower demands require both flow and hydraulic head at the turbine; navigation operations require sufficient flow depths in the channels and flows for the locks; recreation aspects require minimum water level fluctuations in the reservoir and sufficient depth for boats in rivers; ecological considerations require maintaining water quality, selective withdrawal of water from certain zones of storage for temperature maintenance downstream, providing sufficient amounts of aeration for required quantity of dissolved oxygen and at the same time avoiding supersaturation of water with gases such as nitrogen, which can cause gas bubble disease in fish and other biota (Mattice, 1991). These requirements in general become constraints on flow and reservoir storage. In the following some general guidelines as recommended by the U.S. Army Corps of Engineers (USACE) in meeting the demands are outlined.

10.2.2   Reservoir Operation Rules

The USACE's (2003) HEC-ResSim computer program [HEC5 (USACE, 1982)] is designed to simulate system behavior including hydropower under a set of operating rules. The storage index levels for each reservoir along with downstream flow conditions, anticipated inflows, and a view toward reaching equilibrium for the system as a whole dictate how the releases should be made. This approach eliminates much of the leeway in determining the releases; however, this strategy helps in minimizing flooding and empties the system as quickly as possible. Such a strategy should be reconsidered in a situation wherein flood control is not a major issue. In the following the operating rules are presented.

Reservoir Releases

Reservoir releases can be based on the following criteria: (1) channel capacity at the dam; (2) rate of change of release so that current period release cannot deviate from previous period release by more than a specified amount unless the reservoir is in flood surcharge operation; (3) storage not exceeding the top of conservation pool; (4) downstream control point flooding potential; (5) meeting specified target levels; (6) exceeding required low flows; (7) reaching various storage values; and (8) hydropower requirements.

Reservoir Filling/Drawing Down

Reservoirs are operated to satisfy constraints at the individual reservoirs to maintain specified flows at downstream control points and to keep the system as a whole in balance.
  Constraints at individual reservoirs:

1. When the level of a reservoir is between the top of the conservation pool and top of the flood pool, releases are made to attempt to draw the reservoir to the top of the conservation pool without exceeding the channel capacity.
2. Releases are made compatible with the storage divisions within the conservation pool.
3. Channel capacity releases (or greater) are to be made prior to the time the reservoir storage reaches the top of the flood pool if forecasted inflows are excessive. The excess flood water is dumped if sufficient outlet capacity is available. If insufficient capacity exists, a surcharge routing is made.
4. Rate of change criterion specifies the maximum difference between consecutive period releases.

Constraints for downstream control points:

1. Releases are not made (as long as flood storage remains) which would contribute to flooding during a predetermined number of future periods. During flooding at a downstream location, there may not be any release for power requirements.
2. Releases are made, where possible, to exactly maintain downstream releases at channel capacity for flood operation or for minimum desired or required flows for conservation operation.

System Balancing

To keep the entire system in balance, priority is given to making releases from the reservoir with the highest index level; if one of two parallel reservoirs has one or more reservoirs upstream, whose storage should be considered in making the releases, an upstream reservoir release is permitted only if its index level is greater than both the levels of the downstream reservoir and the combined reservoir equivalent index level. The combined equivalent index level corresponds to the sum of the reservoirs' storage at the appropriate index levels. For example, if reservoirs A, B, and C have index level 5 at storage 400, 300, and 200 (in kiloacre feet say) the combined equivalent index 5 corresponds to the storage of 900. For an elaborate discussion the reader is referred to the HEC5 user's manual (USACE, 1982). In the following, a general problem formulation that accommodates many of the facets of the above operating policy guidelines is presented.

10.2.3 Problem Formulation

A general mathematical programming formulation for the reservoir-operations problem may be stated as
Problem 10.1:

$$\text{minimize: deviations} = \text{F}(Sd_{11}^+,\ Qd_{11}^+,\ Ed_{11}^+,\ Sd_{11}^-,\ Qd_{11}^-,\ Ed_{11}^-,\ldots) \tag{10.1}$$

Subject to:

Real Constraints

Reservoir continuity

$$S(i,t+1) + R(i,t) = S(i,t) + I(i,t) \quad \text{for } i=1,\ldots,\text{NRES};\ t=1,\ldots,T \tag{10.2}$$

Reach routing

$$Q(i,t) = \text{g}[P(i,t), \text{TF}(i,t), Q(j_i), R(j_i)] \quad \text{for } i=1,\ldots,\text{NREACH};\ t=1,\ldots,T \tag{10.3}$$

Hydropower

$$R(i,t) = A(i,t) + B(i,t) \quad \text{for } i=1,\ldots,\text{NPLANT};\ t=1,\ldots,T \tag{10.4}$$

$$E(i,t) = \eta(i)\gamma A(i,t)\text{H}\,[S(i,t)] \quad \text{for } i=1,\ldots,\text{NPLANT};\ A=1,\ldots,T \tag{10.5}$$

Bounds on flows

$$\text{LQ}(i) \le Q(i,t) \le \text{UQ}(i) \quad \text{for } i=1,\ldots,\text{NREACH};\ t=1,\ldots,T \tag{10.6}$$

Bounds on storages

$$\text{LS}(i) \le S(i,t) \le \text{US}(i) \quad \text{for } i=1,\ldots,\text{NRES};\ t=1,\ldots,T \tag{10.7}$$

Powerplant capacity

$$E(i,t) \le \text{EP}(i) \quad \text{for } i=1,\ldots,\text{NPLANT} \tag{10.8}$$

Nonnegativity of variables

$$S(i,t), R(i,t), Q(i,t), Sd_{i,t}^-, Sd_{i,t}^+, Qd_{i,t}^-, Qd_{i,t}^+, Ed_{i,t}^-, Ed_{i,t}^+ \ge 0 \tag{10.9}$$

Goal Constraints

$$S(i,t) + Sd_{i,t}^- - Sd_{i,t}^+ = \text{TS}(i,t) \quad \text{for } i=1,\ldots,\text{NRES};\ t=1,\ldots,T \tag{10.10}$$

$$Q(i,t) + Qd_{i,t}^- - Qd_{i,t}^+ = \text{TQ}(i,t) \quad \text{for } i=1,\ldots,\text{NREACH};\ t=1,\ldots,T \tag{10.11}$$

$$E(i,t) + Ed_{i,t}^- - Ed_{i,t}^+ = \text{TE}(i,t) \quad \text{for } i=1,\ldots,\text{NPLANT};\ t=1,\ldots,T \tag{10.12}$$

in which $S(i,t)$ = the storage at the beginning of period; $I(i,t)$ and $R(i,t)$ = inflow and release during period t for reservoir i, respectively; NRES is the number of reservoirs;

$T =$ the operating horizon; $Q(i, t) =$ the flow in reach i written as some function g; $P(i, t)$ and TF $(i, t) =$ precipitation and tributary flow to reach i for period t, respectively; $A(i, t) =$ flow for hydropower from reservoir i for period t; $B(i, t) =$ nonpower release from reservoir i for period t; $\eta(i) =$ efficiency of plant i; $\gamma =$ specific weight of water; $\mathrm{H}[S(i, t)] =$ head over the turbine, a function of reservoir storage and tailwater level; $\mathrm{EP}(i) =$ plant capacity for the ith power plant; NPLANT $=$ the number of power plants; $ji =$ the set of control stations contributing flow for reach i; NREACH $=$ the number of reaches; $\mathrm{LQ}(i)$ and $\mathrm{UQ}(i) =$ lower and upper bounds for flow in reach i, respectively; $\mathrm{LS}(i)$ and $\mathrm{US}(i) =$ lower and upper bounds for storage in reservoir i, respectively; Sd_i^-, Sd_i^+, and Qd_i^-, Qd_i^+, and Ed_i^-, $Ed_i^+ =$ the slack and surplus deviational variables for storage, flow, and power respectively; $\mathrm{TS}(i, t) =$ the storage target for reservoir i at time t; $\mathrm{TQ}(i, t) =$ the flow target for reach i at time t; and $\mathrm{TE}(i, t) =$ power target for plant i at time t. The storage goal constraints provide for recreation and hydropower; the flow goal constraints provide for water supply, navigation, instream flow and irrigation; the hydropower goal constraints provide for target power production at each plant.

Hydropower

Eschenbach et al. (2001) provide a detailed description of Riverware optimization decision support software based on a goal programming formulation. Linearization is used to solve it as a linear program. They also include a description of Tennessee Valley Authority's use of the software in managing their reservoirs to determine optimal hydropower schedules. The hydropower aspect has been considered in various forms, such as assigning a target storage for power production; maintaining a fixed head; linearization by Taylor series about an iterate; induced separability/subsequent linearization; division of storage into known intervals and choosing constant heads for each interval with the selection of intervals aided by integer variables or by dynamic programming state transition; optimal control strategy and direct nonlinear optimization. Comprehensive reviews are given in Labadie (2004), Yeh (1985), Wurbs (1991), and Wunderlich (1991). Martin (1987) and Barritt-Flatt and Cormie (1988) provide detailed formulations. Successive linear programming (Palacios-Gomez et al., 1982; Martin, 1987; Tao and Lennox 1991) and separable programming (Can et al., 1982; Ellis and ReVelle, 1988) are widely adopted for solution. Pardalos et al. (1987) have offered a strategy to convert an indefinite quadratic objective into tight lower bounding separable convex objective which can be solved efficiently. If the hydropower constraints are dropped or linearized, then using linear routing schemes in (Equation 10.3) Problem 10.1 becomes a linear program (Loganathan and Bhattacharya, 1990; Changchit and Terrell, 1989). Martin (1995) presents a well-detailed real system application to the Lower Colorado River Authority district. Moore and Loganathan (2002) present an application to a pumped storage system.

Trigger Volumes for Rationing

Thus far the discussion has paid more attention to floods. However, water shortages do occur and how to cope with them is a critical issue. Lohani and Loganathan (1997) offer a procedure for predicting droughts on a regional scale. In the case of a reservoir one should know at what low storage levels the regular releases should be curtailed toward an impending drought (Shih and ReVelle, 1995). If the available storage plus forecasted inflow is less than V_{1p}, the first-level trigger volume, level 1 rationing is initiated. If the anticipated storage for the next period is less than V_{2p}, the second level trigger volume, level 2 rationing, which is severer than level 1 rationing, is initiated. The objective is not to have any rationing at all by prudently utilizing the storage to satisfy the regular demands. With this objective, the V_{1p} and V_{2p} will be chosen optimally by preserving storage whenever possible to hedge against future shortfalls.

Shih and ReVelle (1995) propose optimal trigger volumes V_{1p} and V_{2p} to initiate two levels of rationing of water during drought periods as described by the following constraints.

$$S_{t-1} + I_t \leq V_{1p} + M\delta_{1t} \tag{10.13}$$

$$S_{t-1} + I_t \geq V_{1p} - M(1 - \delta_{1t}) \tag{10.14}$$

$$S_{t-1} + I_t \leq V_{2p} + M\delta_{2t} \tag{10.15}$$

$$S_{t-1} + I_t \geq V_{2p} - M(1 - \delta_{2t}) \tag{10.16}$$

$$R_t = (1 - \alpha_1)D^*\delta_{1t} + (\alpha_1 - \alpha_2)D * \delta_{2t} + \alpha_{2D} \tag{10.17}$$

$$V_{1p} \geq (1 + \beta_1)V_{2p} \tag{10.18}$$

$$\delta_{1t} \leq \delta_{2,t+1} \text{ with } \delta_{1t}, \delta_{2t} \in 0 \text{ or } 1 \text{ binary} \tag{10.19}$$

in which V_{1p}, V_{2p} = cutoff storage values for levels 1 and 2 rationing; α_1, α_2 = percentages of demand D provided during rationing levels 1 and 2, respectively; β_1 = suitably chosen percentage level, say 20%. When $\delta_{1t} = 1$, from Equation 10.14 there is no rationing; when $\delta_{1t} = 0$, and $\delta_{2t} = 1$ from Equations 10.13 and 10.16 there is level 1 rationing. When $\delta_{2t} = 0$, there is level 2 rationing and $R_t = \alpha_{2D}$. Note that when $\delta_{2t} = 0$, we must have $\delta_{1t} = 0$ by Equation 10.14. When $\delta_{1t} = 1$, δ_{2t} has to be 1 and $R_t = D$. Constraint 10.19 says that level 1 rationing must precede level 2 rationing. The objective is to maximize $\sum \delta_{1t}$ over t.

Linear Release Rule

There have been attempts to follow the rule curve in the sense of relating a fixed amount of storage to a specific time period. For example, the release rule given by ReVelle et al. (1969), also called linear decision rule, is

$$R_t = S_{t-1} - b_t \tag{10.20}$$

in which S_{t-1} = storage at the end of $t - 1$; b_t = decision parameter for period t; R_t = release for period t. This rule says that from the end of $t - 1$ period storage S_{t-1}, save b_t amount and release the remaining during period t as R_t regardless of inflow, I_t, which will add to b_t. It is seen from Equation 10.20 that the end period of period t storage S_t is the sum of the left-over storage b_t from the previous period plus the inflow. That is,

$$S_t = b_t + I_t \tag{10.21}$$

With the aid of Equation 10.21 each storage variable S_t can be replaced by the corresponding decision variable b_t which will be optimally determined for each t. If t represents a month, $b_1 = b$ (January) is fixed and repeats in an annual cycle in replacing S_1, S_{13}, S_{25}, and so on. From Equation 10.21 it is seen that the single decision variable b_1 along with the known I_1, I_{13}, and I_{25} replaces the three variables S_1, S_{13}, and S_{25}. Therefore, this substitution leads to a few decision variables. These twelve b_t values help to regulate a reservoir. Refinements to this type of release rule are given in Loucks et al. (1981) (also see Karamouz et al., 2003, and Jain and Singh, 2003).

Chance Constraint

Instead of using the predicted value of the inflow, its quantile may be used at a particular probability level. For example, consider the probability that storage at the end of time period t denoted by S_t exceeding some maximum storage SMAX must be less than or equal to a small value, say 0.1. That is,

$$P[S_t \geq \text{SMAX}] \leq 0.1 \tag{10.22}$$

Substituting Equation 10.21 in Equation 10.22 we obtain

$$P[I_t \geq \text{SMAX} - b_t] \leq 0.1 \tag{10.23}$$

with its deterministic equivalent given by

$$\text{SMAX} - b_t \geq i_{0.9}(t) \tag{10.24}$$

in which $i_{0.9}(t) = $ cutoff value at the 90% cumulative probability level for period t.

Firm Yield, Storage Capacity, and Maximization of Benefits

As mentioned before, firm yield and storage capacity to meet fixed demands are important parameters. The determination of firm yield requires maximize [minimum $R(i,t)$] for specified reservoir capacity. The required storage capacity is obtained by minimizing capacity to satisfy the demands. The optimal capacity therefore will be determined by the critical period with the most adverse flow situation requiring maximum cumulative withdrawal as dictated by the continuity constraint (Equation 10.2). Of course, a large demand that cannot be supported by the inflow sequence would trigger infeasibility. While the objective in Equation 10.1 minimizes the deviations from targets, one may choose to maximize the benefits from the releases including hydropower, and recreation. Such a formulation can yield a problem with nonlinear objective and linear constraint region. Can et al. (1982) discuss strategies to linearize the objective function so that the problem can be solved as a linear program.

In the simulation program HEC5, by employing the index level discretization scheme for storage capacity, releases compatible with the storage are made, which enhances operations within the conservation zone. This scheme also minimizes flood risk at a reservoir by having its storage compatible with the others. The program attempts to find near optimal policies by cycling through the allocation routines several times. This approach has a tendency to place the releases at their bounds. Practical considerations in reservoir operations include a well spelt out emergency preparedness plan and the operation and maintenance of electrical, mechanical, structural, and dam instrumentation facilities by the operating personnel located at or nearest to the dam. These practical operating procedures at the dam site are covered in the standing operating procedures guide (U.S. Department of Interior, 1985). Because loss of lives and property damages are associated with any faulty decisions with regard to reservoir operations, all model results should be subjected to practical scrutiny.

10.3 Water Distribution Systems Optimization

10.3.1 Introduction

Water distribution systems constitute one of the largest public utilities. Optimal water distribution system design remains intriguing because of its complexity and utility. Comprehensive reviews of optimization of water distribution systems are given in Boulos et al. (2004) and Mays (2000). Bhave (1991) and Walski (1984) provide thorough details on formulating water distribution network problems. The pipe network problems have feasible regions that are nonconvex. Also, the objective function is multimodal. These two aspects make the conventional (convex) optimization methods to result in a local optimum sensitive to the starting point of the search. In this section, a standard test problem from the literature is considered. The pipe network is judiciously subjected to an outer search scheme that chooses alternative flow configurations to find an optimal flow division among pipes. An inner linear program is employed for the design of least cost diameters for the pipes.

The algorithm can also be employed for the optimal design of parallel expansion of existing networks. Three global search schemes, multistart-local search, simulated annealing, and genetic algorithm, are discussed. Multistart-local search selectively saturates portions of the feasible region to identify the local minima. Simulated annealing iteratively improves the objective function by finding successive better points and to escape out of a local minimum it exercises the Metropolis step, which requires an occasional acceptance of a worse point. Genetic algorithm employs a generation of improving solutions as opposed to the other two methods that use a single solution as an iterate. Spall (2003) provides comprehensive details on search methods.

10.3.2 Global Optimization

It is not uncommon to find real life problems that have cost or profit functions defined over a nonconvex feasible region involving multiple local optima and the pipe network optimization problem falls into this group. While classical optimization methods find only a local optimum, global optimization schemes adapt the local optimum seeking methods to migrate among local optima to find the best one. Of course, without presupposing the nature of local optima, a global optimum cannot be guaranteed except to declare a relatively best optimum. Detailed reviews are given in Torn and Zilinskas (1987) and Rinnooy Kan and Timmer (1989). Consider Problem P0 given by:

Problem P0:

$$\text{Minimize} \quad f(x)$$
$$\text{Subject to:} \quad g_i(x) \geq 0 \quad \text{for } i = 1, 2, \ldots, m$$

Let the feasible region be $X = \{x | g_i(x) \geq 0\}$. A solution x^1 is said to be a local optimum if there exists a neighborhood B around x^1 such that $f(x^1) \leq f(x)$ for all $x \in B(x^1)$. A solution x^g is said to be a global optimum if $f(x^g) \leq f(x)$ for all $x \in X$.

10.3.3 Two-Stage Decomposition Scheme

In this section, a two-stage decomposition with the outer search being conducted among feasible flows, and an inner linear programming to optimally select pipe diameters and hydraulic heads is employed. The following formulation retains the same general inner linear programming framework when addressing multiple loadings, pumps, and storage tanks as given in Loganathan et al. (1990). However, for the sake of clarity and the nature of the example problem these elements are suppressed. Consider a pipe network comprised of N nodes. Let S be the set of fixed head nodes. Let $\{N-S\}$ be the set of junction nodes and **L** be the set of links. $Q_{(i,j)}$ is the steady state flow rate through link $(i,j) \in \mathbf{L}$. Let $L_{(i,j)}$ denote the length of link $(i,j) \in \mathbf{L}$ and $D_{(i,j)}$ be its diameter which must be selected from a standard set of discrete diameters $D = \{d_1, d_2, \ldots, d_M\}$. In the present formulation it is assumed that each link (i,j) is made up of M segments of *unknown* lengths $x_{(i,j)m}$ (decision variable) but of *known* diameter d_m for $m = 1, 2, \ldots, M$ (Karmeli et al., 1968). Let $C_{(i,j)m}$ be the cost per unit length of a pipe of diameter d_m. Let r_k be the path from a fixed head node (source) to demand node, k. Let **P** be the set of paths connecting fixed head nodes and basic loops. Let there be P_l loops and b_p denote the head difference between the fixed head nodes for path p connecting them; and it is zero corresponding to loops. Let H_s be the fixed head and $H_k^{\min}$ be the minimum head at node $k \in \{N-S\}$. Let q_i be the supply at node i which is positive; if it is demand, it is negative. Let $J_{(i,j)m}$ be the hydraulic gradient for segment m (partial length of a link with diameter d_m) which is given by

$$J_{(i,j)m} = K[Q_{(i,j)}/C]^{1.85} d_m^{-4.87} \tag{10.25}$$

in which $K = 8.515(10^5)$ for $Q_{(i,j)}$ in cfs and d_m in inches; $C =$ Hazen-Williams Coefficient. The pipe network problem may be stated as follows:

Problem P1:

Minimize $\quad f(x) = \sum_{(i,j)} \sum_{m=1}^{M} C_{(i,j)m} x_{(i,j)m}$ $\qquad$ (10.26)

Subject to: $\quad \sum_j Q_{(i,j)} - \sum_j Q_{(j,i)} = q_i \quad$ for $i \in \{N\!-\!S\}$ $\qquad$ (10.27)

$$H_s - H_k^{\min} - \sum_{(i,j) \in r_k} \pm \sum_m J_{(i,j)m} x_{(i,j)m} \geq 0 \quad \text{for } s \in S \text{ and } k \in \{N\!-\!S\}$$
$\qquad$ (10.28)

$$\sum_{(i,j) \in p} \pm \sum_m J_{(i,j)m} x_{(i,j)m} = b_p \quad \text{for } p \in P$$
$\qquad$ (10.29)

$$\sum_m x_{(i,j)m} = L_{(i,j)} \quad \text{for } (i,j) \in L$$
$\qquad$ (10.30)

$$x_{(i,j)} \geq 0$$
$\qquad$ (10.31)

In Problem P1 pipe cost objective function (Equation 10.26) is minimized; constraint 10.27 represents steady state flow continuity; constraint 10.28 is the minimum head restriction; constraint 10.29 represents the sum of head losses in a path which is zero for loops; constraint 10.30 dictates that sum of segment lengths must equal link length; constraint 10.31 is the nonnegativity on segment lengths. The decision variables are: $Q_{(i,j)}$, $J_{(i,j)m}$, and $x_{(i,j)m}$. The following two-stage strategy Problem P2 is suggested for the solution of Problem P1.

Problem P2:

$$\underset{Q_{(i,j)}}{\text{Min}} \left[\underset{x \in X}{\text{Min}} \; f(x) \right]$$
$\qquad$ (10.32)

in which $Q_{(i,j)}$ are selected as perturbations of the flows of an underlying near optimal spanning tree of the looped layout satisfying constraint 10.27 and X is the feasible region made up of constraints 10.28 through 10.31. It is worth noting that the optimal layout tends to be a tree layout. Deb (1973) shows that a typical pipe cost objective is a concave function that attains its minimum on a boundary point resulting in a tree solution (also see Deb and Sarkar, 1971). Gessler (1982) and Templeman (1982) also argue that because optimization has the tendency to remove any redundancy in the system the optimal layout should be a tree. However, a tree gets disconnected even when one link fails. Loganathan et al. (1990) have proposed a procedure that yields a set of near optimal trees that are to be augmented by loop-forming pipes so that each node has two distinct paths to source. It is suggested that the flows of the optimal tree be considered as initial flows that are to be perturbed to obtain flows in all pipes of the looped network. It is observed that the inner Problem P3 of Problem P2 given by

Problem P3:

$$\underset{x \in X}{\text{Min}} \; f(x), \text{ for fixed flows}$$
$\qquad$ (10.33)

is a linear program that can be solved efficiently. To choose the flows in the outer problem of P3, multistart-local search and simulated annealing are adopted. To make the search efficient, first a set of flows corresponding to a near optimal spanning tree of the network is found. The flows in the looped network are taken as the perturbed tree link flows by

$$Q_{(i,j)}(\text{loop}) = Q_{(i,j)}(\text{tree}) + \sum \pm \Delta Q_{p(i,j)} \qquad (10.34)$$

in which the sum is taken over loops "p" which contain link (i,j) with positive loop change flow ΔQ used for clockwise flow.

10.3.4 Multistart-Local Search

Multistart-local search covers a large number of local minima by saturating the feasible region with randomly generated starting points and applying a local minimization procedure to identify the local optima. A suitable generating function is needed. Because flows in the looped network are generated by perturbing the optimal tree flows, a probability density function for loop change flows with mean zero is a good choice. However, other parameters of the density function must be adjusted by experimentation for the problem under consideration. The algorithm is summarized as follows:

Procedure MULTISTART

Step 0: (*Initialization*) Select a suitable density function (pdf) with known parameters and assign number of seed points to be generated, NMAX; determine an estimate of global optimal value f_g; assign a large value for the incumbent best objective function value, BEST; set $n = 1$.

Step 1: (*Iterative Local Minimization*) Until $n = $ NMAX
do: Generate a point using the pdf that serves as perturbation flow in Equation 10.34. Apply the minimizer Problem P3. Update incumbent BEST.

Step 2: (*Termination*) If $|f_g - BEST| < tolerance$, report BEST and the optimal solution and stop. Otherwise, revise parameter values for the generating function, and/or global value, f_g and tolerance. Set $n = 1$; go to Step 1.

END Procedure.

10.3.5 Simulated Annealing

Simulated annealing is an iterative improvement algorithm in which the cost of the current design is compared with the cost of the new iterate. The new iterate is used as the starting point for the subsequent iteration if the cost difference is favorable. Otherwise, the new iterate is discarded. It is clear that such an algorithm needs help to get out of a local optimum to wander among local minima in search of the global minimum. It uses the Metropolis step to accomplish this. The Metropolis step dictates that a worse point (higher objective value for a minimization problem) be accepted from the current point with a user specified probability "pR." The algorithm accepts the next iterate from the current iterate with probability one if it is a better point (lower objective value).

From the given initial point, the corresponding objective function value, f_0, is calculated. Then, a point is randomly generated from a unit hypersphere with the current iterate at its center. This random point specifies a random direction along which a new point is generated by taking a user specified step size of α from the current iterate. A new objective function value (f_1) is evaluated at the point. The new point is accepted with the probability

p given by:

$$pR = 1 \quad \text{if } \Delta f = f_1 - f_0 \leq 0 \qquad\qquad (10.35)$$
$$= \exp(-\beta \Delta f / f_0^\theta) \quad \text{if } \Delta f > 0$$

where β and θ are user-specified parameters. From the first condition of Equation 10.35 it is seen that the method readily accepts an improving solution if it is found. From the second condition, it is apparent that an inferior point (i.e., with worse objective function value) becomes an acceptable solution with probability $\exp(-\beta \Delta f / f_0^\theta)$ when an improving solution is not available. Bohachevsky et al. (1986) recommend selecting the parameter β such that the inequality $0.5 < \exp(-\beta \Delta f / f_0^\theta) < 0.9$ holds, which implies that 50–90% of the detrimental steps are accepted. The procedure continues until a solution that is within a *tolerance* level from the user-specified (estimated) global optimal value is obtained. The algorithm is stated as follows (Loganathan et al., 1995).

Procedure ANNEALING

   Step 0: (*Initialization*) Let f_g be the user-specified (estimated) global optimal value (which may be unrealistic): α a step size, and β and θ acceptance probability parameters. Let the vector of loop change flows be $\varepsilon = \{\Delta Q_1, \Delta Q_2, \ldots, \Delta Q_{P_l}\}$. Let ε^0 be the arbitrary feasible starting point of dimension P_l.
   Step 1: (*Local Minimization*) Set $f_0 = f(\varepsilon^0)$. If $|f_0 - f_g| < tolerance$, stop.
   Step 2: (*Random Direction*) Generate P_l (number of loops) independent standard normal variates, $Y_1, \ldots, Y_{P_l}$ and compute components of unit vector $\mathbf{U}$: $U_i = Y_i/(Y_1^2 + \cdots Y_{P_l}^2)^{1/2}$, for $i = 1, \ldots, P_l$.
   Step 3: (*Iterative Local Minimum*) Set $\varepsilon^1 = \varepsilon^0 + \alpha \mathbf{U}$. Apply minimizer Problem P3. If ε^1 is infeasible, return to Step 2. Otherwise, set $f_1 = f(\varepsilon^1)$ and $\Delta f = f_1 - f_0$.
   Step 4: (*Termination*) If $f_1 > f_0$, go to step 5. Otherwise, set $\varepsilon^0 = \varepsilon^1$ and $f_0 = f_1$. If $|f_0 - f_g| < tolerance$, save ε^1 and stop. Otherwise go to Step 2.
   Step 5: (*Acceptance–Rejection*) ($f_1 > f_0$): Set $pR = \exp(-\beta \Delta f / f_0^\theta)$. Generate a uniform 0–1 variate v. If $v \geq pR$, go to Step 2. Else $v < pR$ and set $\varepsilon^0 = \varepsilon^1$, $f_0 = f_1$, and go to Step 2.
   END Procedure.

   The success of the procedure depends upon the selection of the parameters α (step size), β (exponential probability parameter), and θ (exponent in the exponential probability function), which must be adjusted as necessary. If β is too large (pR is too small) too many function evaluations are needed to escape from a local minimum; if β is too small (pR is close to unity), an inefficient search results by accepting almost every inferior point generated. The parameter α controls the search radius around ε^0: α should be chosen such that the distance is sufficiently long to prevent from falling back to the same local minimum. Also, estimates for f_g and *tolerance* should be updated based on the performance of the algorithm. Goldman and Mays (2005) provide additional applications of simulated annealing to water distribution systems.

10.3.6 Extension to Parallel Expansion of Existing Networks

Another important aspect that must be considered is the expansion of existing networks. In the present formulation parallel expansion of existing pipes is considered. Adaptation of Problem P2 for existing networks is as follows. For existing pipes, the pipe diameters are fixed and parallel links may be required for carrying additional flow in order not to increase the head loss. The flow from node i to node j, for a parallel system, is denoted by $Q_{(i,j)} = Q_{(i,j),O} + Q_{(i,j),N}$, in which subscripts O and N indicate Old and New, respectively.

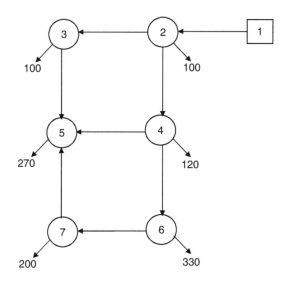

FIGURE 10.2   Two-loop example network.

TABLE 10.1   Node Data

Node i	Elevation (m)	Minimum Head (m)	Demand (m^3/h)
1	210	0	−1120
2	150	30	100
3	160	30	100
4	155	30	120
5	150	30	270
6	165	30	330
7	160	30	200

It is suggested that Problem P2 be solved with the restriction that an existing pipe should not be undersized. From the solution to Problem P2 the head loss for each pipe can be obtained by multiplying the optimal segment lengths, $x_{(i,j)m}$'s and the corresponding head gradients $J_{(i,j)m}$'s. Because optimal head loss, friction parameter, and existing pipe size are known, the flow in an existing link and therefore the flow for the parallel pipe can be computed (Wylie and Streeter, 1985). The parallel pipe in turn is decomposed into discrete diameter pipes in series.

10.3.7   Analysis of an Example Network

In this section, an example network (Alperovits and Shamir, 1977) using the data given therein is considered (see Figure 10.2). The data for the minimization problem P2 are the minimum and maximum heads and demand at each node, length and Hazen-Williams coefficient C for each link, and the unit cost for different pipe diameters. These data for Alperovits and Shamir network are given in Tables 10.1 and 10.2. Each link has length 1000 m and a Hazen-Williams coefficient of 130. For known flow rates, Problem P3 is a linear program. The decision variables are the unknown segment lengths of known diameters. The network consists of seven nodes, eight links (see Figure 10.2), and fourteen different candidate diameters as shown in Table 10.2. Because there are only fifteen possible spanning trees for the network, the global optimal tree solution is easily obtained (Loganathan et al., 1990). These flows are {1120, 370, 650, 0, 530, 200, 270, 0} in links {(1,2), (2,3), (2,4), (4,5), (4,6), (6,7), (3,5), (7,5)}, respectively, which are systematically perturbed to obtain the flows for the looped network.

TABLE 10.2   Diameter/Cost Data

Diameter (in.)	Unit Cost ($/m)	Diameter (in.)	Unit Cost ($/m)
1	2	12	50
2	5	14	60
3	8	16	90
4	11	18	130
6	16	20	170
8	23	22	300
10	32	24	550

TABLE 10.3   Optimal Flows for Different Methods (Flow Rate m^3/h)

Link	Loganathan et al. (1990)	Global Search	Global Search with Min. Flow
(1,2)	1120.0	1120.0	1120.0
(2,3)	360.0	368.33	368.0
(2,4)	660.0	651.67	652.0
(3,5)	260.0	268.33	268.0
(4,6)	534.0	530.69	531.0
(6,7)	204.0	200.69	201.0
(7,5)	4.0	0.97	1.0
(4,5)	6.0	0.69	1.0
Total Cost	412,931	403,657	405,381

TABLE 10.4   Optimal Solution without Minimum Flow

Link	Diameter (in.)	Length (m)	Hydraulic Gradient (m/m)	Head Loss (m)
(1,2)	18	1000.00	0.0068	6.76
(2,3)	10	792.54	0.0150	11.95
	12	207.46	0.0062	1.29
(2,4)	16	1000.00	0.0044	4.40
(4,5)	1	1000.00	0.0190	18.84
(4,6)	14	303.46	0.0058	1.75
	16	696.54	0.0030	2.09
(6,7)	8	10.32	0.0150	0.15
	10	989.68	0.0049	4.85
(3,5)	8	97.70	0.0250	2.43
	10	902.30	0.0084	7.57
(7,5)	1	1000.00	0.0100	10.00

To implement the procedure MULTISTART, a probability distribution function with a high probability mass near the origin (loop change flow vector is zero) accounting for the core tree link flows, is needed. The generated random points are used within Problem P3 to obtain the optimal designs. Because the normal distribution allows for positive or negative perturbations and high probability mass close to the mean, it is chosen as the standard distribution for each loop flow change variable ΔQ_p. A standard deviation of 6.5 is used both for ΔQ_1 and ΔQ_2 and the maximum number of points generated NMAX is set at 200. The optimal point $\varepsilon = (\Delta Q_1, \Delta Q_2) = (1.95, 0.69)$ is further refined by gradient search to obtain $\varepsilon = (\Delta Q_1, \Delta Q_2) = (1.67, 0.69)$ producing a cost of 403,657.94 units. The optimal link flows after ΔQ change flows applied to the optimal tree link flows are given in Table 10.3. The procedure ANNEALING is implemented next. When the procedure locates the optimum, $\varepsilon = (\Delta Q_1, \Delta Q_2) = (2.25, 1.22)$, it circles around the neighborhood and does not escape. The parameters are $\beta = 1$, $\theta = 1$, $\alpha = 2.5$, $f_g = 405,000$, and NMAX $= 200$. Further refinement as well as the gradient search yields the same optimum point (1.67, 0.69).

The global optimal tree solution made up of links $\{(1, 2), (2, 3), (2, 4), (3, 5), (4, 6), (6, 7)\}$ has a cost of \$399,561 (Loganathan et al., 1990). Using the minimum possible diameter of 1 in. for two loop-forming links $\{(4, 5), (5, 7)\}$ at a unit cost of \$2/meter length we obtain the global minimum cost \$403,561. The proposed algorithm yields \$403,657, verifying global optimality. It is noted that for this solution a minimum diameter of 1 in. is imposed for

TABLE 10.5   Energy Heads without Minimum Flow

Node (i)	Elevation (m)	Minimum Head (m)	Head (m)
1	210.00	210.00	210.00
2	150.00	180.00	203.24
3	160.00	190.00	190.00
4	155.00	185.00	198.84
5	150.00	180.00	180.00
6	165.00	195.00	195.00
7	160.00	190.00	190.00

TABLE 10.6   Optimal Solution with Minimum Flow

Link	Diameter (in)	Length (m)	Hydraulic Gradient (m/m)	Head Loss (m)
(1,2)	18	1000.00	0.0068	6.76
(2,3)	10	795.05	0.0150	11.97
	12	204.95	0.0062	1.27
(2,4)	16	1000.00	0.0044	4.40
(4,5)	1	951.65	0.0200	18.81
	2	48.35	0.00068	0.03
(4,6)	14	300.46	0.0058	1.73
	16	699.54	0.0030	2.11
(6,7)	8	9.01	0.0150	0.13
	10	990.99	0.0049	4.87
(3,5)	8	99.12	0.0250	2.46
	10	900.88	0.0084	7.54
(7,5)	1	488.53	0.0200	9.65
	2	511.46	0.00068	0.35

TABLE 10.7   Energy Heads with Minimum Flow

Node (i)	Elevation (m)	Minimum Head (m)	Head (m)
1	210.00	210.00	210.00
2	150.00	180.00	203.24
3	160.00	190.00	190.00
4	155.00	185.00	198.84
5	150.00	180.00	180.00
6	165.00	195.00	195.00
7	160.00	190.00	190.00

all the pipes in the looped layout as previously done by the other authors. The results are given in Tables 10.3 through 10.5. A minimum flow of constraint $1 \, \text{m}^3/\text{h}$, in addition to the minimum diameter of 1 in., was implemented and the results are given in Tables 10.6 and 10.7, with the cost of $405,301.

10.3.8   Genetic Algorithm

Savic and Walters (1997) applied genetic algorithms for the same problem. Deb (2001) presents an authoritative description of genetic algorithms (GA). In GA solutions are represented by strings called chromosomes. A locus is the location of a gene (bit in binary) on a chromosome. An allele represents the set of values that can be assigned to a gene. Good solutions in terms of their objective function value are identified from an initial set of solutions. Multiple copies of good solutions are made while eliminating the weak solutions keeping the population size constant (Selection). The selected good solutions are combined to create new solutions by exchanging portions of strings (Crossover). Note this process retains the population size. To enhance diversity in the populations a bit may be changed (Mutation). Using the new population from crossover and mutation operators, the cycle is restarted with the selection operator. Deb (2001) makes the following key points. The three operators progressively result in strings with similarities at certain string positions (schema). A GA with its selection operator alone increases the density of solutions with above average value while reducing the variance; mutation increases the variance. The crossover operator has the tendency to decorrelate the decision variables. The combined effect of selection and crossover is to enable hyperplanes of higher average fitness values to be sampled more frequently.

Handling constraints is an issue in direct search methods. Variable elimination by constraint substitution for equality constraints, penalty function methods including biasing feasible solutions over infeasible solutions, hybrid methods, and methods based on decoders are procedures for accommodating constraints. Hybrid methods combine the classical pattern search methods within the GA. Bazaraa et al. (2006) and Reklaitis et al. (1983) provide a thorough analysis of penalty function methods. Savic and Walters (1997) use a penalty for constraint violation. They report a cost of \$419,000 with a minimum diameter of 1 in. and without using split pipes.

10.4 Preferences in Choosing Domestic Plumbing Materials

10.4.1 Introduction

In the United States, about 90% of drinking water home plumbing systems are copper pipes. Pinhole leaks in copper plumbing pipes are reported in several parts of the country. Figure 10.1 shows the distribution of pinhole leaks over a five-year period, from 2000 to 2004. For a majority of homeowners, their home is the most valuable asset. A decision has to be made on whether to continue to repair or replace the system. Loganathan and Lee (2005) have proposed an optimal replacement time for the home plumbing system exploiting an economically sustainable optimality criterion. Dietrich et al. (2006) present results of their study related to material propensity to leach organic chemicals, metals, and odorants; ability to promote growth of nitrifying bacteria; and their impact on residual disinfectant for five polymeric and four metallic pipes. Typically, copper, PEX (cross-linked polyethylene), and CPVC (chlorinated polyvinyl chloride) pipes are used. Stainless steel is also being considered. The material-related attributes of price, corrosion resistance, fire retardance, health effects, longevity, re-sale value of home, and taste and odor are considered in making a selection. Consumer preferences toward these attributes are assessed with the aid of a focus group.

10.4.2 Analytical Hierarchy Process

Marshutz (2001) reports that copper accounts for 90% of new homes, followed by PEX (cross linked polyethylene) at 7%, and CPVC (chlorinated polyvinyl chloride) at 2%. All materials have relative advantages and disadvantages. One material may be easy to install but may not be reliable while another material may be corrosion resistant but may have health or taste and odor problems. It is generally recognized that comparing several attributes simultaneously is complicated and people tend to compare two attributes at a time. The analytic hierarchy process (AHP) is used to determine the preference for attributes by pair-wise comparison. Assessing pair-wise preferences is easier as it enables to concentrate judgment on taking a pair of elements and compare them on a single property without thinking about other properties or elements (Saaty, 1990). It is noted that elicited preferences may be based on the standards already established in memory through a person's experience or education. Based on Saaty (1980) the following steps are adopted in performing the analytical hierarchy process.

Step 0: (*Identify attributes*) For plumbing material the following seven attributes are considered: Price—includes cost of materials and labor for installation and repair; Corrosion

TABLE 10.8 Standard Numerical Scores

Preference Level	Numerical Score, $a(i,j)$ 1–9 Scale
Equally preferred	1
Equally to moderately preferred	2
Moderately preferred	3
Moderately to strongly preferred	4
Strongly preferred	5
Strongly to very strongly preferred	6
Very strongly preferred	7
Very strongly to extremely preferred	8
Extremely preferred	9

TABLE 10.9 Pair-Wise Preference Weight Matrix [General]

	Attribute 1	Attribute 2	...	Attribute n
Attribute 1	$w1/w1$	$w1/w2$	...	$w1/wn$
Attribute 2	$w2/w1$	$w2/w2$	...	$w2/wn$
...	...	...	...	...
Attribute n	$wn/w1$	$wn/w2$	...	wn/wn
Sum	$X/w1$	$X/w2$	...	X/wn

$[X = (w1 + w2 + \cdots + wn)].$

TABLE 10.10 Pair-Wise Preference Weight Matrix

	P	C	F	H	L	R	T
P	1	0.20	0.25	0.14	0.20	0.50	0.33
C	5	1	5	0.33	1	6	1
F	4	0.20	1	0.17	0.33	1	0.25
H	7	3	6	1	4	8	3
L	5	1	3	0.25	1	2	0.33
R	2	0.17	1	0.13	0.50	1	0.50
T	3	1	4	0.33	3	2	1
Sum	27.00	6.57	20.25	2.35	10.03	20.50	6.42

P: price, C: corrosion resistance, F: fire retardance, H: health effects, L: longevity, R: resale value of home, and T: taste and odor.

resistance—dependability of material to remain free of corrosion; Fire retardance—ability of material to remain functional at high temperatures and not to cause additional dangers such as toxic fumes; Health effects—ability of material to remain inert in delivering water without threatening human health; Longevity—length of time material remains functional; Resale value of home—people's preference for a particular material including aesthetics; and Taste and odor—ability of material to deliver water without imparting odor or taste.

Step 1: (*Use standard preference scores*) A scale (1–9) of pair-wise preference weights is given in Table 10.8 (Saaty, 1980).

Step 2: (*Develop pair-wise preference matrix*) In AHP, instead of directly assessing the weight for attribute, i, we assess the relative weight $a_{ij} = wi/wj$ between attribute i and j. As shown in Tables 10.9 and 10.10, each participant is asked to fill in a 7×7 attribute matrix of pair-wise preferential weights. An example is given in Table 10.10. In Table 10.10, row H and column P, the entry of 7 implies that health effects are very strongly preferred in comparison to price in the ratio of 7:1. Row H for health effects overwhelms all other attributes with the entries staying well above 1. In row P and column C, the cell value of 0.2 indicates corrosion resistance is strongly preferred to price in the ratio of 5:1.

Step 3: (*Evaluate re-scaled pair-wise preference matrix*) A rescaled preference matrix is generated by dividing each column entry in Table 10.10 by that column's sum, yielding Table 10.11. The last column "Average" contains average weights for each row and shows the ranking of the attributes. Table 10.12 shows the ordered relative ranking of the attributes.

TABLE 10.11   Rescaled Pair-Wise Weight Matrix

Attribute	P	C	F	H	L	R	T	Average
P	0.04	0.03	0.01	0.06	0.02	0.02	0.05	0.03
C	0.19	0.15	0.25	0.14	0.10	0.29	0.16	0.18
F	0.15	0.03	0.05	0.07	0.03	0.05	0.04	0.06
H	0.26	0.46	0.30	0.43	0.40	0.39	0.47	0.38
L	0.19	0.15	0.15	0.11	0.10	0.10	0.05	0.12
R	0.07	0.03	0.05	0.05	0.05	0.05	0.08	0.05
T	0.11	0.15	0.20	0.14	0.30	0.10	0.16	0.17
Sum	1.00	1.00	1.00	1.00	1.00	1.00	1.00	1.00

TABLE 10.12   Relative Ranking of Attributes

Attribute	Weight
Health effects	0.38
Corrosion resistance	0.18
Taste and odor	0.17
Longevity	0.12
Fire resistance	0.06
Resale value	0.05
Price	0.03

TABLE 10.13   Material Pair-Wise Matrix and the Associated Rescaled Matrix for the Attribute Price

		Price						
Material	Mat. A	Mat. B	Mat. C		Mat. A	Mat. B	Mat. C	Average
Mat. A	1.000	0.333	2.000	Mat. A	0.222	0.200	0.333	0.252
Mat. B	3.000	1.000	3.000	Mat. B	0.667	0.600	0.500	0.589
Mat. C	0.500	0.333	1.000	Mat. C	0.111	0.200	0.167	0.159
Sum	4.500	1.667	6.000	Sum	1.000	1.000	1.000	1.000

Step 4: (*Evaluate preferences*) Corresponding to each attribute, pair-wise weight matrix and the associated rescaled matrix for three hypothetical pipe materials A, B, and C are obtained. Results for the price attribute are shown as Table 10.13. This procedure is repeated for the other attributes for the three materials. Table 10.14 shows the results. We obtain the final ranking of the three materials by multiplying the pipe material preference matrix given in Table 10.14 and the attribute preference vector given as the last column of Table 10.11. The results are shown in Table 10.15. Mat. C is the most preferred material with a preference score of 0.460, followed by Mat. A with a score of 0.279. Mat. B and Mat. A have rather close scores of 0.261 and 0.279, respectively. A consistency check can be performed following Saaty (1980) as given in Step 5. Participants were advised to reassess the pair-wise weights if the consistency check failed.

Step 5: (*Perform consistency check*) The calculated maximum eigenvalue for the pair-wise weight matrix, $\lambda max = n$. If it is different from n, we have inconsistencies in our weight assignments. Saaty (1980) defines a consistency index as C.I. $= (\lambda max - n)/(n - 1)$. Instead of the eigenvalue, it is possible to manipulate the matrices to get a different form of the consistency index. Let the average ratio AR $= 1/n \sum_{i=1}^{n} (i$th row of [A]{ave W})/(ith row {ave W}) in which [A] = matrix of pair-wise preference weights in Table 10.10, {ave W} = vector of average rescaled weights in the last column of Table 10.11. The newly formed consistency index is C.I. $= [AR - n]/(n - 1)$. If the ratio of C.I. to R.I. (random index given in Table 10.16) is less than 0.1, the weights should be taken as consistent. Table 10.16 contains the random index values calculated from randomly generated weights as a function of the pair-wise matrix size (number of criteria).

Table 10.17 contains the attribute ranking for 10 participants. From Table 10.17, it is seen that the participants rank health effects the highest, followed by taste and odor, corrosion

TABLE 10.14   Average Ranking for the Materials and Attributes

	P	C	F	H	L	R	T
Mat. A	0.252	0.164	0.444	0.328	0.250	0.297	0.250
Mat. B	0.589	0.297	0.111	0.261	0.250	0.164	0.250
Mat. C	0.159	0.539	0.444	0.411	0.500	0.539	0.500
Average	0.03	0.18	0.06	0.38	0.12	0.05	0.17

TABLE 10.15   Final Preference Matrix

Material	Preference
Mat. A	0.279
Mat. B	0.261
Mat. C	0.460

TABLE 10.16   Random Index (R.I.)

Matrix Size	1	2	3	4	5	6	7	8	9	10	11	12	13	14	15
R.I.	0	0	0.58	0.90	1.12	1.24	1.32	1.41	1.45	1.49	1.51	1.48	1.56	1.57	1.59

TABLE 10.17   Participants' Ranking of Attributes

Participant/Attribute	1	2	3	4	5	6	7	8	9	10
P	0.034	0.161	0.025	0.040	0.200	0.063	0.310	0.202	0.048	0.087
C	0.182	0.037	0.107	0.061	0.090	0.091	0.158	0.111	0.112	0.256
F	0.060	0.056	0.078	0.085	0.070	0.119	0.044	0.057	0.188	0.044
H	0.385	0.211	0.395	0.304	0.270	0.450	0.154	0.369	0.188	0.367
L	0.120	0.097	0.101	0.132	0.090	0.120	0.154	0.067	0.112	0.062
R	0.054	0.020	0.027	0.127	0.050	0.098	0.103	0.064	0.033	0.029
T	0.165	0.419	0.267	0.250	0.230	0.059	0.078	0.131	0.318	0.155

P: price, C: corrosion resistance, F: fire retardance, H: health effects, L: longevity, R: resale value of home, and T: taste and odor.

resistance, and longevity; price, resale value, and fire resistance are showing the lower ranks of the list. Participants 2 and 9 rate taste and odor above health. Participant 5 has the closest preferences among price, health, and taste and odor. Participant 7 has the highest preference for price. These results indicate that health, and taste, and odor may be a surrogate for the purity of water that dominates preferences for a plumbing material. The mental stress resulting from frequent leaks is also an issue. While copper may have remained a relatively inert carrier of water, the corrosion leaks have forced consumers to consider alternatives including other materials, the use of corrosion inhibitors such as phosphate, and lining the interior of the pipe.

10.5   Stormwater Management

10.5.1   Introduction

Urban runoff is a carrier of contaminants. As discussed in Section 10.1.1, the total maximum daily load (TMDL) study should not only estimate the maximum permissible load per time but also should allocate it. Runoff contributes to nonpoint source pollution and a distributed pollutant control strategy is needed. The best management practices (BMPs) are both nonstructural measures and structural controls employed to minimize contaminants carried by the surface runoff. The BMPs also help to serve to control the peak flows due to development. Therefore, the decision should be comprised of optimal location and level of treatment at a selected location.

10.5.2   Problem Formulation

Zhen et al. (2004) suggest the following general framework.

$$\text{Minimize total annual cost} = \sum_{i=1}^{I} C(\mathbf{d_i})$$

$$\text{subject to:}\quad L_j \leq L\max_j \quad \text{for } j = 1, 2, \ldots, J$$

$$\mathbf{d_i} \in S_i$$

in which $C(\mathbf{d_i}) = $ annual cost of implementing a BMP at level d_i at location i; $\mathbf{d_i} = $ decision vector pertaining to the dimensions of the BMP; $L_j = $ annual contaminant load at the designated check point j; $I = $ total number of potential BMP locations; $J = $ total number of check points; and $S_i = $ feasible set of BMPs applicable at location i. The decision vector is given by $\{\mathbf{T_i}; \mathbf{H_i}\}$ in which $\mathbf{T_i}$ is the vector of possible detention times compatible with the BMP dimension vector (height) $\mathbf{H_i}$ at location i. Typically, for a choice of the decision vectors the load, L_j at location j will have to be obtained from a simulation model. Zhen et al. (2004) use the scatter search (Glover, 1999) to find near optimal solutions. Kuo et al. (1987) use the complex search (Box, 1965; Reklaitis et al., 1983) to obtain near optimal solutions.

Loganathan and Sherali (1987) consider the probability of failure of a BMP (also see Loganathan et al., 1994). They consider the risk of failure and cost of the BMP as two objectives within a mutiobjective optimization framework. They generate a cutting plane based on pair-wise tradeoff between objective functions provided by the decision maker at each iterate. It is shown that the iterates are on the efficient frontier and any accumulation point generated by the algorithm is a best compromise solution. The method is applicable even when the feasible region is nonconvex. Lee et al. (2005) consider a production function of release rate and storage for a certain percent pollutant removal. They recommend using an evolutionary optimizer for the solution. They also provide a review of progress in optimizing the storm water control systems. Adams and Papa (2000) provide analytical probabilistic models for estimating the performance of the storm water control systems. They include optimization strategies for these systems.

Mujumdar and Vemula (2004) present an optimization-simulation model for waste load allocation. They apply the genetic algorithm along with an appealing constraint handling strategy due to (Koziel and Michalewicz, 1998, 1999). The decoder method proposed by Koziel and Michalewicz (1999) is based on homomorphous mapping between the representation space of $[-1, 1]^n$ n-dimensional cube and the feasible region, X in the decision space. Because the feasible region may be nonconvex, it may have several points of intersection between a line segment $\mathbf{L}$ and the boundary of the feasible region X. The method permits selecting points from the broken feasible parts of the line segment $\mathbf{L}$. To identify these break points, the following procedure is used. A feasible point $\mathbf{r}$ is chosen. A line segment $\mathbf{L} = \mathbf{r} + \alpha\,(\mathbf{s} - \mathbf{r})$ for $0 \leq \alpha \leq 1$, starting with $\mathbf{r}$ and ending with a boundary point $\mathbf{s}$ of the search space S encompassing the feasible region is considered. Such a defined line segment $\mathbf{L}$ involves a single variable α when substituted into each of the constraints, $g_i(x) \leq 0$. The domain of α is partitioned into v subintervals of length $(1/v)$. Each subinterval is explored for the root $g_i(\alpha) = 0$. The subintervals are chosen in such a manner that there can be at most one root within each subinterval. The intersection points between the line segment and the constraints $g_i(x) \leq 0$ can thus be found.

The problem is as follows. A discharger m (source) removes a fractional amount $x(i, m, n)$ of pollutant n to control water quality indicator i at check point l. For discharger m after removing a fractional load $x(i, m, n)$ of pollutant n, the concentration is $c(m, n)$. This input concentration $c(m, n)$ for $m = 1, 2, \ldots, M$ and $n = 1, 2, \ldots, N$ and the effluent flow at m

denoted as $q(m)$ for $m = 1, 2, \ldots, M$ serve as forcing functions in the water quality simulation model QUAL2E (Brown and Barnwell, 1987) to determine the concentration $c(i, l)$ of water quality indicator i (dissolved oxygen) at the check point l. The discharger would like to limit $x(i, m, n)$ and the pollutant controlling agency would like to control $c(i, l)$ for all check points l. The problem is given below:

Maximize Level of satisfaction $= \lambda$

subject to: $[c(i, l) - c_{\mathrm{L}}(i, l)]/[c_{\mathrm{D}}(i, l) - c_{\mathrm{L}}(i, l)]^{a(i, l)} \geq \lambda$

$$[x_{\mathrm{M}}(i, m, n) - x(i, m, n)]/[x_{\mathrm{M}}(i, m, n) - x_{\mathrm{L}}(i, m, n)]^{b(i, m, n)} \geq \lambda$$

$$c_{\mathrm{L}}(i, l) \leq c(i, l) \leq c_{\mathrm{D}}(i, l)$$

$$x_{\mathrm{L}}(i, m, n) \leq x(i, m, n) \leq x_{\mathrm{M}}(i, m, n)$$

$$0 \leq \lambda \leq 1$$

in which $c_{\mathrm{D}}(i, l) = $ desirable concentration level such as the dissolved oxygen, $c_{\mathrm{L}}(i, l) = $ permissible safe level concentration, $x_{\mathrm{M}}(i, m, n) = $ technologically possible maximum pollutant removal, $x_{\mathrm{L}}(i, m, n) = $ minimum pollutant removal, $a(i, l)$ and $b(i, m, n) = $ positive parameters, and $\lambda = $ compromise satisfactory level.

The decision variables $x(i, m, n)$ are determined with the aid of the genetic algorithm. They use QUAL2E and genetic algorithm optimization packages PGAPack (Levine, 1996) and GENOCOP (Koziel and Michalewicz, 1999) to solve the problem. Using the GA generated population of pollutant reduction values $x(i, m, n)$, QUAL2E determines the corresponding concentration of water quality indicator i at check point l, $c(i, l)$. The feasibility is checked by GENOCOP. They point out that the selection of the initial feasible point in constraint handling could affect the final solution. They recommend using the incumbent best solution as the initial point.

10.6 Groundwater Management

10.6.1 Introduction

As rain falls over land, a part of it runs over land and a part of it infiltrates into the ground. There is void space between soil grains underground and water and air occupy that space. Groundwater flows through the tortuous paths between the solid grains. Because of the extremely small diameter, the tortuous nature of the path, and the nature of the hydraulic head, the groundwater velocity is very small. Therefore, if the groundwater storage is mined, it cannot be replenished in a short time. Driscoll (1986) reports that there has been a 400-ft decline in groundwater table in Phoenix, Arizona, in the last 50 years. Perennial rivers carry the base flow resulting from groundwater storage even when there is no rain. In the United States 83.3 billion [$83.3(10^9)$] gallons of fresh groundwater and 1.26 billion gallons of saline groundwater are used, whereas 262 billion gallons of fresh and 61 billion gallons of saline surface water are used (Hutson et al., 2004). Declining groundwater storage and contaminated groundwater are two major problems.

Willis and Yeh (1987), Bear and Verruijt (1987), and Bedient et al. (1999) provide comprehensive details on groundwater flow and contaminant transport modeling. The groundwater problem involves detailed numerical modeling. There is an added complexity because of the complex nature of the underground domain. The groundwater velocity is determined by hydraulic conductivity which is not known deterministically. Chan Hilton and Culver

(2005) offer a comprehensive review of groundwater remediation problem. They also propose a strategy within the framework of simulation-optimization to consider the uncertainty in the hydraulic conductivity field. Following Bedient et al. (1999) the groundwater remediation may involve one of the following: (1) removal of source by excavation, (2) containment of source by barriers and hydraulic control, (3) reducing the mass of the source by pump and treat, bioremediation, soil vapor extraction, and natural attenuation.

Chan Hilton and Culver (2005) propose the following simulation-optimization approach to groundwater remediation problems. The procedure differs from the previous studies in accommodating the uncertainty in the hydraulic conductivity field. There are three ways in which the hydraulic conductivity field is incorporated: (1) Keep the same field throughout the simulation-optimization study. In the genetic algorithm framework all strings are subjected to the same field in every generation. (2) Use multiple fields within each generation and each such realization is applied to all strings for that generation. Calculate the average fitness value over all these realizations for each string. (3) Obtain a hydraulic conductivity field for a generation. Apply the same field to all the strings in that generation. For the next generation, obtain a new hydraulic conductivity field. The fitness value is calculated applicable to that conductivity field. Chan Hilton and Culver use the third type and call it the robust genetic algorithm.

10.6.2   Problem Formulation

The groundwater remediation problem formulation is as follows. There are N sampling points for checking contaminant concentration (computational nodes) and W potential extraction wells. The problem is to optimally decide which extraction wells should be used such that the contaminant concentration, $C(j)$ can be controlled; hydraulic head, $h(j)$, above a minimum can be maintained; extraction flow rates can be set within chosen bounds. The mathematical formulation is

Minimize   Total Cost = Pumping cost + Treatment cost + Well cost

Subject to: $C(j) \leq C\max$ for $j = 1, 2, \ldots, N$

$h(j) \geq h(\min)$ for $j = 1, 2, \ldots, N$

$Q(\min) \leq Q(i) \leq Q(\max)$ for $I = 1, 2, \ldots, W$

in which Pumping cost $= C$pump $(c1/\eta)\, Q(i)\, [H\text{datum} - h(i) + \mu\, \text{Pads}]\, T$dur, Cpump $=$ energy cost, $c1 =$ unit conversion, $\eta =$ pump efficiency, $Q(i) =$ extracted flow at well i, Hdatum $=$ depth to datum, $\mu =$ technological coefficient for adsorber, Pads $=$ pressure required for the adsorber, Tdur $=$ remediation period in appropriate units; Treatment cost $=$ $f[C\text{ave},\ Q\text{tot}] + C\text{ads}\ N\text{ads}$, $Q\text{tot} = \sum_{i=1}^{W} Q(i)$, $C\text{ave} = \sum_{i=1}^{W} \frac{Q(i)}{Q\text{tot}} \left[\frac{C(i,T\text{beg}) + C(i,T\text{end})}{2} \right]$, $C(i,T\text{beg}) =$ concentration of contaminant at well i at the beginning time, Tbeg, $C(i,T\text{end}) =$ concentration of contaminant at well i at the ending time, Tend, Tdur $= T$end $- T$beg, and $C\text{ave} =$ average influent concentration for the adsorber, $Q\text{tot} =$ total flow, $C\text{ads} =$ cost of an adsorber, $N\text{ads} =$ number of adsorbers $=$ ceiling[$Q\text{tot}(T\text{cont})/V\text{pore}$], ceiling(.) $=$ rounds up to nearest multiple of 1, $V\text{pore} =$ pore volume of an adsorber, $T\text{cont} =$ required contact time.

The above stated problem is solved using simulation–optimization approach by the genetic algorithm. The concentration $c(j)$, and head $h(j)$ constraints are incorporated through a multiplicative penalty function as

Fitness = Total cost $* [1 + \text{pen}C\ \text{Viol}C + \text{pen}H\ \text{Viol}H]$

in which pen$C =$ penalty for violation in concentration $C(j)$ and pen$H =$ penalty for violation in head $h(j)$. For the flow $Q(i)$ constraint, it is assumed that if a well is active, it will extract flow at a fixed rate. Therefore, it becomes a binary variable and is incorporated into the decision variable string in the genetic algorithm.

The entire procedure is as follows. (1) *Initialize:* Generate the initial population of strings with age(i) $= 1$ for the ith string. (2) *Selection:* For a realization of the hydraulic conductivity field, obtain the fitness value for each string, Fitness(i). Rank order the strings by their Fitness values with the highest rank assigned to the string with the best Fitness value. Let $r(i)$ be the rank for string i. Define modified rank fitness measure

$$r\text{fit}(i) = \frac{[\min\{\text{age}(i), \text{age}T\} - 1]r\text{prev}(i) + r(i)}{\min\{\text{age}(i), \text{age}T\}} \quad \text{if age}(i) > 1$$

$$r\text{fit}(i) = 0.9 \ r(i) \quad \text{if age}(i) = 1$$

in which age$T$ = chosen value greater than 1 (set at 11), $r\text{prev}(i) = r\text{fit}(i)$ from the previous generation. List the strings by their $r\text{fit}(i)$ values; assign new $r\text{new}(i)$ ranks to these strings. These become $r\text{prev}(i)$ in the next generation. Retain the highest $r\text{new}(i)$ strings as the *elite group*. These are directly passed to the next generation without crossover and mutation. Apply *tournament selection* to the population in which two strings are compared and the string with better rank fitness is retained for crossover. The tournament selection is repeated with replacement until the desired population size is reached. (3) *Crossover:* Crossover is performed between two strings to create a new string. This new string's age(i) $= 0$. If there is no crossover due to specified probability, the first string proceeds into the next generation with its age unchanged. (4) *Mutation:* If mutation causes a new string to attain the same bits as its parent string, its age is reset to the original string's age. If not, its age(i) $= 0$. (5) *New generation:* The strings passed on to the next generation are assigned age(i) $=$ age(i) $+ 1$. (6) *Termination check:* If there is convergence to lower costs, stop. If not, go to Step (2), selection. The robust GA retains strings that perform well over a number of different hydraulic conductivity fields. In the noisy GA, a number hydraulic conductivity fields are used within the same generation and the fitness value is taken as the average over the realizations. The comparison indicates that the robust GA yields comparable results to the noisy GA while using a smaller number of fitness evaluations.

10.7 Summary

In this chapter, selected mathematical programming formulations related to water resources systems are presented. The overarching theme has been to expose certain underlying problem structures such as in reservoir operations and water distribution systems optimization, the need to assess public preference toward material attributes as in plumbing systems, and the necessity to combine process simulation and optimization such as in stormwater and groundwater management problems. The evolutionary optimization techniques provide a good approach. The constraint handling in these techniques is an important issue; guaranteeing the nature of optimality is also an issue. Often, a zero-one decision has to be made. This decision requires consideration of political and public preferences and implications over a long term. Both the analyst and the decision maker are interested in the alternatives and the solution behavior with changing constraints as the result of incorporating regulatory issues and societal preferences.

Acknowledgments

The writer gratefully acknowledges the support provided by the National Science Foundation under the grant DMI-0329474, American Water Works Association Research Foundation,

Weston Solutions, Inc., Virginia Department of Environmental Quality, and Virginia Water Resources Research Center. He also wishes to thank Professors Marc Edwards, Andrea Dietrich, Darrell Bosch, Sharon Dwyer, Mr. Bob Buglass, Dr. Paolo Scardina, Dr. A. K. Deb, Dr. Hwandon Jun, Mr. Juneseok Lee, and Mr. Owais Farooqi.

References

1. Adams, B.J. and Papa, F., _Urban Stormwater Management Planning with Analytical Probabilistic Models_, John Wiley, New York, 2000.
2. Alperovits, E. and Shamir, U., Design of Optimal Water Distribution Systems, _Water Resources Research_, 13, 885–900, 1977.
3. American Society of Civil Engineers, _Civil Engineering Guidelines for Planning and Designing Hydroelectric Developments_, ASCE, New York, 1989.
4. American Society of Civil Engineers, Sustainability Criteria for Water Resources Systems, _Project M-4.3_, WRPM Division (Task Committee Sustainability Criteria) and UNESCO International Hydrological Programme, 1998.
5. Barritt-Flatt, P.E. and Cormie, A.D., Comprehensive Optimization Model for Hydroelectric Reservoir Operations, in Labadie, L.W., et al. (eds.) _Computerized Decision Support System for Water Managers_, ASCE, New York, 1988.
6. Bazaraa, M.S., Sherali, H.D., and Shetty, C.M., _Nonlinear Programming: Theory and Algorithms_, John Wiley, New York, 2006.
7. Bear, J. and Verruijt, A., _Modeling Groundwater Flow and Pollution_, D. Reidel, Boston, 1987.
8. Bedient, P.B., Rifai, H.S., and Newell, C.J., _Groundwater Contamination_, Prentice Hall, Upper Saddle River, 1999.
9. Berry, L., Water: The Emerging Crisis? L. Berry (Issue editor) _Water Resources Update_, Universities Council on Water Resources, 102, 7–9, Winter, 1996.
10. Bhave, P.R., _Analysis of Flow in Water Distribution Networks_, Technomic Publishers, Lancaster, 1991.
11. Bohachevsky, I.O., Johnson, M.E., and Stain, M.L., Generalized Simulated Annealing for Optimization, _Technometrics_, 28(3), 209–217, 1986.
12. Boulos, P.F., Lansey, K.E., and Karney, B.W., _Comprehensive Water Distribution Analysis Handbook for Engineers and Planners_, MWH Soft, Inc., Pasadena, 2004.
13. Box, M.J., A New Method of Constrained Optimization and a Comparison with Other Methods, _Computer Journal_, 8, 42–51, 1965.
14. Brown, L.C. and Barnwell, T.O., Jr., The Enhanced Stream Water Quality Models QUAL2E and QUAL2E-UNCAS: Documentation and User Manual, _Report No. EPA/600/3/87/007_, U.S. Environmental Protection Agency, Athens, GA, 1987.
15. Can, E.K., Houck, M.H., and Toebes, G.H., Optimal Real-time Reservoir Systems Operation: Innovative Objectives and Implementation Problems, _Technical Report 150_, Purdue University Water Resources Research Center, 1982.
16. Chan Hilton, A.B. and Culver, T., Groundwater Remediation Design under Uncertainty Using Genetic Algorithms, _Journal of Water Resources Planning and Management_, 131(1), 25–34, January/February 2005.
17. Changchit, C. and Terrell, M.P., CCGP Model for Multiobjective Reservoir Systems, _Journal of Water Resources Planning and Management_, 115(5), 658–670, 1989.
18. Chiras, D.D., Reganold, J.P., and Owen, O.S., _Natural Resources Conservation_, 8th ed., Prentice Hall, Upper Saddle River, 2002.

19. Deb, A.K. and Sarkar, A.K., Optimization in Design of Hydraulic Network, *Journal of the Sanitary Engineering Division*, 97(SA2), 141–159, 1971.

20. Deb, A.K., Least Cost Pipe Network Derivation, *Water and Water Engineering*, 77, 18–21, 1973.

21. Deb, K., *Multi-Objective Optimization Using Evolutionary Algorithms*, John Wiley, New York, 2001.

22. Dellapenna, J.W., Is Sustainable Development a Serviceable Legal Standard in the Management of Water? *Water Resources Update*, Universities Council on Water Resources, 127, 87–93, February 2004.

23. Delleur, J.W., The Evolution of Urban Hydrology: Past, Present, and Future, *Journal of Hydraulic Engineering*, 129(8), 563–573, August 2003.

24. Dietrich, A. et al., Plumbing Materials: Costs, Impacts on Drinking Water Quality, and Consumer Willingness to Pay, *Proceedings of 2006 NSF Design, Service, and Manufacturing Grantees Conference*, St. Louis, Missouri, 2006.

25. Driscoll, F.G., *Groundwater and Wells*, Johnson Filtration Systems, St. Paul, MN, 1986.

26. Eschenbach, E., Magee, T., Zagona, E., Goranflo, M., and Shane, R., Goal Programming Decision Support System for Multiobjective Operation of Reservoir Systems, *Journal of Water Resources Planning and Management*, 127(2), 108–120, March/April 2001.

27. Ellis, J.H. and ReVelle, C.S., A Separable Linear Algorithm for Hydropower Optimization, *Water Resources Bulletin*, 24(2), 435–447, 1988.

28. Feldman, A.D., HEC Models for Water Resources System Simulation: Theory and Experience, *Advances in Hydroscience*, 12, 297–423, 1981.

29. Gessler, J., Optimization of Pipe Networks, *International Symposium on Urban Hydrology, Hydraulics and Sediment Control*, University of Kentucky, Lexington, KY, pp. 165–171, 1982.

30. Glover, F., Scatter Search and Path Relinking, *HCES-01-99, Working paper series*, Hearin Center for Enterprise Science, The University of Mississippi, 1999.

31. Goldman, F.E. and Mays, L.W., Water Distribution System Operation: Application of Simulated Annealing, in Mays, L.W. (ed.) *Water Resource Systems Management Tools*, McGraw Hill, New York, 2005.

32. Heintz, T.H., Applying the Concept of Sustainability to Water Resources Management, *Water Resources Update*, Universities Council on Water Resources, 127, 6–10, February 2004.

33. Hutson, S.S., Barber, N.L., Kenny, J.F., Linsey, K.S., Lumia, D.S., and Maupin, M., Estimated Use of Water in the United States in 2000, *U.S. Geological Survey Circular 1268*, U.S. Geological Survey, Reston, VA, 2004.

34. Jain, S.K. and Singh, V.P., *Water Resources Systems Planning and Management*, Elsevier, Boston, 2003.

35. Karamouz, M., Szidarovszky, F., and Zahraie, B., *Water Resources Systems Analysis*, Lewis Publishers, Boca Raton, FL, 2003.

36. Karmeli, D., Gadish, Y., and Myers, S., Design of Optimal Water Distribution Networks, *Journal of Pipeline Division*, 94, 1–10, 1968.

37. Koziel, S. and Michalewicz, Z., A Decoder-Based Evolutionary Algorithm for Constrained Parameter Optimization Problems, *Proceedings of the 5 Parallel Problem Solving from Nature*, T. Back, A.E. Eiben, M. Schoenauer, and H.-P. Schwefel (eds.), Amsterdam, September 27–30, Lecture Notes in Computer Science, Springer-Verlag, New York, 231–240, 1998.

38. Koziel, S. and Michalewicz, Z., Evolutionary Algorithms, Homomorphous Mappings, and Constrained Parameter Optimization, *Evolutionary Computation*, 7(1), 19–44, 1999.

39. Kranz, R., Gasteyer, S.P., Heintz, T., Shafer, R., and Steinman, A., Conceptual Foundations for the Sustainable Water Resources Roundtable, *Water Resources Update*, Universities Council on Water Resources, 127, 11–19, February 2004.

40. Kuo, C.Y., Loganathan, G.V., Cox, W.E., Shrestha, S., and Ying, K.J., *Effectiveness of BMPs for Stormwater Management in Urbanized Watersheds*, Bulletin 159, Virginia Water Resources Research Center, 1987.

41. Labadie, J.W., Optimal Operation of Multireservoir Systems: State of the Art Review, *Journal of Water Resources Planning and Management*, 130(2), 93–111, March/April 2004.

42. Lee, J. and Loganathan, G.V., Consumer Preferences Towards Plumbing Material Attributes, *Virginia Water Central*, Virginia Water Resources Research Center, August 2006.

43. Lee, J.G., Heaney, J.P., and Lai, F.-H., Optimization of Integrated Urban Wet-Weather Control Strategies, *Journal of Water Resources Planning and Management*, 131(4), 307–315, July/August 2005.

44. Levine, D., User's guide to the PGAPack parallel genetic algorithm library, Tech. Rep. ANL-95/18, Argonne National Lab, 1996.

45. Loganathan, G.V., Optimal Operating Policy for Reservoir Systems, in Kenneth D. Lawrence and Gary R. Reeves (eds.), *Applications of Management Science: Engineering Applications*, Vol. 9, JAI Press, Greenwich, CT, 1996.

46. Loganathan, G.V. and Bhattacharya, D., Goal-Programming Techniques for Optimal Reservoir Operations, *Journal of Water Resources Planning and Management*, 116(6), 820–838, 1990.

47. Loganathan, G.V., Greene, J.J., and Ahn, T., A Design Heuristic for Globally Minimum Cost Water Distribution Systems, *Journal of Water Resources Planning and Management*, 121(2), 182–192, 1995.

48. Loganathan, G.V., Sherali, H.D., and Shah, M.P., A Two-Phase Network Design Heuristic for the Minimum Cost Water Distribution Systems under Reliability Constraints, *Engineering Optimization*, 15, 311–336, 1990.

49. Loganathan, G.V. and Lee, J., Decision Tool for Optimal Replacement of Plumbing Systems, *Civil Engineering and Environmental Systems*, 22(4), 189–204, December 2005.

50. Loganathan, G.V. and Sherali, H.D., A Convergent Cutting-Plane Algorithm for Multiobjective Optimization, *Operations Research*, 35(3), 365–377, May–June 1987.

51. Loganathan, G.V., Watkins, E.W., and Kibler, D.F., Sizing Stormwater Detention Basins for Pollutant Removal, *Journal of Environmental Engineering*, 120(6), 1380–1399, 1994.

52. Lohani, V.K. and Loganathan, G.V., An Early Warning System for Drought Management Using the Palmer Drought Index, *Water Resources Bulletin*, 33(6), 1375–1386, 1997.

53. Loucks, D.P. and Gladwell, J., *Sustainability Criteria for Water Resources Systems*, Cambridge University Press, Cambridge, 1999.

54. Loucks, D.P., Quantifying System Sustainability Using Multiple Risk Criteria, in Bogardi, J.J., and Kundzewicz, Z. (eds.), *Risk, Reliability, Uncertainty, and Robustness of Water Resources Systems*, Cambridge University Press/UNESCO, Cambridge, 2002.

55. Loucks, D.P., Stedinger, J.R., and Haith, D.A., *Water Resources Systems Planning and Analysis*, Prentice-Hall, Englewood Cliffs, 1981.

56. Marshutz, S., Hooked on Copper, *Reeves Journal*, March 2001.

57. Martin, Q.W., Optimal Daily Operation of Surface-Water Systems, *Journal of Water Resources Planning and Management*, 113(4), 453–470, 1987.

58. Martin, Q.W., Optimal Reservoir Control for Hydropower on Colorado River, Texas, *Journal of Water Resources Planning and Management*, 121(6), 438–4446, 1995.

59. Material Performance (Supplement), *Cost of Corrosion Study Unveiled*, July 2002.

60. Mattice, J.S., Ecological Effects of Hydropower Facilities, in Gulliver, J.S. and Arndt, R.E.A. (eds.), *Hydropower Engineering Handbook*, McGraw Hill, New York, 1991.

61. Mays, L.W. and Tung, Y.K., *Hydrosystems Engineering and Management*, McGraw Hill, New York, 1992.

62. Mays, L.W., *Water Distribution Systems Handbook*, McGraw Hill, New York, 2000.

63. McMahon, G.F. and Farmer, M.C., Reallocation of Federal Multipurpose Reservoirs: Principles, Policy, and Practice, *Journal of Water Resources Planning and Management*, 130(3), 187–197, July/August 2004.

64. Moore, C. and Loganathan, G.V., Optimizing Hydropower Operation in a Pumped-Storage System, *ASCE Environmental and Water Resources Institute (EWRI) Annual Conference*, May 19–22, 2002, Roanoke, Virginia.

65. Mujumdar, P.P. and Vemula, V.R.S., Fuzzy Waste Load Allocation Model: Simulation-Optimization Approach, *Journal of Computing in Civil Engineering*, 18(2), 120–131, April 2004.

66. National Research Council, *New Directions in Water Resources: Planning for the U.S. Army Corps of Engineers*, National Academy Press, Washington, DC, 1999.

67. Palacios-Gomez, Lasdon, L., and Engquist, M., Nonlinear Optimization by Successive Linear Programming, *Management Science*, 28(10), 1106–1120, 1982.

68. Pardalos, P.M., Glick, J.H., and Rosen, J.B., Global Minimization of Indefinite Quadratic Problems, *Computing*, 39, 281–291, 1987.

69. Reklaitis, G.V., Ravindran, A., and Ragsdell, K.M., *Engineering Optimization*, John Wiley, New York, 1983.

70. ReVelle, C.S., Joeres, E., and Kirby, W., The Linear Decision Rule in Reservoir Management and Design 1: Development of the Stochastic Model, *Water Resources Research*, 5(4), 767–777, 1969.

71. Rinnooy Kan, A.H.G. and Timmer, G.T., Global Optimization, Chapter IX in G.L. Nemhauser et al. (eds.), *Handbooks in OR & MS*, Vol. I, Elsevier Science Publishers B.V. (North Holland), New York, 1989.

72. Saaty, T.L. *The Analytic Hierarchy Process*, McGraw-Hill, New York, 1980.

73. Saaty, T.L. How to Make a Decision: The Analytic Hierarchy Process, *European Journal of Operational Research*, 48, 9–26, 1990.

74. Savic, D.A. and Walters, G.A., Genetic Algorithms for Least-Cost Design of water Distribution Networks, *Journal of Water Resources Planning and Management*, 123(2), 67–77, 1997.

75. Shih, J.S. and ReVelle, C.S., Water Supply Operations During Drought: A Discrete Hedging Rule, *European Journal of Operational Research*, 82, 163–17, 1995.

76. Smith, E.T., Water Resources Criteria and Indicators, *Water Resources Update*, Universities Council on Water Resources, 127, 59–67, February 2004.

77. Smith, E.T. and Lant, C., Water Resources Sustainability, *Water Resources Update*, Universities Council on Water Resources, 127, 1–5, February 2004.

78. Spall. J.C., *Introduction to Stochastic Search and Optimization: Estimation, Simulation, and Control*, John Wiley, New York, 2003.

79. Tao, T. and Lennox, W.C., Reservoir Operations by Successive Linear Programming, *Journal of Water Resources Planning and Management*, 117(2), 274–280, 1991.

80. Templeman, A.B., Discussion of Optimization of Looped Water Distribution Systems by Quindry et al., *Journal of the Environmental Engineering Division*, 108(EE3), 599–602, 1982.

81. Torn, A. and Zilinkas, A., *Global Optimization*, Lecture Notes in Computer Science 350, Springer-Velag, New York, 1987.

82. U.S. Army Corps of Engineers, *HEC-5: Simulation of Flood Control and Conservation Systems*, User's Manual, Hydrologic Engineering Center, Davis, CA, 1982.

83. U.S. Army Corps of Engineers, *HEC-ResSim: Reservoir System Simulation*, User's Manual, Hydrologic Engineering Center, Davis, CA, 2003.

84. U.S. Dept. of Interior, *Standing Operating Procedures Guide for Dams and Reservoirs*, U.S. Government Printing Office, Denver, CO, 1985.

85. Walski, T., *Analysis of Water Distribution Systems*, Van Nostrand Reinhold Company, New York, 1984.

86. Warnick, C.C., Mayo, H.A., Carson, J.L., and Sheldon, L.H., *Hydropower Engineering*, Prentice-Hall, Englewood Cliffs, 1984.

87. Wylie, E.B. and Streeter, V.L., *Fluid Mechanics*, McGraw Hill, New York, 1985.

88. Willis, R. and Yeh, W.W., *Ground Water Systems Planning and Management*, Prentice Hall, Englewood Cliffs, 1987.

89. World Commission on Environment and Development, *Our Common Future*, Oxford University Press, Oxford, 1987.

90. Wunderlich, W.O., System Planning and Operation, in Gulliver, J.S. and Arndt, R.E.A. (eds.), *Hydropower Engineering Handbook*, McGraw Hill, New York, 1991.

91. Wurbs, R.A., Optimization of Multiple Purpose Reservoir Operations: A Review of Modeling and Analysis Approaches, *Research Document No. 34*, U.S. Army Corps of Engineers, Hydrologic Engineering Center, Davis, CA, 1991.

92. Yeh, W.W.-G., Reservoir Management and Operations Models: A State of the Art Review, *Water Resources Research*, 21(12), 1797–1818, 1985.

93. Younos, T., *Water Research Needs in Virginia*, Virginia Water Resources Research Center, 2005. Available at http://www.vwrrc.vt.edu/publications/recent.htm (accessed in April 2006).

94. Zhen, X.-Y., Yu, S.L., and Lin, J.-Y., Optimal Location and Sizing of Stormwater Basins at Watershed Scale, *Journal of Water Resources Planning and Management*, 130(4), 339–347, July/August 2004.

11

Military Applications

Jeffery D. Weir and
Marlin U. Thomas
*Air Force Institute of Technology**

11.1 Introduction

Operations Research (OR) has evolved through the need for analytical and analysis methods for dealing with decisions involving the use and utilization of critical resources during military operations. One could argue that OR has been around as long as standing armies and navies. History is replete with examples of military "geniuses" who used mathematics and science to outwit and overcome their enemy. In more recent history, one need only look at Thomas Edison helping the United States Navy overcome the stifling effect of German U-boats during World War I or Lanchester's modeling of attrition rates to see OR being applied to the military. It was not until the late 1930s, however, that OR was officially recognized as a scientific discipline that could help the military. At that time, British scientists were called upon to apply their scientific methods not to improving the technical aspects of the recent discoveries in Radio Detection and Ranging (RADAR) but rather the best way to use it operationally within the military. The field of OR grew rapidly from this point and many of the advances in the field can be directly tied to their applications in the military. Evidence of this can be found in the articles that appeared in the early issues of OR, documenting the interest and applications of OR methods following World War II and throughout the three decades that followed. It is easy to argue that military OR is the beginning and actual foundation of the OR profession which has since contributed to successful efforts in gaining and sustaining world peace, and further established the foundation for the field of OR and management science as it is known today.

The purpose of this chapter is to introduce and provide an overview of military OR. We will start in Section 11.2 with a brief background on its history. There is a fairly rich literature on the field including some excellent surveys that have appeared in several books

---

* The views expressed in this chapter are those of the authors and do not reflect the official policy or position of the United States Air Force, the Department of Defense, or the United States Government.

during the past 10 years. Of particular note here is the chapter by Washburn (1994). He gives an excellent summary of the measures of effectiveness, game theory applications, and combat modeling applied in wartime applications. Our focus will be on the methodological aspects of military OR. In Section 11.3, we discuss current military applications. Although the first known applications of OR might be considered the application of experimental designs for establishing optimal machine gun firing methods during World War II by the British, the initial methodological developments were through mathematical programming (Dantzig, 1963). In Section 11.3, these and other methods and applications are summarized. In Section 11.4, we provide some discussion on the current trends that are prevalent in military OR.

11.2 Background on Military OR

Before and during World War II the military recognized the value of OR and began setting up OR groups in the different branches. The British were the first to set up these groups and in fact by 1941 they had OR groups in all three branches of their military: Army, Air Force, and Navy. P. M. S. Blackett, credited by most as the father of operations research, stood up the British Army's first group, which as mentioned earlier dealt with air defense and the best use of radar. Blackett also stood up the British Navy's first OR section at their Costal Command. The naval groups were concerned with U-boats, as they were in World War I, but also started looking at convoy operations and protection of ships from aircraft. The Air Force OR groups in Britain were more focused on air-to-air and bombing tactics. About this same time, OR was growing in the United States military.

Several incidents have been credited with transferring OR from Great Britain to the United States. In 1940, Henry Tizzard from Britain's Committee for the Scientific Study on Air Defense traveled to the United States to share some of the secret developments Great Britain had made in sonar, radar, and atomic energy. Although there was no direct mention of OR in this technical interchange, there was mention of assigning scientists to the various commands within the United States. Captain Wilder D. Baker, in charge of antisubmarine studies for the Atlantic Fleet, was influenced by "Scientists at the Operational Level," a paper written by P. M. S. Blackett, and asked John Tate of the National Defense Research Committee to provide some scientists for his studies. It was at this time that Phillip M. Morse was brought into operations research. Finally, the commander of the United States' Eighth Bomber Command requested scientists be assigned to his command after conferring with B. G. Dickens of the ORS Bomber Command of Great Britain. This led to the formation of what became known as operations analysis in the U.S. Army Air Forces. How much these events actually contributed to the formation of OR units in the United States is debatable, but shortly after its entry into World War II, the U.S. Navy had two OR groups and the Army Air Forces had its first of many (McCloskey, 1987).

Before and during World War II, operations researchers proved the value of applying the scientific method to the day-to-day operations of the military and not just to weapon system development and strategy. The exploits of these early researchers are chronicled in many articles and books, too numerous to name (see, for example, Koopman's (1980) *Search and Screening*, E. S. Quade's (1964) *Analysis for Military Decisions*, and any issue of *Operations Research* dated in the 1950s). There is one, however, that can be used to demonstrate how OR modeling during this time was quickly outpacing our ability to solve many of the mathematical formulations and another that can be used to highlight the differences between the beginnings of OR and how it is practiced today.

Saul Gass in his article "The First Linear-Programming Shoppe" (2002a) details how the development of the computer industry and OR were linked. Of importance to this chapter, however, is how that link came about. In 1947, Project SCOOP (Scientific Computation Of Optimal Programs) was formed by the U.S. Air Force. The members of Project SCOOP included at some time during its eight-year existence, George B. Dantzig, Saul I. Gass, Murray Geisler, Leon Goldstein, Walter Jacobs, Julian Holley, George O'Brien, Alex Orden, Thomas L. Saaty, Emil D. Schell, Philip Wolfe, and Marshall K. Wood. Its main objective was to develop better answers to the problem of programming Air Force requirements. Wood and Dantzig defined programming as "the construction of a schedule of actions by means of which an economy, organization, or other complex of activities may move from one defined state to another, or from a defined state toward some specifically defined objective" (Wood and Dantzig 1949, p. 15).

This definition led Dantzig to name his rectangular optimization model linear programming (LP). In addition to research into LP and the simplex method, Project SCOOP researched inequality systems by the relaxation method and zero-sum two-person games by the fictitious play method. Project Scoop was the impetus for the U.S. National Bureau of Standards contracting the first Standards Eastern Automated Computer (SEAC) and purchasing the second production unit of the UNIVAC. During its eight-year span, Project SCOOP went from solving a 9×77 Stigler's diet problem in 120 person-days on a hand-operated desk calculator to solving an 18×34 gasoline blending LP in 20 minutes on an IBM 701. While Project SCOOP and other works were outpacing technology, there was also a shift beginning to happen in military OR (Gass, 2002).

After World War II many of the scientists that had helped the war effort were returning to their original lives in academia and were being offered jobs in industry. Businesses, seeing the success of OR in the military, began using these techniques to improve their own operations and the field of management science was beginning to evolve. In 1951, Morse and Kimball published *Methods of Operations Research*, originally written in 1946 as a classified report for the U.S. Navy's Operations Evaluation Group, in hopes to share information with the scientific population at large about the successes of OR before and during World War II. Morse and Kimball saw that many of the successes in OR were not widely known either due to classification or just a general lack of free flowing information. Their hope was that by compiling this information the field of OR could be expanded in the military and be applied to nonmilitary operations as well. Of note in this book is how Morse and his peers performed OR during World War II.

In the early stages of OR in the military, practitioners were focused on improving tactics and the effective use of newly invented equipment developed by scientists to enable the military to counter or overcome enemy techniques and equipment. Military analysts used close observation and mathematical modeling of processes and their salient properties to improve desired outcomes. After the war, the size of the nation's militaries was decreasing both because the global war was over and many of the tactics had been so improved that smaller forces could have the same effect as larger ones. Beasley in his OR notes discusses how the probability to attack and kill a U-boat went from 2% to 3% in 1941 to over 40% by 1945. Bothers (1954) explains how the Eighth Air Force used operations analysis to increase the percentage of bombs that fell within 1000 feet of their aimpoint from less than 15% in 1942 to over 60% by 1944.

11.3   Current Military Applications of OR

The rest of this chapter will outline selected general application areas for OR within the military. The goal is to describe how military organizations use common OR techniques and to provide some specific examples where it is useful. It is not a complete review of current literature, as that would require an entire book. On their Web site, http://www.mors.org, the Military Operations Research Society of the United States has a list of several excellent references regarding OR and the military. When talking about the application of OR to military problems, it is useful to look at what level the tools or techniques are being applied. These levels are similar to those of businesses. At the topmost level are strategic (corporate) decisions that are made for the good of a nation. The bottom level is comprised of tactical decisions. These decisions influence how small units or individuals will engage for short durations. Linking these levels is the operational level. At this level, decisions are made to achieve the strategic goals. It determines the when, where, and for what purpose forces will be employed. While each level of war is unique, many of the OR tools can be applied at more than one of these levels.

11.3.1 Linear and Integer Programming

Linear and integer programming are used throughout the military in a variety of ways. Most of their uses are at the strategic and operational levels. At the operational level, LPs and integer programs (IP) have been used to allocate operating room space, route unmanned air vehicles, and even to optimize rehabilitation and restoration of damaged land at training facilities. Its most frequent use, however, is at the strategic level in the area that it was originally designed by Dantzig to help, that being the planning of large-scale operations.

Often called campaigns, these operations look at possible future scenarios for major wars, that is, country on country conflicts. These models provide decision-makers with an analytical tool for determining the impacts of budget changes, force structures, target allocations, munitions inventories, and combat attrition rates on a nation's war-fighting capability. Typically, the scenarios analyzed use all of a nation's military forces, air, ground, and sea, and are modeled over a large time-period, for example, months or years. These large-scale operations can also be humanitarian in nature such as natural disaster relief. Typically, the decision variables used reflect combinations of platform, weapon, target, profile, loadout, and weather. A platform refers to the type of system delivering the weapon, for example, a tank, aircraft, or naval vessel. The weapon refers to what type or types of munitions or aid are being delivered, for example, food pallets, large equipment, flares, bombs, guided munitions, or torpedoes. The profile refers to the manner in which the platform will attack a target or deliver its cargo. For aircraft, this might include altitude, angle of attack, or landing area. The loadout refers to the number of weapons used against or delivered to the target. A submarine might use a volley of two torpedoes against a single target, or a large cargo aircraft might deliver bags of grain and a truck to haul them to a different location. Finally, target and weather are self-explanatory and refer to the target to be attacked or location of aid delivery and the weather expected at the target area. Associated with each of the variables are several parameters.

It is beyond the scope of this chapter to outline all of the parameters that could be associated with these decision variables, but a few bear mentioning as they will be mentioned later in this chapter. With respect to war-like operations, one parameter of interest is how much damage to the target can be expected from choosing the decision variable. This is referred to as damage expectancy (DE) or probability of kill (PK) and takes into account all of the characteristics used to describe the decision variable as well as those of the target. Another is the probability that the platform making the attack described by the variable will survive and be available for subsequent attacks. The complement of this is called the attrition rate. How these are modeled is discussed later in this chapter and is usually done at the tactical level.

11.3.2 Network Optimization

Most of the applications for network optimization within the military are no different from those of major corporations or small cities. Like corporations, they are concerned with building reliable communication networks to include local area networks (LANs) and wide area networks (WANs), logistics networks for supply chain management, and the exploitation of network structures and formulations for LPs and IPs. As the military must build and maintain self-sufficient bases, like municipalities they work with electrical grids, water and sewer supply, and other public utility networks. There are, however, some aspects of network optimization that are unique to the military. One area of research that is unique to the military (and possibly law enforcement) deals with the interdiction of an enemy's network.

When dealing with the interdiction of an enemy's network, the military considers both physical networks, such as command and control networks, and more esoteric networks such as clandestine and terrorist networks. Regardless, the goal for these types of network operations is the same, disruption. Many of the techniques used in optimizing networks can also be used to identify vulnerabilities within networks. The most obvious of these techniques is the max-flow min-cut. By modeling the max flow of an enemy's network, bottlenecks can be identified for potential exploitation. Besides looking at arcs, one might also be interested in the nodes of a network. In this instance a facility location problem may be useful to identify key nodes that are in contact with many other nodes or the use of graph theoretic centrality measures might be useful to exploit.

11.3.3 Multiple Criteria Decision-Making

Like nonlinear programming, multicriteria decisionmaking within the military has followed from LP and IP development. While not the only techniques for multicriteria decisionmaking, goal programming and multiple objective linear programming are used in conjunction with large-scale campaign models to account for various objectives. Two examples of these techniques follow from the U.S. Strategic Command's Weapon Allocation Model (WAM) and the U.S. Air Force Space Command's (AFSPC) Aerospace Integrated Investment Study (ASIIS).

WAM is a large-scale integer goal-programming model. It allocates weapons to targets. Several goals need to be met by the allocation. These include maximizing damage expectancy, ensuring specific weapons cover specific targets, and goals for damage expectancy within varying target categories. These goals are user defined and can be specified either as a goal to be obtained or as a hard constraint. Once the user identifies the goals, they are given priority and the model uses preemptive goal programming. That is, the model runs with the highest priority goal remaining as the objective function and if the goal is obtained, that goal is added to the model as a hard constraint. If a goal cannot be achieved it will be adjusted to the level of best achievement and added to the model as a constraint. This technique was chosen over weighting the priorities because there was no single decision maker from whom to solicit weights. ASIIS does support a single decision maker and therefore uses a weighted goal programming approach.

AFSPC is responsible for training and equipping the U.S. space force. They use the ASIIS model to aid in making decisions with regard to how to balance their spending on space assets and maintain a capable force. ASIIS is a large-scale integer program that uses *elastic* constraints (Brown et al., 2003) and has a linear objective function that penalizes deviations in the *elastic* constraints. This is a long-range planning tool that looks at an epoch of time, usually 25–30 years. There are four *elastic* constraints or goals in ASIIS: (1) do not exceed the authorized funding levels in any given year, (2) do not exceed total funding over the epoch, (3) minimize capability gaps, and (4) minimize underspending.

The first goal ensures that AFSPC does not spend more money on its investments than is authorized by the government in any given year. The second goal ensures that if in any year more money is needed, that money can come from a future or past year in the epoch. The third goal tries to purchase or fund those programs that will ensure that AFSPC can meet its current and future mission requirements with as little capability gap as possible based on the money and technology available. The fourth goal was added to the model to let the model choose redundant systems for better capability coverage. This was needed due to the way the capability gaps are calculated in the third goal. The ASIIS model uses absolute deviations from the goals and therefore, weights for each goal both prioritize the goals and ensure that they are of equal scale. For example, the units for money and capability gaps are several orders of magnitude different. As there is a single decision maker in charge of AFPSC, these weights can be solicited directly from the decision maker to ensure their priorities are met.

11.3.4   Decision Analysis

There has been much debate in the last 5 years as to what is decision analysis (DA). Keefer et al. (2002) define DA as a normative, not descriptive, systematic quantitative approach to making better decisions. Specifically, for an application to be considered DA, it must "explicitly analyze alternatives for a decision problem using judgmental probabilities and/or subjectively assessed utility/value functions." This definition excludes multicriteria decisionmaking and the analytic hierarchy process. In the military, DA applications fall into two categories: decision trees and value focused thinking (VFT).

VFT has been used extensively in the military to help senior decision makers gain insight into decisions that have competing objectives. Of the ten articles summarized by Keefer et al. in their survey of DA literature from 1990 to 2001, eight used some form of VFT. The models described in these articles were usually very large in size and scope. The largest effort described was the U.S. Air Force's 2025 (Jackson et al., 1997; Parnell et al., 1998) study that included over 200 subject matter experts working over a year to develop a 134-attribute model to score 43 future concepts using weights based on six different scenarios. Not all VFT models are at the strategic level, however. The Croatian Army (Barković and Peharda, 2004) used a VFT model to determine what rifle to acquire to replace its aging stores and a VFT model was developed to help U.S. psychological operations (PSYOP) detachment commanders choose the best strategy for the use of PSYOP, where PSYOPs are defined as operations that use selected information to influence an audience's behavior (Kerchner et al., 2001). Along with the VFT model, the military also use decision trees.

The application of decision trees is most useful in the military when there is a single objective to consider. When faced with a variety of options, the decision maker can then choose the best option based on the expected value of the objective. An example of this in the military was the use of a decision tree to choose the best allocation of aircraft to targets by Griggs et al. (1997). In this example, an IP was used to build the allocations with varying weather states and the plan with the best expected value was found by solving a decision tree with the IP solutions at the varying weather states.

11.3.5   Stochastic Models

One of the earliest quantitative approaches to combat modeling was by F. W. Lanchester (1916) who proposed the study of the dynamics of two opposing forces, say Red versus Blue, through a set of differential equations. Let $R(t)$ and $B(t)$ be the number of combatants remaining for each force at a time $t > 0$ and assume that:

$$\frac{\mathrm{d}}{\mathrm{d}t} R(t) = -bB(t), \qquad R(t) > 0$$

$$\frac{\mathrm{d}}{\mathrm{d}t} B(t) = -aR(t), \qquad B(t) > 0$$

where a and b are constants representing the respective lethality rates of kill against each other. It can be shown from these equations that

$$a[R(0)]^2 > b[B(0)]^2 \Rightarrow \text{Red wins}$$

$$b[B(0)]^2 > a[R(0)]^2 \Rightarrow \text{Blue wins}$$

implying that the force with the greatest strength will first deplete the number of combatants of the weaker force. Strength is given by the product of the attrition rate and the square of the initial force. This has become known as the *Lanchester square law*. A simpler linear law follows by assuming that the attrition rates are directly proportional to the number of opposing units. Many variations of the attrition conditions for this model have been studied including that by Brackney (1959) and probabilistic extensions by Morse and Kimball (1951) and further work by Smith (1965), Kisi and Hirose (1966), and Jain and Nagabhushanam (1974).

The earlier studies of the Lanchester models focused on the development of closed form results for force survival probabilities and expected duration of combat. A complete treatment of these models can be found in Taylor (1980, 1983). Bhat (1984) formulated a continuous parameter Markov chain model for this force-on-force combat scenario. We assume the two forces to have given fixed initial levels $R(0) = R$ and $B(0) = B$. The states of the process are outcomes of the 2-tuple $\langle R(t), B(t) \rangle$ with transitions among the finite number of states satisfying the Markov property. For example, for $R = 2$ and $B = 3$ the state space is

$$\Omega = \{(2,3), (2,2), (2,1), (2,0), (1,3), (1,2), (1,1), (1,0)\}$$

The survival probabilities can be derived from the Kolmogorov equations and for a Red win with j survivors it can be shown that,

$$P_{i,0} = b_{i,1} \int_0^\infty P_{i,1}(t)\mathrm{d}t$$

for the duration of combat, T

$$P(T > t) = \sum_{i,j>0} \sum P_{i,j}(t)$$

and

$$E[T] = \int_0^\infty P(T > t)\mathrm{d}t.$$

Another area of important military applications of OR is in search and detection. The basic problem is to try to locate a hidden object in a region A through examination of an area $a \in A$. This problem appears in many military settings; for example, A could be a region of water for antisubmarine operations, a weapon site in a region of desert, or a failed component in a piece of equipment. Various solution approaches have been taken, the earliest, of course, being an exhaustive search which, of course, has limitations due to costs for searching, that is, in time, costs, or other effort, and the criticality of the objective. Another popular method applied during World War II is the random search that can be formulated as a nonlinear programming problem. Let p_i denote the probability of a target being in $A_i \subseteq A$, $i = 1, \ldots, n$, and B is the amount of available search effort. The problem is then to find a solution to

$$\text{min:} \qquad \sum_{i=1}^{n} p_i e^{-b_i/A_i}$$

$$\text{subject to:} \quad \sum_{i=1}^{n} b_i = B$$

More information on methods and approaches for dealing with search and detection can be found in Washburn (2002).

11.3.6 Simulation

Simulation has become one of the most widely used OR techniques in the military. Many of the processes that were modeled with mathematical programming or that had difficult to find closed formed solutions using probabilistic models are now being modeled with simulations. Today's more powerful computers allow practitioners to better model the chance outcomes associated with many real world events and help describe how systems with many such events react to various inputs. Within the military, Monte Carlo, discrete event, and agent-based simulations are being used at all levels.

The most recognizable type of simulation in the military is probably the "war game." The "war game" gets its fame from movie and television, but actually has the same purpose as mathematical programming at the strategic level. This type of discrete event simulation allows decision makers to try out different strategies, force structures, and distributions of lethality parameters and see their affect on the outcome of a given scenario. Unlike their deterministic counterparts, these simulations show decision makers a distribution on the output parameters of interest and not just the most likely. This provides senior military officials insight into the inherent risks associated with war. Like their deterministic counterparts, these models too are aggregated at the strategic level, but can also be used at the operational and tactical level when disaggregated.

While senior military officials do not need to know what every soldier, sailor, or airman is doing on the battlefield, there are leaders that do. At the operational level, simulations focus more on small groups and allow commanders to dry run an operation. This type of simulation is not used for planning purposes necessarily, but to provide insight to commanders so that they can see how different decisions might play out on the battlefield. By identifying decisions that will likely result in bad outcomes, commanders can design new techniques, tactics, and procedures to mitigate those bad outcomes. These same types of simulations when run with even smaller groups can be useful at the tactical level.

At the tactical level, leaders of small units can rehearse actions, trying out different strategies for a single scenario. An army unit might simulate patrolling a village and try different tactics for repelling an ambush. Airmen can try several different approaches for bombing a target and sailors for enforcing an embargo. In these cases, the simulation acts as a training tool allowing the military to gain experience in what can happen during an operation without the loss of life and with substantial monetary savings when compared to live exercises. There is more value to these simulations thought than just training.

As is discussed in the section on LPs and IPs, many models require parameter estimates. As the funding of militaries decreases and the cost of weapons increases, simulation can help to estimate these parameters. During World War II, OR sections within bombing units would gather data on thousands of bomb drops to try and estimate the effectiveness of a particular tactic and try to find ways to improve them. Today's militaries do not have thousands of bomb drops to gather data and have turned to simulation for help. With more powerful computers, simulations that mirror reality better can be built at the individual weapon level. These simulations model the physics behind the weapon and its delivery vehicle allowing researchers to generate the thousands of "bomb drops" they cannot afford. Like their predecessors, these researchers are able to uncover problems with tactics as well as design flaws and produce better effects. These simulations can also help militaries learn about their enemies.

The same simulations that improve one's own tactics and weapons can also be used to model an enemy's weapons. These simulations can be used to provide the distributions used in simulations at the tactical level or higher. For example, when an aircraft is approaching a guarded target, a simulation can be used to build a distribution on the probability that the aircraft will survive the attack. By using engineering data and modeling the known physics of the system guarding the target, the simulation provides both a training tool to try to maximize the likelihood of surviving a mission, and a parameter estimate of the enemy system that can be used in larger simulations. While all of these types of simulation are discrete event, they are not the only type of simulation being used by the military today.

Agent-based simulations (ABS) are the newest simulation tool for the military. Originally developed by those who study the evolution of (complex) systems, such as biologists, AI researchers, sociologists, and psychologists, agent-based simulation is used to predict future events as well as past processes. It is not a computational approach to rational action such as game theory, but rather attempts to model autonomous and heterogeneous agents capable of adapting and learning. Most military operations require that individuals interact with their environment and other individuals. Before ABS, these interactions were modeled based on physics vice first principles of behavior (assuming one knows what these are). That is to say, most simulations before ABS were based on kinetics. As the military's missions continue to grow to include more peacekeeping and humanitarian relief, kinetic models are less useful.

For these ever-increasing missions, militaries need to be able to model the reaction of populations to their presence. Like the tactical discrete event simulations, these ABS models can help units train for these new missions. For example, what is the best tactic for crowd control? Should you call in more reinforcements, fire a warning shot, use non-lethal weapons? ABS models can show how each of these choices affects different cultures and are affected by different environments. These types of insights help commanders identify changing situations before they get out of control. While this type of modeling is still in its infancy, it shows promise for future military operations other than war.

11.3.7 Metrics

Once a military operations researcher has learned the techniques and tools described in this chapter and others and applies them to a problem or process, they must then try to build what Morse and Kimball (1970) call a "quantitative picture" for the decision maker. This quantitative picture allows for comparisons both within an operation and between competing alternatives that may be used for an operation. Morse and Kimball (1970) called these comparators "constants of the operation." Today we know them as metrics.

The two most common categories of metrics are measures of performance (MOP) and measures of effectiveness (MOE). MOPs are used to answer the question, "Are we doing things right?" MOEs answer the question, "Are we doing the right things?" Examples of military MOPs are casualties, bombs dropped, ton miles of cargo moved, or square miles searched in a given time period. Observed actual values of MOPs can be compared to theoretical values if calculable and show how well a unit is doing or where possible improvements may be made. However, even when a unit is operating at or near its theoretical limits, it still may not be effective. MOEs provide the commander or senior decision maker insight into how well an operation or groups of operations are achieving the overall objective. Going back to the example of ton miles of cargo moved, while a unit may be moving goods at its capacity, if it is moving the wrong type of goods or the goods are consistently late, then that unit is not effective.

There are countless examples in the literature of the use of metrics. Morse and Kimball (1970) devote a whole chapter of their book to the use of metrics and give many examples from their time with the Navy. Dyson (2006) recalls examples of metrics he used while assigned to the Operational Research Section (ORS) of the British Royal Air Force's Bomber Command in 1943. While these are historical examples, they help provide today's military operations researchers with foundations for developing the "art" of building the quantitative picture of tomorrow's battles. To see that this "art" is still being practiced today one need only look at recent publications. In the book by Perry et al. (2002), the Navy develops measures of effectiveness to verify the hypothesis that network-centric operations will enhance the effectiveness of combat systems. Other examples can be found by searching the Defense Technical Information Center (www.dtic.mil), where articles and technical reports by Schamburg and Kwinn (2005) and West (2005) detail the use of MOPs and MOEs to build a quantitative picture of future Army weapon systems.

11.4   Concluding Remarks

The nature of warfare is changing throughout the world and military OR practitioners will need to change along with it. In the United States, this new generation of warfare is referred to as the 4th generation (Lind et al., 1989). The generations of warfare are concisely defined by Echevarria (2001) as (1) massed manpower, (2) firepower, (3) maneuver, and (4) insurgency. While there is not complete agreement about 4th generation warfare, one thing is clear: insurgency and terrorist attacks have presented the military with new problems to solve.

Since World War II military OR practitioners have been focused on planning and training and not the quick fielding of new technologies. Their modeling has been characterized by the realistic portrayal of combat in simulations. They were often dislocated from the actual problem being solved and have lost sight of their beginnings. The notion of 4th generation warfare is changing this. Increased access to technology, information, and finances has given terrorists and insurgents alike the ability to create chaos at will. The scientific community has been asked to develop and field new technologies rapidly to thwart these new tactics. Military OR practitioners are once again being called back to the field to help the military employ these new technologies effectively against their enemies.

The threat of terrorism is not only affecting military OR practitioners through rapid fielding initiatives, but also through how to plan for and respond to a terrorist attack. While the location and timing of these attacks are not known with certainty, it is still possible to plan for some of them. In the military, this is called contingency planning. This is in contrast to deliberate planning where the location and enemy are known with certainty. During any type of planning, logistics support for personnel and equipment is important. While the military has many systems for assessing their deliberate plans, contingency or crisis planning assessment is lacking (Thomas, 2004). The idea of contingency logistics has been brought to the forefront of research by the recent tsunamis in the Indian Ocean and hurricanes in the United States. While not typically the responsibility of the military, responding to these types of contingencies is becoming more and more common. The military can learn from these operations what is needed to support a contingency. In the future, military OR practitioners will need to develop models both to plan for and assess contingency operations whether they be for war or operations other than war.

Finally, a new field of study, system of systems (SoS), is emerging and OR practitioners seem well placed to help. While there are many definitions for SoS, this chapter will use one from the U.S. Defense Acquisition University's Defense Acquisition Guidebook: "[SoS] engineering deals with planning, analyzing, organizing, and integrating the capabilities of a mix of existing and new systems into a system of systems capability greater than the sum of the capabilities of the constituent parts." The notion is that there are few standalone systems and by considering during the design phase how one system will interact with others, we can get better results in the end. As these SoS cut across many disciplines and OR practitioners are used to interacting with multiple disciplines it seems logical that we can help in this new field. Since the military purchases many high-cost systems it seems prudent that military OR practitioners begin to embrace this new field and see how they can help save their countries money and improve military capabilities. For an example of how the military is already beginning to use SoS, see Pei (2000).

References

1. Barković, M. and I. Peharda, 2004. The value-focused thinking application: Selection of the most suitable 5.56 mm automatic rifle for the Croatian Armed Forces, presented at Croatian Operations Research Society 10th International Conference on Operations Research, Trogir Croatia, Sept. 22–24, 2004.
2. Bhat, U. N., 1984. *Elements of Applied Stochastic Processes*, 2nd ed., Wiley, New York.
3. Beasley, J. E., "OR Notes," Downloaded from http://people.brunel.ac.uk/~mastjjb/jeb/or/intro.html.
4. Blackett, P. M. S., 1942. "Scientists at the Operational Level," reprinted in Blackett, P. M. S. *Studies of War* (Oliver and Boyd, Edinburgh, 1962), Part II, Chapter 1, pp. 171–176.
5. Brackney, H., 1959. "The Dynamics of Military Combat," *Operations Research*, 7(1):30–44.
6. Brothers, L. A., 1954. "Operations Analysis in the United States Air force," *Operations Research Society of America*, 2(1):1–16.
7. Brown, G. G., R. F. Dell, H. Heath, and A. M. Newman, 2003. "How US Air Force Space Command Optimizes Long-Term Investment in Space Systems," *Interfaces*, 33(4):1–14.
8. Dantzig, G. B., 1963. *Linear Programming and Extensions*, Princeton University Press, Princeton, NJ.
9. Defense Acquisition Guidebook, Version 1.6 (07/24/2006), http://akss.dau.mil/dag/.
10. Dyson F., 2006. "A Failure of Intelligence," Essay in *Technology Review* (Nov.–Dec. 2006).
11. Echevarria II, A. J., 2001. Fourth-Generation War and Other Myths, monograph from the Strategic Studies Institute of the U.S. Army War College. Available at http://www.strategicstudiesinstitute.army.mil/pubs/display.cfm?pubID=632, 8 Dec 2006.
12. Gass, S. I., 2002. "The First Linear-Programming Shoppe," *Operations Research*, 50(1):61–68.

13. Griggs, B. J., G. S. Parnell, and L. J. Lehmkuhl, 1997. "An Air Mission Planning Algorithm Using Decision Analysis and Mixed Integer Programming," *Operations Research*, 45:662–676.
14. Jackson, J. A., G. S. Parnell, B. L. Jones, L. J. Lehmkuhl, H. W. Conley, and J. M. Andrew, 1997. "Air Force 2025 Operational Analysis," *Military Operations Research*, 3(4):5–21.
15. Jain, G. C. and A. Nagabhushanarn, 1974. "Two-State Markovian Correlated Combat," *Operations Research*, 22(2): 440–444.
16. Keefer, D. L., C. W. Kirkwood, and J. L. Corner, 2002. Summary of Decision Analysis Applications in the Operations Research Literature, 1990–2001, Technical Report, Department of Supply Chain Management, Arizona State University, Nov. 2002.
17. Kerchner, P. M., R. F. Deckro, and J. M. Kloeber, 2001. "Valuing Psychological Operations," *Military Operations Research*, 6(2):45–65.
18. Kisi, T. and T. Hirose, 1966. "Winning Probability in an Ambush Engagement," *Operations Research*, 14(6):1137–1138.
19. Koopman, B. O., 1980. *Search and Screening: General Principles with Historical Application*. Pergamon, New York.
20. Lanchester, F. W. 1916. *Aircraft in Warfare: The Dawn of the Future Arm*, Constable and Co., London.
21. Lind, W. S., K. Nightengale, J. F. Schmitt, J. W. Sutton, and G. I. Wilson, 1989. "The Changing Face of War: Into the Fourth Generation," *Marine Corps Gazette*, October 1989, pp. 22–26.
22. McCloskey, J. F., 1987. "U.S. Operations Research in World War II," *Operations Research*, 35(6):910–925.
23. Morse, P. M. and G. E. Kimball, 1970. *Methods of Operations Research*. Peninsula Publishing, San Francisco, CA, 1970.
24. Parnell, G. S., H. W. Conley, J. A. Jackson, L. J. Lehmkuhl, and J. M. Andrew, 1998. "Foundations 2025: A Value Model for Evaluating Future Air and Space Forces," *Management Science*, 44:1336–1350.
25. Pei, R.S., 2000. "Systems-of-Systems Integration (SoSI)—A Smart Way of Acquiring Army C4I2WS Systems," Proceedings of the Summer Computer Simulation Conference, (2000), pp. 574–579.
26. Perry, W., R. Button, J. Bracken, T. Sullivan, and J. Mitchell, 2002. Measures of Effectiveness for the Information-Age Navy: The Effects of Network-Centric Operations on Combat Outcomes, RAND. Available electronically at www.rand.org/publications/MR/MR1449/.
27. Quade, E. S., ed., 1964. *Analysis of Military Decisions*. Amsterdam, the Netherlands.
28. Schamburg, J. B. and M. J. Kwinn, Jr., 2005. *Future Force Warrior Integrated Analysis Planning*, Technical Report. United States Military Academy Operations Research Center of Excellence. DSE-TR-0542, DTIC#: ADA434914
29. Smith, D. G., 1965. "The Probability Distribution of the Number of Survivors in a Two-Sided Combat Situation," *OR*, 16(4):429–437.
30. Taylor, J., 1980. *Force on Force Attrition Modeling*, Military Applications Section of the Operations Research Society, Baltimore, MD.
31. Taylor, J., 1983. *Lanchester Models of Warfare*, Vol. 1 and 2, Operations Research Society of America, Baltimore, MD.
32. Thomas, M. U., 2004. "Assessing the Reliability of a Contingency Logistics Network," *Military Operations Research*, 9(1):33–41.
33. Washburn, A., 1994. Military Operations Research, in S. M. Pollock et al. (eds.), *Handbooks in OR & MS. Vol. 6, Operations Research and the Public Sector*, Elsevier Science, North Holland.

34. Washburn, A. R., 2002. *Topics in O.R.: Search and Detection*, 4th ed., Military Applications Section, INFORMS.

35. West, P. D., 2005. *Network-Centric System Implications for the Hypersonic Interceptor System*, Technical Report. United States Military Academy Operations Research Center of Excellence. DSE-TR-0547, DTIC#: ADA434078.

36. Wood, M. K. and G. B. Dantzig, 1949. Programming of interdependent activities I, general discussion. Project SCOOP Report Number 5, Headquarters, U.S. Air Force, Washington, D.C. Also published in T. C. Koopmans, ed. *Activity Analysis of Production and Allocation*. John Wiley & Sons, New York, 1951, 15–18, and *Econometrica* 17 (3&4) July–October, 1949, pp. 193–199.

12

Future of OR/MS Applications: A Practitioner's Perspective

P. Balasubramanian
Theme Work Analytics

As per Maltz [1], Simeon-Denis Poisson developed and used the Poisson distribution in the 1830s to analyze inherent variations in jury decisions and to calculate the probability of conviction in French courts. There have been many other initiatives prior to the 1930s similar to Poisson's, highlighting the use of quantitative techniques for decision support. However, it is widely acknowledged that operations research (OR) started as a practice in the 1930s. It owes its origin to the needs of Great Britain during World War II to enhance the deployment effectiveness of combat resources and the support facilities. Its subsequent growth is credited to the achievements during the war period. The year 1937 can be considered to be the base year as this was the year the term operational research (OR) was coined [2]. It is not even a century old and compared to other functional areas (such as production, marketing, financial management, or human resource development) or to scientific disciplines (like mathematics, physics, or chemistry) it is a very young field. Yet, there are many practitioners of OR who began questioning its utility, applicability [3], and even its survivability within four decades of its existence. Fildes and Ranyard [4] have documented that 96% of Fortune 500 companies had in-house OR groups in 1970 but the numbers dwindled by the 1980s.

Ackoff [5] fueled this debate in 1979 with the assertion that "The future of Operational Research is Past." In 1997, Gass [6] proposed that in many lines of business, issues pertaining to fundamental activities have been modeled and OR has saturated. And this debate seems to be in a continuous present tense in spite of the growth in the number of OR practitioners (more than 100,000), applications, and OR/MS journals. (The 1998 estimates are 42 primary OR journals, 72 supplementary journals, 26 specialist journals, 66 models of common practices, 43 distinct arenas of application, and 77 types of related theory development, according to Miser [7].) The fundamental issue has been the gap between developments on the theoretical front versus the application of OR models and concepts in practical circumstances. Consequently, the career growth of OR professionals in corporations beyond middle management positions has become difficult.

Rapid growth in affordable computing power since the 1970s and the advent of Internet access on a large scale have rekindled this debate during the current decade, however, with one significant difference. While the earlier issues of theory versus application remain, the need for inventing newer algorithms to locate optima is also questioned now. If exhaustive enumeration and evaluation can be done in seconds and at a small cost, even for large or complex problems, why would one want to invent yet another superior search algorithm?

These issues can be embedded into the larger context of the relevance and utility of OR in the present century, considered to be the growth period of Information Age. Will OR as a discipline survive in this digital economy? Will there be newer contexts in the emerging market place where OR would find relevance? Does the discipline need to undergo a paradigm shift? What is the future of OR practitioners?

12.1   Past as a Guide to the Future

I propose to start with the early days of OR in the British armed forces during World War II [7]. The original team at Royal Coastal Command (RCC) consisted of Henry T. Tizard, Professor P. M. S. Blackett, and A. V. Hill. They were specialists in chemistry, physics, and anatomy, respectively. They worked with many others in the military operations with backgrounds in a variety of other disciplines. They were asked to study the deployment of military resources and come up with strategies to improve their effectiveness. They were to help RCC in protecting the royal merchant ships in the Atlantic Ocean from being decimated by the German U-Boats; in enhancing the ability of the royal airforce to sight U-Boats in time; and in increasing the effectiveness of patrol of the coast with a given aircraft fleet size.

Time was of the essence here. Solutions had to be found quickly, often in days or weeks but not in months. Proposed solutions could not be simulated for verification of effectiveness, yet any wrong solution could spell doom for the entire program.

Blackett and the team relied on available but limited data not only to study each of the stated problems but also to give it a focus. They were inventive enough to relate concepts from mathematics (the ratio of circumference to area of a circle decreases as the radius increases) to suggest that the size of a convoy of merchant ships be increased to minimize damage to them against enemy attacks. They were lateral thinkers who asked for painting the aircrafts white to reduce their visibility against a light sky so that the aircraft could sight the U-Boats before the latter detected their presence. They converted the coast patrol problem into one of effective maintenance of aircraft. They documented the before and after scenarios so that the effectiveness of the solutions proposed could be measured objectively. They interacted with all stakeholders throughout the project to ensure that their understanding was correct, meaningful solutions were considered, and final recommendations would be acceptable. It helped their cause that they had the ears of RCC leaders so that their proposals could be tried with commitment.

In other words, the team was interdisciplinary to facilitate innovative thinking. It consisted of scientists who could take an objective, data-based approach for situation analysis, solution construct, and evaluation of solution effectiveness. They were capable modelers as they knew the concepts and their applicability. Above all, they had the management support from the start, they understood the time criticality of the assignment, and they were tackling real world problems.

Should we ever wonder why they succeeded?

Compare this to

Mitchell and Tomlinson's [8] enunciation of six principles for effective use of OR in practice in 1978. The second principle calls for OR to play the role of a Change Agent.

and

Ackoff's assertion in 1979 [5] that "American Operations Research is dead even though it has yet to be buried" (for want of interdisciplinary approach and due to techniques chasing solutions and not the other way round).

or

Chamberlain's agony [9] even in 2004 that OR faced the challenge of "to get the decision maker to pay attention to us" and to "get the practitioner and academics to talk to each other."

My detailed study of innumerable OR projects (both successful and failed) reveals four discriminators of success, namely

1. Problem formulation in an inclusive mode with all stakeholders
2. An appropriate model (that recognizes time criticality, organization structure for decision making, and ease of availability of data)
3. Management support at all stages of the project
4. Innovative solution.

These are to be considered the Four Legs of OR. Absence of any one or more of these legs seem to be the prime cause for any failed OR project. Presence of all four legs gives an almost complete assurance of the project's success. Besides, they seem to be time invariant. They hold good when applied to the pre-Internet era as well as the current period. (A review of the Franz Edelman Prize winning articles from 1972 to 2006 lends strength to this assertion.) Let us explore these in detail.

12.1.1   Leg 1: Problem Formulation in an Inclusive Mode

This is a lesson learned early in my career as an OR practitioner.

The governor of the state is the chancellor of the state-run university in India. The then governor of Maharashtra called the director-in-charge of my company, Mr. F. C. Kohli, to discuss a confidential assignment. As a young consultant I went with Mr. Kohli and met the governor in private. The governor, extremely upset about the malpractices in the university examination system wanted us to take up a consulting assignment and to submit a report suggesting remedial action. He wished to keep the study confidential and not to involve any of the university officials whose complicity was suspected. Mr. Kohli intervened to state that it would be impossible to do such a study focusing on malpractices without involving the key stakeholders. Much against his belief, the governor finally relented to let Mr. Kohli try.

We went and met the vice chancellor of the university and briefed him about the assignment. It was the turn of the VC to express his anguish. He berated the government in general for not giving the university an appropriate budget to meet the growing admission and examination needs of the university during the past decade. Mr. Kohli then suggested that the scope of the study be enhanced to cover the functioning of the examination system in entirety (and not the malpractices alone) and was able to get the VC to agree to participation.

As a consequence we went back to the governor and impressed upon him the need to enlarge the scope. He consented.

Identifying all stakeholders and getting their commitment to participate from the beginning is an essential prerequisite for the success of any OR assignment. That this has to be a part of the project process is often lost among the OR practitioners.

Stakeholder involvement enables scoping the project appropriately, facilitates problem construction in a comprehensive manner, and eases the path for solution implementation.

Above all it recognizes and accepts the decision making role assigned to stakeholders and creates an inclusive environment. It is a natural route to forming a multidisciplinary team.

12.1.2 Leg 2: Evolving Appropriate Models

This has turned out to be the Achilles' heel of OR professionals.

Stated simply, a model is not reality but a simplified and partial representation of it. A model solution needs to be adapted to reality. The right approach calls for qualitative description, quantification, and model building as a subset and finally qualitative evaluation and adaptation of the model solution. This realization is reflected in the approach of Professor Blackett in 1938. Yet it has been lost in many a subsequent effort by others.

Many a problem of OR practice, in the pre-Internet era, can be traced to the practitioner lamenting about the quality or reliability of data and nonavailability of data at the desired microlevel at the right time as per model needs. This is ironical as the role of the practitioner is to have a firm grip on data quality, availability, and reliability at the early stages so that an appropriate model is constructed. In other words, the relationship between model and data is not independent but interdependent.

Shrinking time horizons for decisions is the order of the day. Model-based solutions need to be operationalized in current time. Prevention of spread of communicable diseases before they assume epidemic proportions is critical to societal welfare when dealing with AIDS, SARS, or avian flu. Models that depend on knowing the spread rate of a current episode cannot do justice to the above proposition (while they are good enough to deal with communicable diseases that are endemic in a society and do flare up every now and then).

I was involved in a traffic management study in suburban commuter trains in Mumbai in the early 1980s. The intent was to minimize the peak load and flatten it across different commuting time periods. We evolved an ingenious way of using a linear programming (LP) model with peak load minimization as the objective and scheduling the work start time for different firms as decision variables. And we obtained optimality that showed reduction of peak load by 15%. This admirable solution, however, could never be implemented as it did not reflect the goal of thousands of decision makers in those firms. Instead their goals were to maximize interact time with their customer and supplier firms as well as with their branch offices spread out in the city.

Having learnt this lesson, I included the study of organization structure and decision making authority as an integral aspect of any subsequent OR project. When our group came up with the optimal product mix for a tire company, where I was the head of systems, we ensured that the objective function reflected the aspirations of stakeholders from sales, marketing, production, and finance functions; the constraint sets were as stated by them and the decision variables were realistic from their perspective. The optimal solution, along with sensitive analysis was given as an input to their fortnightly decision committee on production planning, resulting in a high level of acceptance of our work.

Models are context sensitive and contexts change over time. Hence data used to derive model parameters from a given period turn out to be poor representers with the passage of time. The model itself is considered unusable at this stage. The need to construct models where model parameters are recalculated and refreshed at regular intervals is evident. Assessing and establishing the effectiveness of a model continuously and revising the model at regular intervals depend on data handling capabilities inbuilt into the solution.

Data collection, analysis, hypothesis formulation, and evaluation in many instances is iterative (as opposed to the belief that they are strictly sequential). Andy Grove, then the CEO of Intel, was afflicted with prostate cancer in the early 1990s. He was advised to go for surgery by eminent urologists. His approach to understanding the situation and constructing a solution different from their suggestion is documented in Ref. [10]. The use of data and data-based approach and the iterative nature of finding a solution are exemplified here.

12.1.3 Leg 3: Ensuring Management Support

Operations research functionaries within a firm are usually housed in corporate planning, production planning and scheduling, management information systems (MIS), or industrial engineering (IE) departments or in an OR cell. They tend to be involved in staff functions. They are given the lead or coordination responsibility in productivity improvement assignments by the top management. As direct responsibility to manage resources (machines, materials, money, or workforce) rests with functional departments such as production, purchase, finance, and HRD the onus to form an all inclusive stakeholder team rests with the OR practitioner.

Most organizations are goal driven and these are generally short term. For listed companies the time horizon can be as short as a quarter and rarely extend beyond a year. The functional departments mirror this reality with monthly goals to be met.

Many OR assignments originate with top management support. As the assignment progresses, the OR practitioner has a challenging task of including the interest of all stakeholders and ensuring their continued participation vis-à-vis keeping the support and commitment of top management. These two interests can diverge at times as the former's goal is to protect the interest of a particular resource team while management, concerned with overall effectiveness of all resources, is willing for a tradeoff. Issues and conflicts arise at this stage resulting in inordinate delay in project execution. Management support soon evaporates as a timely solution is not found.

Mitchell and Tomlinson [8] have stressed the Change Agent role of OR. Playing this role effectively calls for many skills such as excellent articulation, unbiased representation of all interests, negotiation, and diplomacy. OR practitioners need to ensure that they are adequately skilled to do justice to these role requirements.

Apart from helping to ensure top management support, possession of these skills will facilitate long-term career growth of the OR professional by switching over to line functions. This is a career dilemma for many OR professionals. While their services are required at all times, their role rarely rises above that of a middle or senior management position. There are no equivalent top management positions such as Chief Finance Officer, Chief Operations Officer, or CEO for them. (Some with IT skills can become Chief Information Officers.) Hence, diversifying their skill base (beyond quantitative) is a precondition for the growth of any OR professional in most organizations.

12.1.4 Leg 4: Innovative Solutions

Andy Grove, in his article about his battle with prostate cancer [10], states that "The tenors always sang tenor, the baritones, baritone, and the basses, bass. As a patient whose life and well-being depended on a meeting of minds, I realized I would have to do some cross-disciplinary work on my own."

The single value most realized out of an interdisciplinary team is the innovativeness of the solution generated. Either the model or the process adopted should result in its emergence. Discussing the practice of OR, Murphy [11] asserts that the "OR tools in and of themselves are of no value to organizations." According to him, the ability to build a whole greater than the sum of the parts "is the missing link." Blackett and his team took full advantage of this. In spite of the use of models to structure and analyze the given problems, they went out of the box when finding solutions. (But returned to quantitative techniques to assess the impact of solutions.)

of the computer to do exhaustive enumeration in a few seconds, thus striking a death knell on many efforts to find the algorithms for optimal solution.

Personal productivity tools such as spreadsheets with programmable cells have enabled wide-scale use of simulation, particularly of financial scenarios. Some optimization routines are also embedded into functions such as internal rate of return (IRR) calculation resulting in extensive use of OR concepts by masses but without the need to know the concepts.

The second phenomenon, namely, the Internet with its ubiquitous connectivity since the mid-1980s, coupled with increase in computing power available per dollar, has impacted on every field of human endeavor dramatically. The field of OR is no exception.

Gone are the days of data issues such as adequacy, granularity, reliability, and timeliness. Instead it is a data avalanche now presenting a data management problem. Montgomery [13] highlights issues relating to extraction of value from the huge volume of data in the Internet era. The arrival of tools such as data warehousing and data mining has helped the storage and analysis aspects. Yet the interpretation and cause–effect establishment challenges remain.

Many practitioners have wondered if IT developments have overshadowed the practice of OR. They also believe that OR applications in industry have reached a saturation point with the modeling of all conceivable problems. Some other developments are considered even more profound.

For example, for decades, inventory management has been a business area that has received enormous attention from the OR community. Models have been built and successfully used to consider deterministic and stochastic demand patterns, supply and logistics constraints, price discounts, and the like. The focus has remained in determining the optimal inventory to hold. Contrast this with the developments in this era of supply chain. Supply chain management (SCM) has been largely facilitated by the relative ease and cost effective manner in which disjoint databases, processes, and decision makers can be linked in real time and by real time collection and processing of point of sale (POS) data. Coupled with the total quality management (TQM) principles it has also given rise to the new paradigm that inventory in any form is evil and hence needs to be minimized. No one is yet talking about the futility of inventory models, at least at the manufacturing plant level, but that managerial focus has shifted to just in time (JIT) and minimal finished goods (FG) inventory is real.

The Internet era has also squeezed out intermediaries between manufacturers and consumers and has usurped their roles. Issues such as capacity planning and product mix optimization (essential in made to stock scenario) have yielded place to direct order acceptance, delivery commitment, and efficient scheduling of available capacity (made to order scenario). Dell's success has been credited [14] to eliminating FG inventory as well as to minimizing inventory held with its suppliers. They treated inventory as a liability as many components lose 0.5% to 2.0% of value per week in the high tech industry.

OR expertise was called for in improving the forecasting techniques earlier. Forecasts were fundamental to many capacity planning and utilization decisions. With the ability to contact the end user directly, firms have embraced mass customization methodologies and choose to count each order than forecast. In other words, minimal forecasting is in.

As the Internet has shrunk distances and time lines and as it has facilitated exponential growth in networking, the planning horizon has shrunk from quarters to days or even hours. Decisions that were considered strategic once have become operational now. The revenue management principle that originated in the airline industry is a prime example. The allocation of seats to different price segments was strategic earlier. It is totally operational now and is embedded into the real time order processing systems. This practice has spread to many other service industries such as hospitality, cruise lines, and rental cars (a beneficial side effect of strategic decisions becoming operational is the reduced need for assumptions and worrying about the constancy aspects of assumptions).

When the look-ahead period was long, the need to unravel the cause and effect relationship between different variables was strong. Regression and other techniques used to determine this dependency have always been suspected of throwing up spurious correlations. This has resulted in low acceptance of such techniques for certain purposes, particularly in the field of social sciences. In the Internet age, the emphasis is not in unraveling but to accept the scenario in totality, build empirical relationships without probing for root causes, and move ahead with scenario management tasks.

Does it mean that many revered OR techniques of the past are becoming irrelevant in this ICE (information, communication, and entertainment) age? Is there a change in paradigm where the brute force method (applied at a low cost) of computers and the Internet is sweeping to dust all optimization algorithms? Do we see a trend to embed relevant OR concepts into many day to day systems that make every commoner an OR expert?

Or is there another side to this coin? Given that OR has the best set of tool kits to cope with complexity and uncertainty and facilitates virtual experimentation, as seen by Geoffrion and Krishnan [15], is this a door opening for vast new opportunities in the digital era?

I have chosen to highlight some situations already encountered in a limited way untill now but are likely to be widespread in future. These are forerunners of possible opportunities for OR professionals but are by no means exhaustive.

12.3   Emerging Opportunities

Gass [6] opined in 1997 that the future challenges for OR will come from real time decision support and control problems such as air traffic control, retail POS, inventory control, electricity generation and transmission, and highway traffic control. This prediction is in conformity with our findings that many strategic decisions have become operational in the Information Age. Data collection at source, instantaneous availability of data across the supply and service chains, and shrinking planning horizons have meant that fundamental tasks of OR such as cause–effect discovery, simulation of likely scenarios, evaluation of outcomes, and optimal utilization of resources, all need to be performed on the fly. It implies that for routine and repeatable situations OR faces the challenge of automating functions such as modeling, determining the coefficient values, and selecting the best solutions. In other words, static modeling needs to yield its place to dynamic modeling. I contend that this is a whole new paradigm. OR professionals, both academicians and practitioners alike, have to invent newer methodologies and processes, and construct templates to meet this challenge.

Gawande and Bohara [16], discussing the issue faced by U.S. Coast Guard, exemplify the dilemma faced by many developed nations. As economic development reaches new highs, so will concerns for ecology and preservation. Hence, many policy decisions have to wrestle with the complex issue of balancing between incentives versus penalties. According to Gawande and Bohara, maritime pollution laws dealing with oil spill prevention have to decide between penalties for safety (and design) violations of ships and penalties for pollution caused. This issue can be resolved only with OR models that reflect an excellent understanding of maritime laws, public policy, incentive economics, and organizational behavior. In other words, bringing domain knowledge along with the already identified interdisciplinary team into the solution process and enhancing the quantitative skill sets of the OR practitioner with appropriate soft skills is mission critical here. I assert that structuring of incentives vis-à-vis penalties will be a key issue in the future and OR practitioners with enhanced skill sets and domain knowledge will play a significant role in helping to solve such issues.

Another dilemma across many systems relates to investments made for prevention as opposed to breakdown management. Replacement models of yesteryear chose to balance between cost of prevention against cost of failure. The scope of application of these principles will expand vastly in this new age, from goods to services. They were difficult to quantify for the services sector earlier. One of the significant developments in the digital era has been the sharpening of our ability to quantify the cost of an opportunity lost or poor quality of service. Further, a class of situations has emerged where cost of failure is backbreaking and can lead to system shutdown. Aircraft and spaceship failures in flight, bugs in mission critical software systems, and spread of deadly communicable diseases among humans are good examples. Even a single failure has dramatic consequences. Innovative solutions need to be found. Redundancy and real time alternatives are possible solutions with cost implications. OR experts with innovation in their genes will find a great opportunity here.

The spread of avian flu is a global challenge threatening the lives of millions of birds, fowls, and humans. Within the past 5 years it has spread from the Southeast Asian subcontinent to Africa and Eastern Europe. The losses incurred run into millions of dollars whenever the H5N1 virus is spotted in any place. The United States alone has budgeted 7 billion dollars to stockpile drugs needed to tackle an epidemic. Policies to quarantine and to disinfect affected population, to create disjoint clusters of susceptible population, have been implemented in some countries. Innovative measures to monitor, test, and isolate birds and fowls that move across national borders through legitimate trade, illegitimate routes, and natural migration are needed in many countries [17]. Formulating detection, prevention, and treatment measures calls for spending millions of dollars judiciously. Cost effectiveness of each of these measures and their region-wise efficacy can be ensured by application of OR concepts at the highest levels of government in every country.

Almost a decade ago, Meyer [18] discussed the complexities in selecting employees for random drug tests at Union Pacific Railroad. The issue of selection is entwined with concerns for protection of employee privacy, confidentiality, as well as fairness. Post 9/11, similar situations have occurred in the airline industry with passengers being selected randomly for personal search. Modern society will face this dilemma more and more across many fields, from work places to public facilities to even individual homes. For the good of the public at large, a select few need to be subject to inconvenience, discomfort, and even temporary suspension of their fundamental rights. (The alternative of covering the entire population is cost prohibitive.) How does one be fair to all concerned and at the same time maximize the search effectiveness for a given budget? Or what is an optimal budget in such cases? OR expertise is in demand here.

Market researchers have struggled with the 4Ps and their interrelationships for four decades since McCarthy [19] identified them, namely, Product, Performance, Price, and Promotion. It is the application of OR that has helped to discover the value of each of the 4Ps and relate them in a commercially viable way. The 4Ps have been applied in a statistical sense in product and system design so far. Markets were segmented and products were designed to meet the needs of a given segment with features likely to appeal to either the majority or a mythical ideal customer. This meant that product attributes were subject to either the rule of least common denominator or with gaps in meeting specific customer needs. The Internet era has extended direct connectivity to each customer, and hence the 4Ps as they relate to any individual can be customized. The front end of mass customization, namely, configuring the product with relevant Ps for each customer, understanding his/her tradeoffs, can be done cost effectively only with innovative application of OR concepts.

Thompson et al. [20] have dealt with one aspect of the above challenge, namely the need for product design to reach a new level of sophistication. As costs plummet, every manufacturer wishes to cram the product with as many features as one can. Market research has shown that buyers assign a higher weightage to product capability over its usability at the time of purchase. This results in high expected utility. However, their experienced utility is always lower as many features remain unused or underutilized. This results in feature fatigue and lower customer satisfaction. Hence, manufacturers need to segment the market better, increase product variety, and invest to learn consumer behavior on an ongoing basis.

Similarly, consumer research can scale new heights in the emerging decades. Customer behavior can be tracked easily, whereas customer perception is hard to decipher. Correlating the two and deriving meaningful insights will be the focus of market researchers. OR when combined with market research and behavioral science concepts can play a vital role here.

Communicating with customers and potential customers on time and in a cost effective manner is the dream of every enterprise. As the Internet and phone-based customer reach/ distribution channels have emerged, as customers seek to interact in multiple modes with an organization, optimal channel design in terms of segments, activities within a segment, and balancing of direct versus indirect reach partners will assume criticality. Channel cannibalization or channel conflicts are issues to contend with. Channels such as the Internet, kiosks, and telephone have been around for a decade or more. However, data on channel efficacy is either sparse or too macro. Over time a higher level of sophistication in channel usage will emerge as a key competitive edge. OR has enough tools and concepts to help out in such decisions for different segments in the financial services sector.

Banks, savings and loan associations, insurance firms, and mutual funds constitute the financial services sector. Customer segmentation, product structuring, pricing, and credit worthiness assessment are issues tackled through sophisticated OR applications since the 1970s in this sector. The concept of multi-layering of risk and treating each layer as a distinct market segment is inherent to insurance and reinsurance lines of business [21]. Product portfolio optimization for a given level of capital and risk appetite is mission critical here. The Internet and online systems have enabled firms to gather product, customer segment, and channel specific data on costs, margins, and risks. Extending the product mix optimization applications from the manufacturing area to financial services is an exciting and emerging opportunity. Formulating cost effective programs to minimize identity theft and fraudulent claims without alienating genuine customers are industry challenges where OR professionals can play significant enabler roles.

With reference to physical goods, the back end of mass customization is intelligent manufacturing that can incorporate a feedback loop from the demand side into the manufacturing processes. Modularizing components and subassemblies, delaying final assembly, and linking final assembly line to specific customer orders as well as items on hand are tasks consequent to mass customization. As discussed earlier, the move now is toward minimal forecasting. Capacity management and order management are in focus rather than better forecasting and finished goods inventory control. Simulation techniques have found wide acceptance in shop floor layout, sequencing of production steps, scheduling and rescheduling of operations to meet customer commitments and to make delivery commitments. While optimization held sway when many decisions were strategic, operationalizing the strategic decisions has resulted in simulation systems replacing the optimizers.

FedEx, UPS, and DHL are early movers in the courier transportation industry to have used complex OR algorithms and sophisticated computer systems for routing and tracking packages [22]. Cell phones, RFID devices, and GPS technologies have immense potential to provide solutions to mobility-related issues of products and resources. Dynamic data captured through such means facilitate building optimizing online and real time solutions for logistic and transportation problems. When coupled with context-specific information from GIS databases as well as from enterprise-wide computers, higher levels of complex issues can be solved in an integrated mode. Constructing such solutions is a skill and capability building means for OR professionals, thus facilitating their career growth. Camm et al. report a pioneering GIS application at Proctor & Gamble carried out by OR practitioners [23].

OR being called upon to play the role of a fair judge recurs in the telecom sector too. Systems need to reach a new level of sophistication at the design stage itself. Many systems have facilities that are multifaceted (can carry data, voice, video, etc.). Hence, they end up serving multiple customer segments and multiple products. Then the critical issue is one of design to honor service level agreements (SLAs) with each customer segment and at the same time not to build overcapacity. This turns out to be both a facility design issue and operational policy issue for a given system capacity. Ramaswamy et al. [24] have discussed an Internet dial-up service clashing with the emergency phone services when provided over the same carriers and how OR models have been used to resolve attendant issues.

Law enforcement and criminal justice systems have had a long history of use of quantitative techniques, starting from Poisson in the 1830s [1]. From effective deployment of police forces, determining patrol patterns, inferring criminality, concluding guilt or innocence, deciding on appropriate level of punishment to scheduling of court resources, numerous instances are citied in the literature about the use of OR techniques. However, a virtually untapped arena of vast potential for OR practitioners is attacking cyber crime. This is fertile ground for correlating data from multiple and unconnected sources, drawing inference, and detection and prevention of criminal activities that take place through the Internet.

One of the telling effects of the Internet has been online auctions [25]. It is the virtual space where the buyer and seller meet disregarding time and place separations. One can imagine well the advantage of such auction sites when compared to physical sites. The emerging issues in this scenario call for ensuring fairness in transactions through an objective process while preserving user confidentiality. This is an exciting emerging application area for OR experts with in-depth understanding of market behavior as well as appropriate domain knowledge relating to the product or service in focus. Bidders in such auctions need to choose not only the best prices to offer but also sequence their offer effectively. The emergence of intelligent agents that can gather focused data, evolve user specific criteria for evaluation, help both parties to choose the best path of behavior (price to quote and when) are challenges awaiting the OR practitioner. Gregg and Walizak [26] are early entrants to this arena.

A conventional assumption in any system design has been to overlook the transient state and design for the steady state (in most queuing systems). Today's computing technology permits the analysis of the transient state within reasonable cost. Also it is true that many systems spend a large amount of time in the transient state and hence have the need to design them optimally in this state as well. OR theory has much to contribute in this area.

Developments in life sciences and related fields such as drug discovery, genetics, and telemedicine have opened doors to many opportunities for OR applications. OR experts are playing an increasingly assertive role in cancer diagnosis and treatment. 3D pattern recognition issues rely heavily on statistical and related OR concepts to identify the best fits. It is evident that such opportunities will grow manifold going forward.

The year 2003 is significant in the field of biology and bioinformatics. This was the 50th anniversary of the discovery of the double helical structure of DNA and the completion of the human genome project. While the first event gave birth to the field of genetics, the second is the beginning of the use of genomes in medicine. The current challenge is to utilize the genome data to its full extent and to develop tools that enhance our knowledge of biological pathways. This would lead to accelerated drug discovery. Abbas and Holmes [27] provide enormous insight into the opportunities for OR/MS expertise application in this area. Sequence alignment, phylogenetic tree, protein structure prediction, and genome rearrangement and the like are problems solvable with OR tools of probabilistic modeling, simulation, and Boolean networks.

Similarly, the next revolution in information technology itself awaits the invention of intelligent devices that recognize inputs beyond keyboard tapping and speech. Other means of inputs such as smell, vision, and touch require advancement in both hardware and software. And the software would rely heavily on OR to draw meaningful conclusions.

Industries in the services sector will dominate the economies of both developed and developing countries in the twenty-first century. To answer the challenges in the service sector, there is an emerging *science of service management* advocated by IBM and other global companies. This would open up many opportunities for the OR/MS practitioners (see the article by Dietrich and Harrison [28]). In this handbook, OR/MS applications to the service sector (airlines, energy, finance, military, and water resources) are discussed in detail.

Finally, one of the significant but unheralded contributions of OR to society at large has been the creation of means to handle variability. Previously, most systems and products were designed with either the average in mind or custom developed. OR with its embedded statistical tools and optimization algorithms has enabled a better understanding of variability at multiple levels (variance, skewness, and kurtosis), its associated costs, and has enabled meaningful and commercially viable market segmentation (as in the airlines industry). The twenty-first century demands that we accept diversity as an integral and inevitable aspect of nature in every aspect of our endeavor. We as a society are expected to build systems, evolve policies, and promote thoughts that accept diversity in totality. This can, however, be achieved only in an economic model that balances the associated costs with resultant benefits. OR is the only discipline that can quantify the diversity aspects anywhere and create the economic models to deal with the issues effectively. In that sense its longevity is assured.

References

1. Maltz, M.D., From Poisson to the present: Applying operations research to problems of crime and justice, *Journal of Quantitative Criminology*, 12(1), 3, 1996. http://tigger.uic.edu/~mikem//Poisson.PDF.

2. Haley, K.B., War and peace: The first 25 years of OR in Great Britain, *Operations Research*, 50(1), 82, Jan.–Feb. 2002.

3. Ledbetter, W. and Cox, J.F., Are OR techniques being used?, *Journal of Industrial Engineering*, 19, Feb. 1977.

4. Fildes, R. and Ranyard, J., Internal OR consulting: Effective practice in a changing environment, *Interfaces*, 30(5) 34, Sep.–Oct. 2000.

5. Ackoff, R.L., The future of operational research is past, *Journal of the Operational Research Society*, 30(2), 93, 1979.

6. Gass, S.I., Model world: OR is the bridge to the 21st century, *Interfaces*, 27(6), 65, Nov.–Dec. 1997.

7. Miser, H.J., The easy chair: What OR/MS workers should know about the early formative years of their profession, *Interfaces*, 30(2), 97, Mar.–Apr. 2000.

8. Mitchell, G.H. and Tomlinson, R.C., Six principles for effective OR—their basis in practice, Proceedings of the Eighth IFORS International Conference on Operational Research, Toronto, Canada, June 19–23, 1978, Edited by Haley, K.B., North Holland Publishing Company, 1979.

9. Chamberlain, R.G., 20/30 hindsight: What is OR?, *Interfaces*, 34(2), 123, Mar.–Apr. 2004.

10. Andy Grove, Taking on prostate cancer, *Fortune*, May 13, 1996 www.prostate-cancerfoundation.org/andygrove www.phoenix5.org/articles/Fortune96Grove.html.
11. Murphy, F.H., The practice of operations research and the role of practice and practitioners in INFORMS, *Interfaces*, 31(6), 98, Nov.–Dec. 2001.
12. Smith, B.C. et al., E-commerce and operations research in airline planning, marketing and distribution, *Interfaces*, 31(2), 37, Mar.–Apr. 2001.
13. Montgomery, A.L., Applying quantitative marketing techniques to the Internet, *Interfaces*, 31(2), 90, Mar.–Apr. 2001.
14. Kapuscinski, R. et al., Inventory decisions in Dell's supply chain, *Interfaces*, 34(3), 191, May–Jun. 2004.
15. Geoffrion, A.M. and Krishnan, R., Prospects for operations research in the e-business era, *Interfaces*, 31(2), 6, Mar.–Apr. 2001.
16. Gawande, K. and Bohara, A.K., Agency problems in law enforcement: Theory and applications to the U.S. Coast Guard, *Management Science*, 51(11), 1593, Nov. 2005.
17. Butler, D. and Ruttiman, J., Avian flu and the New World, *Nature*, 441, 137, 11 May 2006.
18. Meyer, J.L., Selecting employees for random drug tests at Union Pacific Railroad, *Interfaces*, 27(5), 58, Sep.–Oct. 1997.
19. McCarthy, E.J., *Basic Marketing: A Managerial Approach*, Homewood, IL: Richard D Irwin Inc., 1960.
20. Thompson, D.V., Hamilton, R.W., and Rust, R.T., Feature fatigue: When product capabilities become too much of a good thing, *Journal of Marketing Research*, XLII, 431, Nov. 2005.
21. Brackett, P.L. and Xia, X., OR in insurance: A review, *Transaction of Society of Actuaries*, 47, 1995.
22. Mason, R.O. et al., Absolutely positive OR: The Federal Express story, *Interfaces*, 27(2), 17, Mar.–Apr. 1997.
23. Camm, J.D. et al., Blending OR MS, judgement and GIS: restructuring P & G's supply chain, *Interfaces*, 27(1), 128, Jan.–Feb. 1997.
24. Ramaswami, V. et al., Ensuring access to emergency services in the presence of long Internet dial-up calls, *Interfaces*, 35(5), 411, Sept.–Oct. 2005.
25. Iyer, K.K. and Hult, G.T.M., Customer behavior in an online ordering application. A decision scoring model, *Decision Sciences*, 36(4), 569, Dec. 2005.
26. Gregg, D.G. and Walizak, S., Auction advisor: An agent based online auction decision support system, *Journal of Decision Support Systems*, 41, 449, 2006.
27. Abbas, A.E. and Holmes, S.P., Bio-informatics and management science: Some common tools and techniques, *Operations Research*, 52(2), 165, Mar.–Apr. 2004.
28. Dietrich, B. and Harrison, T., Serving the services industry, *OR/MS Today*, 33, 42–49, 2006.

Index